Isoenzymes

Isoenzymes

J. HENRY WILKINSON
PH.D., D.SC.(LOND.), F.R.I.C.,
M.R.C.PATH., F.P.S., F.I.BIOL.
Professor of Chemical Pathology,
Charing Cross Hospital Medical School
University of London

SPRINGER-SCIENCE+BUSINESS MEDIA, B.V.

Originally published by Chapman and Hall in 1970
Softcover reprint of the hardcover 1st edition 1970

ISBN 978-0-412-10150-2   ISBN 978-1-4899-6834-0 (eBook)
DOI 10.1007/978-1-4899-6834-0

Contents

zymes in mammalian tissues; malate dehydrogenase isoenzymes in insect development; malate dehydrogenase isoenzymes of the chicken embryo. Isocitrate dehydrogenase: isocitrate dehydrogenase isoenzymes in heart and liver diseases. Glutamate dehydrogenase. Alcohol dehydrogenase: alcohol dehydrogenases of insects; avian alcohol dehydrogenases; alcohol dehydrogenases in mammalian liver; alcohol dehydrogenase polymorphism in plants. Xanthine dehydrogenase. Miscellaneous dehydrogenases

Preface

The term *isoenzyme* is used to describe enzymically active proteins, catalysing the same reaction and occurring in the same species, but differing in certain of their physico-chemical properties. Although the occurrence of enzymes in multiple molecular forms has been recognized for several decades, the application of electrophoretic and, to a lesser extent, chromatographic techniques has led to rapid developments in recent years. Since the discovery of the heterogeneity of lactate dehydrogenase, reported almost simultaneously by Vesell and Bearn and by Wieland and Pfleiderer in 1957, the multiple forms of this enzyme have become the subject of an extensive literature, and the techniques introduced have been applied to investigations of other isoenzymes.

Isoenzymes, especially those of lactate dehydrogenase and alkaline phosphatase, have found many applications as diagnostic agents in clinical chemistry, while others, such as those of cholinesterase and glucose 6-phosphate dehydrogenase, exhibit a biological polymorphism of considerable interest in genetics. Recent additions to our understanding of the chemical structure of the lactate dehydrogenase isoenzymes have indicated their great potential in fields as diverse as immunology, embryology and comparative anatomy. These remarkable developments suggest that the time is opportune for present knowledge to be surveyed, and I have compiled this monograph in the hope that it will be found useful by biochemists, pathologists, clinicians, geneticists and others with interests in enzymology.

After an introductory chapter concerned with definitions and nomenclature there follow chapters dealing with procedures for the separation of isoenzymes and methods for their detection and determination of their activities. The main part of the book is devoted to detailed discussion of multiple forms of individual enzymes. The chapter on lactate dehydrogenase isoenzymes, the longest in the book, includes consideration of their distribution; chemical composition; immunological, chemical and kinetic properties; the role of lactate dehydrogenase isoenzymes in developing tissues, and finally a survey of their diagnostic applications. Other dehydrogenases, aminotransferase (transaminase) isoenzymes, phosphatases, esterases and miscellaneous enzymes are discussed in the remaining chapters.

I am very grateful to my wife, Dorothy, not only for typing the manuscript but also for her patience in coping with numerous changes during its revision. Any errors that remain are mine alone, and I should appreciate hearing about any that have been overlooked.

I wish to thank the authors whose names appear in the legends and the following editors and publishers for permission to reproduce illustrations: The Editor, *Annals of the New York Academy of Sciences* (Plate 3a); Springer-Verlag, Heidelberg (Fig. 5); The Editor, *Science*, and the American Association for the Advancement of Science (Fig. 11 and Plate 4); The Editorial Board, *The Biochemical Journal* (Figs. 8, 9, 10, 12, 17 and 18); The Editorial Board, *Clinical Science* (Figs. 14, 15 and 20); The Secretary, American Society for Microbiology, and the Pergamon Press, Oxford (Fig. 16); Elsevier Publishing Company, Amsterdam (Figs. 19 and 22); and The Editor, *Nature* (Fig. 21). I am also indebted to the Shandon Scientific Company, London, N.W.10, and the E.C. Apparatus Corporation, Philadelphia, for kindly providing Plates 1 and 2 respectively.

I am most grateful to numerous colleagues at Westminster Hospital and elsewhere for their advice and helpful criticism, especially Dr W. G. Haije, Dr M. de Jong, Dr S. J. Holt, Mrs P. C. Barrow, Dr D. W. Moss and Dr S. B. Rosalki, for kindly making available to me material in advance of publication, and Dr Rosalki, Dr D. T. Plummer, Dr B. A. Elliott, Dr Pauline Emerson and Dr Wendy Withycombe for their collaboration in some of the investigations mentioned in the book. Dr Moss, Dr Rosalki, Dr Emerson and Dr Withycombe also assisted by reading and checking several chapters of the manuscript. I am much indebted to the General Editor, Professor Cyril Long, for his suggestion that I should write this monograph, for his advice during its preparation and for reading and criticizing the entire manuscript.

The book was written while I was a member of the staff of the Department of Chemical Pathology, Westminster Medical School, and I should like to take this opportunity of expressing my warmest appreciation for the help and encouragement I received throughout its preparation from Professor N. F. Maclagan.

London, S.W.1   J. H. WILKINSON
March 1965

Preface to the Second Edition

During the five years that have elapsed since this monograph was first published an enormous number of papers dealing with multiple molecular forms of enzymes has appeared in the scientific literature. A major occasion was the conference organized by the New York Academy of Sciences in December 1966, at which a most impressive series of papers was presented. During the quinquennium the emphasis has shifted from the clinical application of isoenzymes to the genetic and structural aspects of enzyme multiplicity. This change is reflected in the contents of the present volume.

New chapters have been added on the chemical nature of isoenzymes, on enzyme multiplicity in the glycolytic pathway and the pentose–phosphate cycle, and on the biological significance of isoenzymes, but the material from the first edition has also been extensively revised. The accretion of so much new knowledge necessitated selection of material for inclusion, but over a thousand new references have been added. I must acknowledge my indebtedness to all investigators in the field whether or not I have quoted their work. I am most grateful to the many research workers with whom I have enjoyed personal conversation or correspondence and to those who kindly sent me reprints.

I particularly wish to thank Drs Dean Arvan, Frank Oski, Donald Moss, Alex Pinsky, Margaret Fletcher and Jim Swale for reading one or more chapters, and Mrs Maryanne Metz and Mr Yukio Fujimoto for their help with some of the investigations and with the preparation of some of the illustrations. To my secretary, Mrs Regina Esposito, who prepared the entire manuscript, I am especially indebted.

I should like to thank the authors whose names appear in the legends and the following editors and publishers who gave permission to reproduce the additional illustrations included in the second edition: Elsevier Publishing Company, Amsterdam (Figs. 11, 12, 55 and 56); The Editorial Board, *Journal of Biological Chemistry* (Figs. 14 and 15); The Editor, *Canadian Journal of Physiology and Pharmacology* (Fig. 17); The Editor, *Endocrinology* (Figs. 19 and 20); The Editor, *Nature* (Figs. 21, 23, 24, 41 and 59); The Editor, *Biochemistry* (Fig. 25); The Editor, *Science* and the American Association for the Advancement of Science (Figs. 26, 30 and 38); S. Karger AG, Basel and New York (Figs. 27,

42, 43 and 49); The Editor, *Annals of the New York Academy of Sciences* (Figs. 28, 48 and 64); The Editor, *Journal of Experimental Zoology* (Fig. 39); The Editor, *Journal of Clinical Pathology* (Fig. 57); and The Editor, *Gut* (Fig. 62).

Once again it is my pleasure to express my appreciation to the General Editor, Professor Cyril Long, for his valuable criticism of the entire typescript. I should also like to thank the publishers for their excellent co-operation. I am greatly indebted to my wife, Dorothy, for her enthusiastic support during the preparation of this book and for her patient acceptance of the loss of some three hundred evenings.

This edition has been prepared during my tenure at the William Pepper Laboratory, University of Pennsylvania, and I should like to record my appreciation of the unfailing support given me by the Director, Dr Howard M. Rawnsley.

Philadelphia, Pa.   J. H. WILKINSON
July 1969

CHAPTER 1

Introduction

The discovery that the lactate dehydrogenase of various human and animal tissues consists of several different fractions which can be separated by electrophoresis (Vesell and Bearn, 1957; Wieland and Pfleiderer, 1957; Sayre and Hill, 1957) led to a spectacular development of interest in the occurrence of enzymes in multiple molecular forms. However, the rather isolated, but nevertheless quite numerous, earlier reports that many different enzymes found in animals, plants and micro-organisms may occur in several distinct forms should not be overlooked.

It has long been the practice for biochemists when describing enzymes to follow the advice of Emil Fischer (1895) in stating the species and organ or tissue of origin, thus recognizing that enzymes catalysing the same chemical reaction and having the same substrate specificity, but derived from different sources, may differ in other properties. However, the discovery, a generation ago, that the various coenzymes, e.g. phosphopyridoxal, coenzyme A, nicotinamide-adenine dinucleotide (formerly known as diphosphopyridine nucleotide), etc., have the same chemical structures irrespective of their biological sources, induced the idea that apoenzymes may also have identical structures (Wieland and Pfleiderer, 1962).

An enzyme thus came to be regarded as a single entity, and it was not until Warburg and his colleagues published their findings about three decades ago that this was seriously questioned. Warburg and Christian (1943) demonstrated that aldolase derived from yeast differed in several respects from that obtained from animal tissues; other investigators have reported the occurrence of pepsin, chymotrypsin, trypsin, xanthine dehydrogenase and lysozyme in a number of different forms.

Differences between the properties of analogous enzymes obtained from the same tissues of different species have been observed on numerous occasions. For example, human salivary and pancreatic α-amylases may be identical, but they differ in solubility, pH optimum and other respects from that obtained from hog pancreas (Meyer, Fischer and Bernfeld, 1947).

Even within the same species, enzymes with similar catalytic properties, but markedly different in other respects, may be found in different tissues. An acid phosphatase which originates in the human prostate behaves quite differently with inhibitors from that in the erythrocytes, though both catalyse the hydrolysis of a range of phosphate esters (Demuth, 1925; Kutscher and Wolbergs, 1935; Abul-Fadl and King, 1949). Other examples of enzymes catalysing closely related reactions are the cholinesterases, which are widely distributed throughout the human body. Acetylcholinesterase, which occurs in erythrocytes and in nervous tissue, readily hydrolyses acetylcholine and acetyl-β-methylcholine, but not benzoylcholine, whereas pseudocholinesterase found in the liver, intestine and heart and also in the blood serum has little or no action upon acetyl-β-methylcholine, but readily hydrolyses benzoylcholine in addition to acetylcholine (Mendel and Rudney, 1943). Thus, although the chemical reaction is the same in both cases, these enzymes differ considerably in their substrate specificities.

Similar metabolic reactions take place in the cells of a wide variety of tissues, and Warburg (1948) suggested that enzymes catalysing the same reaction in different tissues might be organ specific. An example of this was demonstrated by Pfleiderer and Jeckel (1957), who found that rat skeletal muscle lactate dehydrogenase and rat heart muscle lactate dehydrogenase differ in their electrophoretic mobilities and in certain other respects. Meister (1950) had previously shown that crystalline beef heart lactate dehydrogenase contained two electrophoretically distinct proteins, and Neilands (1952) found that both possessed enzymic activity. Thus, an enzyme from a single organ was clearly shown to occur in more than one form. Another example was soon forthcoming, for Krebs (1953) reported the electrophoretic separation of yeast protein into four separate forms, each of which exhibited glyceraldehyde phosphate dehydrogenase activity. It could be argued that these multiple forms of enzymes might arise as a result of their association with different carrier proteins, but the separation of crystalline enzymes into different fractions would appear to eliminate this possibility.

Nomenclature of multiple forms of enzymes
The term 'isozyme' has been proposed by Markert and Møller (1959) to describe different proteins with similar enzymic activity, but the alternative form 'isoenzyme' is perhaps to be preferred. 'Isoenzyme' is now officially recommended by the Standing Committee on Enzymes of the International Union of Biochemistry to describe the multiple enzyme forms occurring in a single species. 'Isozyme', however, is officially regarded as an acceptable alternative (Webb, 1964).

The precise definition of the term presents problems, for, though the electrophoretically distinct lactate dehydrogenase fractions obtained from a single species could without difficulty be described as isoenzymes, the position of the various esterases is much more obscure even when these occur in a single tissue. The acid and alkaline phosphatases are generally considered to be separate enzymes, and so are acetylcholinesterase and pseudocholinesterase. Markert and Møller (1959), however, confined their study to lactate, malate and isocitrate dehydrogenases, all of which have a high degree of substrate specificity. They considered that the multiple molecular forms of alkaline phosphatase, peroxidase and esterase do not qualify for inclusion as isoenzymes because their broad specificities might be attributed to the existence of each as a family of enzymes, the members of which have distinct but overlapping specificities. Recent work, however, suggests that some of these related components might now be classified as isoenzymes.

The classification of the multiple molecular forms of enzymes has also been discussed by Wieland and Pfleiderer (1962). These authors have suggested that *isoenzymes* should be distinguished from *heteroenzymes*, the first term being restricted to those forms which are derived from the same organ and tissue of origin as well as being isodynamic, i.e. having the same catalytic action. Heteroenzymes are also isodynamic, but may be derived from different organs or even from different species. In general, differences between heteroenzymes are more marked than those between isoenzymes, but while there are several examples of heteroenzymes differing considerably in the size of their molecules, e.g. alcohol dehydrogenase from yeast has a molecular weight of 150,000, whereas that from liver has a molecular weight of 73,000 (Dixon and Webb, 1958), others show remarkable similarity, e.g. beef, pig and sheep trypsins (Vithayathil, Buck, Bier and Nord, 1961). It follows that problems of definition can frequently arise, especially when, as in the case of aspartate aminotransferase, the enzyme occurs in different forms in the mitochondria and in the supernatant of the same cell (Eichel and Bukovsky, 1961; Boyd, 1961). Enzymes performing similar biochemical functions may thus occur in several forms not only in different species, or even in different organs of the same animal, but also in different parts of the same cell.

Moreover, certain enzymes can be converted artificially into different forms which still retain the original catalytic properties. Purified yeast hexokinase can be separated by chromatography on DEAE-cellulose columns into at least six different forms, two of which (*a* and *b*) are eluted by a pH gradient with succinate buffers (Group I), while the remainder (*c*, *d*, *e* and *f*) are eluted with a sodium chloride gradient

(Group II) (Kaji, Trayser and Colowick, 1961). Those of Group I can be converted into those of Group II by the action of trypsin or chymotrypsin in the presence of glucose. Since neither the enzymes of Group I nor those of Group II consists of a single protein entity, Wieland and Pfleiderer (1962) do not regard them as isoenzymes, but their individual components might qualify for such a description.

The fact that some isoenzymes differ in their relative affinities for substrate analogues, coenzyme analogues and inhibitors raises the question of the suitability of the prefix 'iso-', but no satisfactory alternative has yet been suggested. The almost unpronounceable 'homoioenzyme' (similar enzyme), however, cannot be accepted. For the purposes of this book the author has adopted the rather broad definition of the term 'isoenzyme' recommended by the Standing Committee on Enzymes (p. 2) and has endeavoured to include the majority of the more important enzymes which occur in multiple forms. The enzymes concerned are summarized in Table 1. During the past five years this task has become increasingly difficult owing to the large number of such enzymes now known. At a recent symposium on isoenzymes it was estimated that some 100 enzymes exist in multiple forms (Vesell, 1968).

Numbering of isoenzymes

Since 1957 when the diagnostic importance of the lactate dehydrogenase isoenzymes was first recognized, it has become established practice to distinguish the various components by numbering them according to their electrophoretic mobilities. Unfortunately two contradictory systems have been evolved. Most European investigators have followed the example of Wieland and Pfleiderer (1957) in assigning number 1 to the isoenzyme with greatest (anodic) mobility, while the majority of American workers have described the slowest component as isoenzyme 1 (e.g. Plagemann, Gregory and Wróblewski, 1960). Consequently there is considerable confusion in the early literature.

Wieme (1962) has appealed for international agreement and has pointed out that it is the usual convention to number α_1- and α_2-globulins in order of decreasing electrophoretic mobility, and in consequence the European system of numbering isoenzymes appears more logical. Rates of migration, however, are dependent upon the pH of the buffer employed, and it could be argued that numbering of isoenzymes on this basis would require the pH to be specified. The use of an acid buffer might lead to the cathodic γ-globulins migrating to a greater extent than albumin. This objection is not on very firm ground, since many enzymes are inactivated in the presence of acid, and electrophoresis is usually carried out with buffers in the pH range 7–9. In these circumstances most enzymes are negatively charged and migrate towards the anode. Accordingly, the Standing Committee on Enzymes of the International Union of Biochemistry has now ruled in favour of the convention of numbering isoenzymes in decreasing order of negative charge, but points out that numbering should be regarded only as a temporary expedient until more is known of the chemical differences between isoenzymes (Webb, 1964).

T A B L E 1 *List of enzymes* occurring in multiple molecular forms*

Enzyme Commission Number	Systematic name	Trivial name
1.1.1.1	Alcohol: NAD oxidoreductase	Alcohol dehydrogenase
1.1.1.8	L-glycerol-3-phosphate: NAD oxidoreductase	α-glycerophosphate dehydrogenase
1.1.1.27	L-lactate: NAD oxidoreductase	Lactate dehydrogenase
1.1.1.37	L-malate: NAD oxidoreductase	Malate dehydrogenase
1.1.1.42	L$_x$-isocitrate: NADP oxidoreductase	Isocitrate dehydrogenase
1.1.1.44	6-phospho-D-gluconate: NADP oxidoreductase	6-phosphogluconate dehydrogenase
1.1.1.49	D-glucose-6-phosphate: NADP oxidoreductase	Glucose-6-phosphate dehydrogenase
1.11.1.7	Donor: H_2O_2 oxidoreductase	Peroxidase
1.2.1.12	D-glyceraldehyde-3-phosphate: NAD oxidoreductase	Glyceraldehyde-3-phosphate dehydrogenase
1.4.1.2	L-Glutamate: NAD oxidoreductase	Glutamate dehydrogenase
1.10.3.1	o-Diphenol: O_2 oxidoreductase	o-Diphenol oxidase (tyrosinase)
1.11.1.6	H_2O_2 : H_2O_2 oxidoreductase	Catalase
2.6.1.1	L-aspartate: 2-oxoglutarate aminotransferase	Aspartate aminotransferase (glutamate-oxaloacetate transaminase)
2.7.1.1	ATP: D-hexose 6-phosphotransferase	Hexokinase
2.7.1.40	ATP: pyruvate phosphotransferase	Pyruvate kinase
2.7.3.2	ATP: creatine phosphotransferase	Creatine kinase

TABLE 1—*continued*

Enzyme Commission Number	Systematic name	Trivial name
2.7.5.1	D-glucose-1,6-diphosphate: glucose-1-phosphate phosphotransferase	Phosphoglucomutase
2.7.7.10	UTP: α-D-galactose-1-phosphate uridylyl transferase	Galactose-1-phosphate uridylyltransferase
3.1.1.1	Carboxylic ester hydrolase	Esterase
3.1.1.2	Aryl ester hydrolase	Esterase
3.1.1.7	Acetylcholine acetyl-hydrolase	Acetylcholinesterase
3.1.1.8	Acylcholine acyl-hydrolase	Cholinesterase (pseudocholinesterase)
3.1.3.1	Orthophosphoric monoester phosphohydrolase	Alkaline phosphatase
3.1.3.2	Orthophosphoric monoester phosphohydrolase	Acid phosphatase
3.1.3.5	5'-ribonucleotide phosphohydrolase	5'-nucleotidase
3.4.1.1	—	Leucine aminopeptidase
3.4.1.1	L-leucyl-peptide hydrolase	Leucine aminopeptidase (arylamidase)
4.2.1.1	Carbonate hydro-lyase	Carbonic anhydrase
4.2.99.1	Hyaluronate lyase	Hyaluronate lyase
—	—	Caeruloplasmin (copper oxidase)

* Enzyme Nomenclature. Recommendations 1964 of the International Union of Biochemistry (1965), Amsterdam: Elsevier.

REFERENCES

ABUL-FADL, M. A. M. and KING, E. J. (1949), *Biochem. J.* **45**, 51.
BOYD, J. W. (1961), *Biochem. J.* **81**, 434.
DEMUTH, F. (1925), *Biochem. Z.* **166**, 162.
DIXON, M. and WEBB, E. C. (1958), *Enzymes*, London: Longmans, Green, p. 479.
EICHEL, H. J. and BUKOVSKY, J. (1961), *Nature, Lond.* **191**, 243.
FISCHER, E. (1895), *Ber. dtsch. chem. Ges.* **28**, 1429.
KAJI, A., TRAYSER, K. A. and COLOWICK, S. P. (1961), *Ann. N.Y. Acad. Sci.* **94**, 798.
KREBS, E. G. (1953), *J. biol. Chem.* **200**, 471.
KUTSCHER, W. and WOLBERGS, H. (1935), *Z. physiol. Chem.* **236**, 237.
MARKERT, C. L. and MØLLER, F. (1959), *Proc. Natl. Acad. Sci., Wash.* **45**, 753.
MEISTER, A. (1950), *J. biol. Chem.* **184**, 117.
MENDEL, B. and RUDNEY, H. (1943), *Biochem. J.* **37**, 53.

MEYER, K. H., FISCHER, E. H. and BERNFELD, P. (1947), *Helv. chim. Acta* **30**, 64.

NEILANDS, J. B. (1952), *J. biol. Chem.* **199**, 373.

PFLEIDERER, G. and JECKEL, D. (1957), *Biochem. Z.* **329**, 371.

PLAGEMANN, P. G. W., GREGORY, K. F. and WRÓBLEWSKI, F. (1960), *J. biol. Chem.* **235**, 2282.

SAYRE, F. W. and HILL, B. R. (1957), *Proc. Soc. exp. Biol. N.Y.* **96**, 695.

VESELL, E. S. (1968), *Ann. N.Y. Acad. Sci.* **151**, 5.

VESELL, E. S. and BEARN, A. G. (1957), *Proc. Soc. exp. Biol. N.Y.* **94**, 96.

VITHAYATHIL, A. J., BUCK, F., BIER, M. and NORD, F. F. (1961), *Arch. Biochem. Biophys.* **92**, 532.

WARBURG, O. (1948), *Wasserstoffübertragende Fermente*, p. 54, Berlin: Verlag Werner Sänger (quoted by Weiland and Pfleiderer, 1962).

WARBURG, O. and CHRISTIAN, W. (1943), *Biochem. Z.* **314**, 149.

WEBB, E. C. (1964), *Nature, Lond.* **203**, 821.

WIELAND, T. and PFLEIDERER, G. (1957), *Biochem Z.* **329**, 112.

WIELAND, T. and PFLEIDERER, G. (1962), *Angew. Chem. Internat. Edn.* **1**, 169.

WIEME, R. J. (1962), *Lancet* **i**, 270.

CHAPTER 2

Techniques for the Separation of Isoenzymes

Two main groups of procedures are available for the separation of isoenzymes, namely electrophoresis and ion-exchange chromatography. Both depend primarily upon the nature and extent of the resultant charge on the protein fractions in the buffer solution used. For analytical purposes several micro-electrophoretic techniques can be used, but when it is required to separate isoenzymes on a preparative scale chromatographic methods may sometimes be more convenient. Before discussing the various procedures which can be employed, brief consideration will be given to some of the factors to be taken into account in the preparation of enzyme solutions.

Preparation of materials for isoenzyme separation
In general, it is advisable to use fresh materials for the study of the relative distribution of the various forms of an enzyme, since certain isoenzymes are frequently less stable than others. The cationic lactate dehydrogenases, for example, disappear on storage at a much faster rate than the anionic forms.

It has been the practice to store serum and tissues in the frozen state, but there are indications that this may not always be the best means of retarding loss of enzymatic activity. Kreutzer and Fennis (1964) reported that over a period of 10 days the cationic lactate dehydrogenase isoenzymes retained their activities at room temperature, but were rapidly inactivated at 0–4° or at −10°. These findings have been confirmed in the writer's laboratory. Owing to bacterial contamination, however, it is not practicable to keep serum for more than a few days at room temperature, and it is probably better to store it at 4°, even though under these conditions there is disproportionate loss of the cationic isoenzymes.

Aspartate transaminase is relatively stable in serum at 4° and at −20°. By contrast, the glucose 6-phosphate dehydrogenase activity of human haemolysates is lost after a few hours at room temperature, while remaining substantially unaffected if the intact erythrocytes are stored at 4° overnight.

Tissues can be stored frozen for quite long periods without significant loss of enzyme activity. The author obtained, from samples of muscle sent by air from London to Philadelphia, lactate dehydrogenase isoenzyme patterns which were identical with those obtained by his colleagues in London (Johnston, Wilkinson, Withycombe and Raymond, 1966). Other investigators have kept muscle specimens frozen for up to two months without detectable change in the lactate dehydrogenase isoenzyme pattern (Kelly, Belorit and Copeland, 1967). Other tissues such as the liver, heart, placenta and various endocrine organs also appear to retain their complement of enzymes when stored frozen for quite long periods. Crude extracts, however, frequently lose activity unless subjected to partial purification. This is partly due to breakdown of the enzyme protein by lysosomal enzymes and other peptidases. Losses may also result from excessive dilution (Vesell, 1962), or exposure to extremes of pH, organic solvents, etc.

Tissues are usually homogenized in a Potter–Elvehjem homogenizer at 0–5°, but electric homogenizers (e.g. Waring Blender, Atomix) may be used for larger-scale operations. Fibrous material and other tissue debris is removed by filtration through cotton gauze, but subsequent centrifugation conditions depend upon the intracellular distribution of the enzyme in question. Solutions of nuclear, mitochondrial and microsomal enzymes are prepared from the appropriate particulate fraction rather than from crude homogenates. Extracts of soluble enzymes should be centrifuged at high speed in a refrigerated centrifuge to remove as much of the particulate matter as possible. It should be borne in mind, however, that some enzymes occur in more than one subcellular fraction.

Space does not permit detailed consideration of the preparation of tissue extracts for isoenzyme investigation, but the stability of the various isoenzymes under the conditions used for their separation must be checked before their observed relative distributions can be accepted as valid.

Electrophoretic techniques

The development of simplified procedures for the zone-electrophoretic separation of serum proteins during the decade 1945–55 provided a ready-made and convenient means for the separation of isoenzymes, once their existence had been recognized. Consequently, there has been little need for the large-scale application of the classical Tiselius moving-boundary technique in this field. Paper electrophoresis has been extensively used in clinical biochemistry for the study of serum protein abnormalities, and this technique was one of the first to be applied to isoenzyme separation. In recent years, however, a number of alternative methods employing a variety of supporting media, e.g. starch grain, cellulose acetate, starch gel, agar gel, polyacrylamide, etc., has been introduced.

General considerations

Although detailed discussion of the general theory underlying the electrophoretic separation of proteins is beyond the scope of the present volume, some of the principal factors involved might advantageously be summarized before considering the special problems associated with enzyme separation. For more details the reader is referred to the excellent accounts by Stern (1956), Durrum (1958), Bier (1959) and Smith (1968).

The ability of a protein molecule to migrate in an electric field depends, as stated above, primarily upon its net electric charge. Since this can be varied, within quite wide limits, by changing the pH of the medium, it is possible for a given protein to travel towards either electrode according to the conditions. In alkaline solution (e.g. pH 8·6)

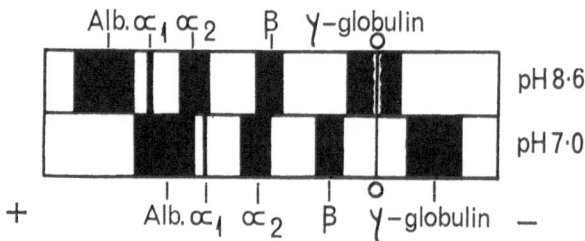

Fig. 1. Electrophoresis of serum proteins on paper: (*a*) in phosphate buffer at pH 7·0; (*b*) in barbitone buffer at pH 8·6.

most protein molecules are negatively charged and travel towards the anode. Of the major serum proteins albumin possesses the greatest charge, and hence the greatest mobility, while)'-globulin is virtually uncharged at this pH, and in the absence of other influences remains near the starting-point. At pH 7·0, however,)'-globulin is positively charged and migrates towards the cathode, whereas albumin still bears a net negative charge and moves in the opposite direction. i.e. towards the anode (Fig. 1).

Although the net electric charge is the most important single factor governing the separation of proteins during electrophoresis, its effect can be modified very considerably by other factors, such as the size and shape of the protein molecule. The nature of the buffer solution may have a profound effect on the rate of migration, for its ionic strength, its viscosity and its chemical properties may exert different effects on different proteins. Moreover, in certain cases buffer ions may react with uncharged groups to form charged complexes, e.g. borate ions combine with sugar molecules in this way.

The rates of migration of protein particles during zone electrophoresis may be modified by the electro-osmotic flow of the solvent. This is due to the effect of charged groups bound in the solid supporting medium. When exposed to an electric field the ionized carboxyl groups of paper, for example, induce positive charges on nearby solvent molecules, which then migrate towards the cathode (Kunkel and Tiselius, 1951). The extent of this movement of solvent can be measured by applying a non-ionized substance, e.g. *N*-dinitrophenylethanolamine or a suitable dye, which moves passively with the buffer and whose presence can easily be detected. Electro-osmotic flow is increased when conditions favour the ionization of the fixed groups, so that the use of alkaline buffers increases the degree of ionization of the carboxyl groups of paper. It is minimized by increasing the ionic strength of the buffer solution.

The actual distances travelled by the various protein fractions in a given time are dependent largely on the strength of the electric field to which they are subjected. As the applied voltage and current are increased, the greater is the heat produced during electrophoresis and, unless this is adequately dissipated, the temperature will rise and there is risk of partial or complete denaturation of the proteins. It follows that for a given apparatus the maximum voltage which can safely be applied is strictly limited.

All the factors which affect the electrophoretic separation of proteins generally apply with equal force to the separation of isoenzymes, but great care must be exercised to avoid inactivation. It is usual to work at 4°, with precautions to avoid local overheating.

Difficulties may also be experienced through the relative instability of certain isoenzymes when removed from the protective influence of other proteins. These can sometimes be overcome by the addition of an enzymically inert protein to the buffer solution, e.g. albumin effectively stabilizes aspartate aminotransferase isoenzymes (Fleisher, Potter and Wakim, 1960). Excessive dilution may also lead to denaturation, and in general the greater the amount of enzyme protein used in the process, the less is the loss of activity during electrophoresis. On the other hand, the more dilute the solution, the better is the separation achieved and, clearly, in defining conditions for enzyme electrophoresis a compromise has to be made. Furthermore, individual isoenzymes may be selectively adsorbed on the walls of the apparatus used, so that it is desirable to employ relatively non-polar materials, e.g. polystyrene, for its construction.

Despite this rather formidable list of factors which has to be taken into account in carrying out enzyme electrophoresis, several elegant techniques which give excellent results have been developed.

Paper electrophoresis

Lactate dehydrogenase was already of considerable diagnostic importance when it became one of the first enzymes known to exist in multimolecular forms. It was natural therefore that much of the impetus during the early stages of isoenzyme investigation should come from the clinical laboratory and, as paper electrophoresis was at that time the principal means for the routine examination of the serum proteins, several of the early isoenzyme separations were carried out by this technique. These have included serum and tissue alkaline phosphatases (Baker and Pellegrino, 1954; Ahmed, Abul-Fadl and King, 1959), human and animal tissue lactate dehydrogenases (Sayre and Hill, 1957; Wieland and Pfleiderer, 1957; Wieland, Pfleiderer and Ortanderl, 1959), plant peroxidases (Jermyn and Thomas, 1954; Garay and Faragó, 1959), serum and tissue aspartate aminotransferases (Pryse-Davies and Wilkinson, 1958; Fleisher *et al.*, 1960) and serum and urinary amylase (Searcy, Ujihira, Hayashi and Berk, 1964; Franzini, 1965).

The isoenzymes of human serum and tissue lactate dehydrogenase have also been separated by continuous paper electrophoresis (Hill, 1958; Rosalki and Wilkinson, 1960; Hill and Meacham, 1961).

Either the vertical or the flat-bed type of apparatus may be used, and although the latter, in which the paper strip is supported horizontally, is perhaps to be preferred on theoretical grounds, equally satisfactory results can be obtained with the vertical or hanging-strip apparatus. In both types the ends of the paper strips dip into buffer solutions in which the platinum electrodes are immersed. The electrodes are usually separated from the main bulk of buffer solution by some sort of partition penetrated by suitable wicks. This arrangement ensures that pH changes resulting from electrolysis of the buffer solution are confined to a small volume immediately surrounding the electrode. The levels of the buffer solutions in all four compartments must be identical, otherwise siphoning through the paper strips might occur. In the horizontal apparatus a Perspex bridge is usually provided to support the paper, while in the vertical type the papers are suspended over a nylon thread in the form of an inverted V.

The paper strips of Whatman No. 1 or 3MM grade are cut to the appropriate size and inserted into the apparatus, where they are wetted with the buffer solution. The buffer solutions used for isoenzyme separation are discussed when considering individual enzymes. After allowing several minutes for equilibration the enzyme preparation is applied at a pencil line previously drawn on the paper. Since under the conditions normally used, enzymes travel a greater distance towards the anode than towards the cathode, the line is drawn somewhat nearer the cathode when using the horizontal apparatus. When using the vertical apparatus, however, the line is drawn at the fold in the paper. The capacity of the wet paper to accommodate the sample is limited: with Whatman No. 1 paper about 3–4 μl./cm. can be applied, while thicker grades can take up to 8–10 μl./cm. Overloading should be avoided, since it leads to lengthwise spreading of the sample, and hence to poor resolution.

The potential applied should not normally exceed 10 v./cm., otherwise there is likely to be a significant rise in temperature which may lead to protein denaturation. High-voltage paper electrophoresis is consequently not readily applicable to enzyme studies. Excessive heat production may also lead to distortion of the protein pattern as a result of evaporation of buffer solution from the paper. In order to obtain good separation of the isoenzyme bands, paper electrophoresis is usually continued for several hours, and it is often convenient to allow the process to run overnight.

The amount of material which can be applied to a paper strip is strictly limited, but when it is required to separate larger quantities the technique of continuous paper electrophoresis may be employed (Haugaard and Kroner, 1948; Grassmann and Hannig, 1950; Durrum, 1958; Hobart and Rose, 1955). The sample is fed continuously at a slow rate regulated by a wick or a mechanically operated syringe to a vertically supported 'curtain' of Whatman 3MM paper through which a buffer solution is allowed to flow evenly and continuously from a reservoir fitted with a constant-feed device supported at the top of the apparatus. An electric potential is applied either at the bottom corners of the 'curtain' or along its vertical edges. The lower edge of the paper is serrated and arranged so that the buffer solution drips from the teeth into suitably placed test tubes. In order to prevent excessive evaporation the whole apparatus is enclosed by Perspex screens. As electrophoresis proceeds, the protein fractions fan out as they descend the curtain from the point source. Thus albumin collects in tubes near the anodic edge, while the γ-globulins are found in tubes near the cathode. When all the enzyme solution has been applied to the curtain the current may be switched off and the buffer removed from the upper reservoir. It is then a simple matter to dry the paper and to stain the pathways taken by the individual protein fractions with a suitable dye. This enables the mobilities of the isoenzymes collected in the test tubes to be related to those of the major protein fractions.

Another technique for continuous electrophoresis of potential value for the separation of isoenzymes is the 'curtain-less curtain' apparatus of Hannig (1961), in which a potential is applied across a thin film of buffer solution flowing at a slow rate between two parallel glass plates separated by about 1 mm. The effluent is collected by means of a number of delivery tubes placed along the lower edge of the plates and extending from the anode to the cathode. These lead to test tubes in which the various fractions are collected. The protein (or enzyme) solution is applied continuously just below the top edge of the plates, and separation of the fractions is effected in much the same manner as in continuous paper electrophoresis.

Starch-block electrophoresis
The use of potato starch grains as an anticonvectant and supporting medium for the zone-electrophoretic separation of proteins was introduced by Kunkel and Slater (1952), and its first application in the isoenzyme field was that of Vesell and Bearn (1957), who separated the isoenzymes of human serum lactate and malate dehydrogenases. Subsequently, human serum, liver and bone alkaline phosphatases were also electrophoretically separated into a number of components with the aid of this medium (Keiding, 1959; Rosenberg, 1959). Starch-block electrophoresis has proved to be a very useful procedure for enabling individual lactate dehydrogenase isoenzymes to be separated in quantities sufficiently large to permit the study of their properties (Wilkinson, Cooke, Elliott and Plummer, 1961; Plummer, Elliott, Cooke and Wilkinson, 1963; Krieg, Rosenblum and Henry, 1967). It has also been employed for the separation of lactate dehydrogenase isoenzymes as an aid to the differential diagnosis of malignant and benign effusions (Richterich, Locker, Zuppinger and Rossi, 1962), and in the study of the peroxidases of *Zea mays* (Galston and McCune, 1961). Though technically simpler than continuous paper electrophoresis, it is too laborious, however, for purely analytical purposes.

After washing with water, potato starch is equilibrated with the buffer solution to be used and separated by decantation. The moist starch is then compressed into a block on a glass plate with the aid of a Perspex former. The dimensions can be varied within wide limits according to the volume of enzyme solution to be applied, but in the writer's laboratory it has been found convenient to use blocks 35 cm. long, 1 cm. thick and 10–25 cm. wide. If necessary, thicker blocks can be used, though in such cases it is difficult to prevent local overheating in the centre of the block. Wicks consisting of several thicknesses of absorbent gauze provide electrical connection with the buffer solution in

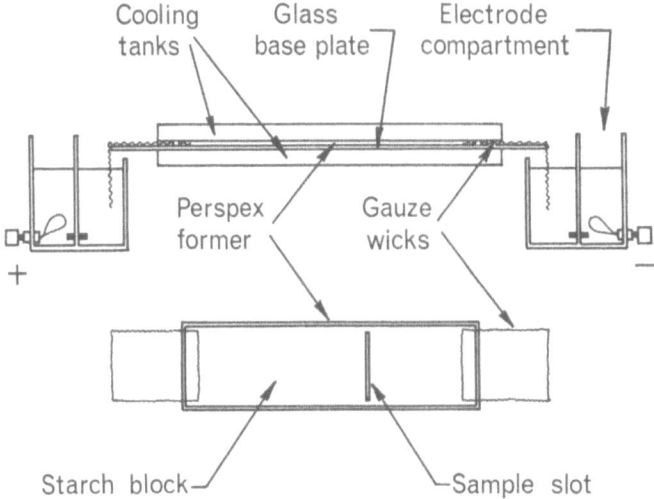

Fig. 2. Apparatus for starch-block electrophoresis, showing electrode assembly and cooling unit.

the Perspex electrode tanks. Each of these contains a platinum loop electrode in a separate compartment connected with the buffer in the rest of the tank by a number of asbestos or gauze wicks. This arrangement prevents pH changes in the neighbourhood of the electrodes from affecting the buffer solution in contact with the block.

In order to minimize temperature changes the glass plate supporting the block is placed on a cooling tank, and a similar tank coated with Parafilm to insulate against electrical short-circuiting is placed on top of the block (Tombs, Cooke, Burston and Maclagan, 1961). Aqueous ethylene glycol (20% w/v), cooled to 0–2 in a refrigerator unit, is circulated through the cooling tanks by means of a pump. The general arrangement of the apparatus is shown in Fig. 2.

Before commencing a run, the current is switched on for about $\frac{1}{2}$–1 hr. to permit equilibration, then, *having first switched off again*, a groove is cut transversely in the block to allow for the introduction of the sample. When using a 35-cm. block with a buffer of pH 8·6, the origin should be about 9 cm. from the cathodic end. The groove, which should not extend nearer than 1 cm. from the edges of the block, is blotted with filter paper and the sample is applied, displaced buffer being removed by blotting. A 1-cm.-thick block can accommodate up to about 2·5 ml. sample per 10 cm. Larger amounts than this tend to spread longitudinally, and consequently lead to poor resolution. The groove is then filled with starch previously equilibrated with buffer. A marker, such as mixed bovine serum albumin and human haemoglobin solution, should be applied to the block alongside the sample. Good separations of the lactate dehydrogenases of normal and pathological human sera are obtained when a potential of 10 v./cm. is applied over a period of 18 hr. With a 35-cm. × 15-cm. block (cross-section area, 15 sq. cm.), the current flow is about 90 mA.

After electrophoresis the positions of the markers can be detected by applying a narrow strip of Whatman 3MM paper along the length of the block over the markers. The paper is then removed, dried at 100° and stained with Light Green or some other protein stain to indicate the positions taken up by the albumin and haemoglobin. Alternatively, bromophenol blue can be incorporated into the marker solution, and the albumin position is then denoted by the lilac-coloured spot. The use of two marker proteins enables the extent of the electro-osmotic flow to be estimated.

The remainder of the block is then cut transversely into 1 cm. (or 0·5 cm.) strips, each of which is quantitatively transferred to a sintered-glass funnel and eluted with a suitable buffer solution. The volume of each combined eluate is measured and a sample taken for enzyme determination.

Cellulose acetate electrophoresis
Cellulose acetate membranes have several advantages over paper as media for electrophoresis, since adsorption of proteins is very much reduced and albumin 'tailing' is practically eliminated (Kohn, 1957*a*, 1957*b*, 1958*a*, 1958*b*). They are chemically homogeneous and can readily be rendered transparent so that stained protein bands can be determined by scanning. Adequate separation of protein bands for analytical purposes can usually be achieved in 1–2 hr., and cellulose acetate has the additional advantage of giving excellent separation of albumin and α_1-globulin. It has been extensively used for the clinical

investigation of lactate dehydrogenase isoenzymes (Wieland *et al.*, 1959; Pfleiderer and Wachsmuth, 1961; Hawkins and Whyley, 1962; Barnett, 1962, 1964; Meade and Rosalki, 1963; Preston, Briere and Batsakis, 1965), and also for the leucine aminopeptidases of human serum and tissues (Smith and Rutenburg, 1963; Meade and Rosalki, 1964), glucose 6-phosphate dehydrogenase (Poznanska and Gore, 1965), creatine kinase (Rosalki, 1965), glyceraldehyde 3-phosphate dehydrogenase (Lebherz and Rutter, 1967) and phosphoglucose isomerase (Nakagawa and Noltmann, 1967). The writer has observed that though glucose 6-phosphate dehydrogenase isoenzymes can be separated by cellulose acetate electrophoresis, better results are obtained when acrylamide gel is used. Cellulose acetate electrophoresis may conveniently be carried out in the Shandon Universal apparatus (Kohn, 1960), or in the Beckman Microzone, Gelman or Millipore cells. Cellulose acetate membrane is cut into 10-cm. × 2·5-cm. strips, on each of which a line is drawn with a soft pencil between the mid-points of each long axis to mark the point of application of the sample. Identification marks and the direction of current flow must also be noted before the strips are wetted. The strips are then floated on the surface of the buffer so that air is displaced from the pores of the membrane. Too rapid immersion in the buffer leads to the trapping of air bubbles. Excess buffer is then removed by blotting the strips with filter paper.

The strips are now ready for insertion into the apparatus, which has previously been charged with the appropriate buffer solution. Kohn (1960) recommends the use of 0·07M-barbitone buffer of pH 8·6 for the separation of serum proteins, and this solution is quite suitable for isoenzyme studies, though of course other buffers may be used with cellulose acetate as with other media. A 10-cm. strip fits conveniently across an 8-cm. gap, and the ends are held in place by means of filter-paper wicks. Up to eight such strips can be accommodated at a time in the apparatus, but it is important to ensure that the lines marking the points of application of the sample should be in line with each other so that the various isoenzyme patterns may subsequently be compared. A ruler is placed across the apparatus to provide a guide for the micropipette used for applying the sample. About 5 μl. is about as much as can be applied to a strip 2·5 cm. wide, but it is important not to allow the sample streak to extend nearer than 0·5 cm. to the margins. The lid of the apparatus is then placed in position and a potential of about 160 v. is switched on. This gives a current flow of about 0·5 mA/cm., which enables satisfactory separation of isoenzymes for analytical purposes to be achieved in about 60 min. Electrophoretic flow can be followed by running a minute trace of albumin stained with bromophenol blue alongside the enzyme preparation. In certain cases, e.g. serum, it is permissible to incorporate the indicator in the sample. A histochemical staining procedure (see Chapter 3) is preferable to elution for the detection of the positions taken up by the isoenzymes.

Difficulty is sometimes experienced due to disproportionate loss of the cationic lactate dehydrogenase isoenzymes, but this can be overcome by incorporating 1% bovine serum albumin into the buffer used to saturate the cellulose acetate strip (Rosalki and Montgomery, 1967).

Cellulose acetate is also available in gel form in blocks and strips ('Cellogel' manufactured by Chemetron of Milan, Italy). It has proved satisfactory for the separation of the isoenzymes of lactate dehydrogenase (Dioguardi, Agostoni and Fiorelli, 1964; Warburton and Waddecar, 1966), alkaline phosphatase (Kitchener, Neale, Posen and Brudenell-Woods, 1965; Posen, Neale, Birkett and Brudenell-Woods, 1967) and glucose 6-phosphate dehydrogenase (Rattazzi, Bernini, Fiorelli and Mannucci, 1967).

Cellogel is supplied moistened with 50% methanol, which must be removed by blotting and soaking in buffer before use. The gel strips can be arranged in the electrophoretic cell so that their ends are immersed in the buffer, but the blocks require paper wicks. The sample (*ca.* 1·5 μl.) can be applied to the surface of the strip, but when the block is used 5–8 μl. are inserted into a slit previously cut with a razor blade. The material can be used with a wide range of buffer solutions, including borate, tris-citrate and barbitone. A voltage of about 150–190 v. is applied, and each run requires 1·5–3 hr. The enzyme can be detected by staining with nitro-blue tetrazolium or tetrazolium MTT (see Chapter 3).

Cellogel appears to have certain advantages over other media in that it is not necessary to prepare the gel immediately before use, and that relatively small amounts of the expensive staining reagents are required.

Agar-gel electrophoresis

Although Arrhenius studied ionic transport in agar gels as long ago as 1901, it was not until nearly fifty years later that protein fractionation by agar-gel electrophoresis was reported (Gordon, Keil and Sebesta, 1949), and the first application of this technique to isoenzyme separation was that of Wieme (1959a). Agar gel has several advantages over paper as a support medium for zone electrophoresis; since it does not adsorb proteins, it can be prepared in a homogeneous state, it is transparent and the separated protein fractions can be determined by direct photometric measurement. It has high resolving power, and agar-gel electrophoresis may readily be adapted for use on an ultramicro-scale. It has the particular advantage that during isoenzyme separation passive diffusion is less than with paper or cellulose acetate. An excellent account of the technique and applications of agar-gel electrophoresis has recently been compiled by Wieme (1965).

Wieme applied the new procedure to the fractionation of mouse- and rat-tissue lactate and sorbitol dehydrogenases and to the lactate dehydrogenases of human pathological sera (Vuylsteek and Wieme, 1958; Wieme, 1959a; Wieme and Demeulenaere, 1959; Wieme and Van Maercke, 1961). Other investigators have also used this technique for the study of the isoenzymes of lactate dehydrogenase from a variety of tissues and species (Ressler and Moy, 1959; Lowenthal, Van Sande and Karcher, 1961; Blanchaer, 1961a; Van der Helm, 1961, 1962; Van der Helm, Zondag, Hartog and Van der Kooi, 1962; Ressler and Joseph, 1962; Bonavita and Guaneri, 1962; Bonavita, Ponte and Amore, 1962; Laursen, 1962; Gerhardt, Clausen, Christensen and Riishede, 1963; Fröhlich, 1965; Wright, Cawley and Eberhardt, 1966; Papadopoulos and Kintzios, 1967). It has also been much used for the fractionation of alkaline phosphatase (Haije and de Jong, 1963), esterases and acid phosphatase (Oort and Willighagen, 1961; Gerhardt *et al.*, 1963; Pla, Papadopoulos and Rosen, 1968), malate dehydrogenase (Lowenthal *et al.*, 1961), aspartate transaminase (Boyd, 1961, 1962); alcohol dehydrogenase (Papenberg, Von Wartburg and Aebi, 1965); creatine kinase (Deul and van Breeman, 1964; Burger, Richterich and Aebi, 1964; Sjövall and Voigt, 1964; Jacob, Heldt and Klingenburg, 1964; Kar and Pearson, 1965; Brownlow and Gammack, 1967); caeruloplasmin (Secchi, Petrella, Lomanto and Gervasini, 1967) and glucose 6-phosphate dehydrogenase (Haywood, Starkweather, Spencer and Zarafonetis, 1968).

For enzyme electrophoresis, Wieme (1959a, 1959b) recommends Difco Special Agar-Noble for the preparation of the gels, but other workers (e.g. Bodman, 1960) have found other grades of agar to be equally satisfactory, though deionized agars sometimes do not form good gels. Kreutzer and Eggels (1965), however, have carried out an extensive survey with different kinds of agar and reported that the basic lactate dehydrogenase isoenzymes do not appear to penetrate certain agar gels so readily during electrophoresis except in the presence of relatively high concentrations of γ-globulin. Best results were obtained with Agar-Noble and agarose, but abnormal mobilities of LD_5 were observed when gels prepared from Behring agar and Ionagar were used.

A 1% solution of agar in warm buffer solution forms a satisfactory gel on cooling. For the separation of proteins generally, it is often recommended that 0·5% sodium azide be incorporated to prevent bacterial contamination, but azide should not be used during enzyme electrophoresis, since it is a potent inhibitor of the activities of certain enzymes. A wide choice of buffer solutions is available (Bodman, 1960), but Wieme (1959*b*) prefers barbitone buffer of pH 8·4, while Ressler and Joseph (1962) use glycine buffer of pH 8·7.

Under certain conditions LD_5 appears to be associated with a negatively charged component of the agar, and consequently migrates towards the anode at a more rapid rate. Ressler, Schulz and Joseph (1963) have demonstrated that this effect can be overcome by either subjecting the gel to preliminary electrophoresis before applying the sample or by increasing the ionic strength of the buffer. The effect is accentuated by using higher concentrations of agar and weaker buffer solutions. Starkweather, Spencer, Schwarz and Schoch (1966) have also recommended the use of a buffer of adequate ionic strength (barbitone buffer 0·1 *I*).

Treatment of the sample with coenzyme (NAD) prior to electrophoresis also increases the mobilities of lactate dehydrogenase isoenzymes (Zondag, 1963; Kreutzer and Jacobs, 1965). Similar effects are observed with NAD analogues (Vesell and Bearn, 1962).

The gel is melted and poured on to clean microscope slides (or larger sheets of glass) supported horizontally so as to produce an even layer about 2 mm. thick completely covering the upper surface of the slide. The enzyme solution (5 μl.) may be applied to a short slit 1 mm. wide cut transversely in the gel, or alternatively, it may be applied by means of a small strip of filter paper (5 × 2 mm.) placed in position on the gel. After about 15 min. most of the enzyme will have diffused into the gel and the paper is removed with forceps. An albumin marker containing bromophenol blue run alongside the enzyme solution provides a measure of the rate of migration, while endosmotic flow can be measured by means of a second marker consisting of dextran stained with *o*-nitroaniline. The slide is placed in a Universal electrophoresis apparatus, and electrical connection between the gel and the buffer is made by paper strips or by agar gel layered on a plastic backing. The lid of the apparatus is now placed in position and the current switched on. When high current densities are used temperature elevation may be prevented by filling the apparatus with light petroleum (Wieme, 1959*b*). Evaporation of the solvent prevents thermal inactivation of the enzymes, but precautions must be taken to minimize the fire risk!

Wieme (1964, 1965) has described a procedure for high voltage electrophoresis which may have applications in the study of isoenzymes.

Starch-gel electrophoresis

Since its introduction by Smithies (1955) the technique of starch-gel electrophoresis has been used more than any other for the separation of isoenzymes (see Table 2).

TABLE 2 *Isoenzymes separated by starch-gel electrophoresis* *

Enzyme	Sources	References
Alcohol dehydro-genase	Mammalian liver Maize	Shaw and Koen (1965, 1967) Beutler (1967) Scandalios (1967)
Alkaline phosphatase	Human and animal serum, bone, liver, placenta, intestine, kidney, erythrocytes Protozoa Insects	Estborn (1959) Moss, Campbell, Anagnostou-Kakaras and King (1961a, 1961b) Boyer (1961) Latner (1961) Paul and Fottrell (1961) Kowlessar, Haeffner and Riley (1961) Moss (1962a, 1962b) Chiandussi, Greene and Sherlock (1962) Hodson, Latner and Raine (1962) Latner and Skillen (1962) Meade and Rosalki (1963) Schneiderman (1967) Fishman, Inglis and Ghosh (1968) Fishman, Inglis, Stolbach and Krant (1968)
Acid phosphatase	Human prostate, semen, serum, erythrocytes, tissues Algae Protozoa	Estborn and Swedin (1959) Keck (1961) Sur, Moss and King (1962) Grundig, Czitober and Schobel (1965) Allen, Allen and Licht (1965) Giblett and Scott (1965) Beckman and Beckman (1967) Lai and Kwa (1968)
Esterases	Human and animal serum and tissues Protozoa	Hunter and Markert (1957) Lawrence, Melnick and Weimer (1960) Laufer (1960, 1961) Allen, S. L. (1961)

TABLE 2—*continued*

Enzyme	Sources	References
	Insects	Hunter and Strachan (1961)
	Root nodules of	Paul and Fottrell (1961)
	Leguminosae	Harris, Hopkinson and Robson (1962)
		Hess, Angel, Barron and Bernsohn (1963)
		Moustafa (1963)
Leucine amino-	Human serum, liver, pan-	Kowlessar, Haeffner and
peptidase	creas, placenta, kidney	Sleisenger (1960)
		Kowlessar *et al.* (1961)
		Smith and Rutenburg (1963)
5´-Nucleotidase	Human serum	Kowlessar *et al.* (1961)
Glucose 6-phosphate	Rat tissues	Tsao (1960)
dehydrogenase	Human and animal	Boyer, Porter and Weilbacher
	erythrocytes, uterus	(1962)
		Trujillo, Walden, O'Neil and Anstall (1965)
		Linder and Gartler (1965)
		Carson, Ajmar, Hashimoto and Bowman (1966)
		Beutler and Collins (1966)
		Stamatoyannopoulos, Yoshida, Bacapoulos and Motulsky (1967)
		Porter (1968)
6-Phosphogluconate	Human erythrocytes	Fildes and Parr (1963)
dehydrogenase		Carson *et al.* (1966)
		Bowman, Carson, Frischer and Garay (1966)
		Parr (1966)
		Gordon, Keraan and Vooijs (1967)
α-Glycerophosphate	Insects	Laufer (1960, 1961)
dehydrogenase	Rat tissues	Tsao (1960)
	Amphibia	Manwell (1966)
Isocitrate dehydro-	Rat tissues	Tsao (1960)
genase	Human serum and tissues	Bell and Baron (1962)
		Baron and Bell (1962)
		Campbell and Moss (1962)
Lactate dehydro-	Human and animal serum,	Markert and Møller (1959)
genase	heart, liver, skeletal	Tsao (1960)
	muscle, kidney, erythro-	Laufer (1960, 1961)
	cytes	Dewey and Conklin (1960)
	Insects	Plagemann, Gregory and Wróblewski (1960)

TABLE 2—*continued*

Enzyme	Sources	References
		Wróblewski and Gregory (1961)
		Blanchaer (1961*b*)
		Latner and Skillen (1961, 1962)
		Markert and Appella (1961)
		Allen, J. M. (1961)
		Kaplan and Cahn (1962)
		Vesell (1962)
		Ressler, Schulz and Joseph (1962, 1963)
		Meade and Rosalki (1963)
Hexokinase	Yeast	Kaji, Trayser and Colowick (1961)
	Human erythrocytes and leucocytes	Eaton, Brewer and Tashian (1966)
Malate dehydro-	Insects	Laufer (1960, 1961)
genase	Rat tissues	Tsao (1960)
		Thorne, Grossman and Kaplan (1963)
		Conklin and Nebel (1965)
		Davidson and Cortner (1967)
		Murphey, Barnaby, Lin and Kaplan (1967)
Aspartate amino-	Human serum	Boyde and Latner (1961)
transferase		
Galactose dehydro-	Mammalian liver	Cuatrecasas and Segal (1967)
genase		
Peroxidase	Maize	Endo (1967)
	Rice	Chu (1967)
Caeruloplasmin	Human serum	Morell and Scheinberg (1961)
		Poulik and Bearn (1962)
		Shreffler, Brewer, Gall and Honeyman (1967)
Creatine kinase	Animal tissues	Dawson, Eppenberger and Kaplan (1967)
		Sjövall and Voigt (1964)
Carbonic anhydrase	Human erythrocytes	Shows (1967)
Arginine kinase	Molluscs	Moreland and Watts (1967)
Glutamate dehydro-	Amphibia	Manwell (1966)
genase		
α-Glucosidase	Yeast	Yau and Lindegren (1967)
Phosphoglucose	Mouse haemolysates	Carter and Parr (1967)
isomerase		

* Owing to the very large number of applications of starch-gel electrophoresis to isoenzyme separation, this list of references is representative rather than complete.

Starch gel possesses a number of advantages over other supporting media, especially in analytical procedures, though it can also be used for small-scale preparative electrophoresis. It gives excellent resolution of the serum proteins, at least twelve fractions being detectable as sharply defined lines when the gel is subsequently stained. This is partly due to the fact that proteins only slowly enter the gel at the point of application – thus the origin becomes effectively much narrower than when other media are used. Another factor is that the high starch concentration employed gives a gel with a pore size comparable with the dimensions of the protein molecules being investigated. The gel thus appears to behave as a molecular sieve, and protein fractionation by starch-gel electrophoresis depends not only upon the charge on the various components but also upon their abilities to penetrate the 'sieve', i.e. upon their sizes and shapes. Thus, human α_2-globulin may be separated into as many as eight sub-fractions (Bearn, quoted by Kunkel and Trautman, 1959). A typical electropherogram obtained with normal human serum is shown in Fig. 3.

Partly hydrolysed 'soluble' starch suitable for preparing gels is available commercially. It is suspended in the buffer solution (12–13 g./100 ml.) and the mixture is heated with vigorous stirring until it forms a translucent but homogeneous viscous liquid. Heating is stopped and pressure reduced until the mixture boils vigorously; this ensures the removal of air bubbles. Suction is disconnected and the liquid is poured into Perspex trays supported on a levelling board. Sufficient starch is placed in each tray so that the meniscus is slightly higher than the edges of the trays. The excess starch is then pressed out while still molten so as to leave a gel of uniform thickness free from air bubbles (Smithies, 1955; Bodman, 1960).

Most separations of serum proteins have been carried out on horizontal gels with borate buffer ($0 \cdot 03\text{M-}H_3BO_3$ and $0 \cdot 012\text{M-NaOH}$) which gives a gel of pH $8 \cdot 48$ (Smithies, 1955), but phosphate or acetate buffers may also be used. When applying the technique to enzyme separations, however, the ionic strength may exert a marked influence on the rate of migration of individual isoenzymes. Ressler *et al.* (1963) have recently shown that with buffers of low ionic strength the cathodic lactate dehydrogenase isoenzyme (LD_5) tends to migrate towards the anode, whereas with buffers of higher ionic strength it migrates to the cathode. The authors attribute this effect to association of the isoenzyme with anionic substances present in the medium, since increase in the concentration of starch in the gel increases the effect of weak buffers, while preliminary electrophoresis of the gel before applying the sample eliminates it. Similar phenomena also occur with agar gels (p. 18). Traces of metals may affect the mobilities of alcohol dehydrogenase isoenzymes (Watts, Donninger and Whitehead, 1961).

Discontinuous buffer systems, in which a different buffer is used in the electrode vessels from that used in the preparation of the gel, may be used in starch-gel electrophoresis (Poulik, 1957). The various components of crude horse-radish peroxidase are resolved much more sharply

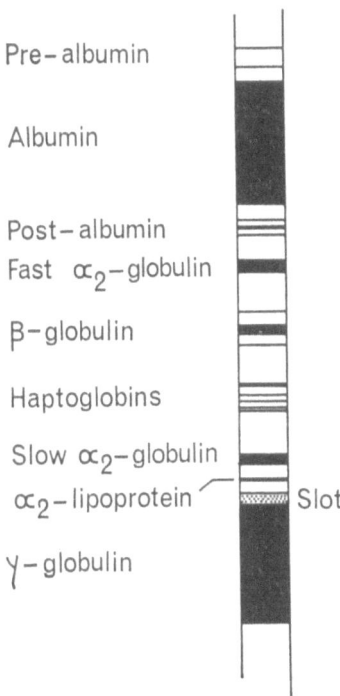

Pre-albumin

Albumin

Post-albumin
Fast α_2-globulin

β-globulin

Haptoglobins

Slow α_2-globulin
α_2-lipoprotein

γ-globulin

Slot

Fig. 3. The proteins of normal human serum separated by starch-gel electrophoresis.

when tris-citrate buffer is used for gel preparation and the electrode vessels are filled with borate buffer than when either solution is used in a continuous system. Several investigators have described the separation of isoenzymes with similar discontinuous systems, e.g. alkaline phosphatases (Moss, 1962a, 1962b; Chiandussi *et al.*, 1962), lactate dehydrogenase (Latner and Skillen, 1961) and esterases (Paul and Fottrell, 1961).

The sample is applied in a narrow slit cut transversely in the gel, and markers can be used as described above for agar gel. The gels may be inserted in the Universal apparatus and electrical connection established by means of filter-paper wicks. The lid is then placed in position and a current of 0·5–2·0 mA/cm. is passed through the gels.

The technique for vertical starch-gel electrophoresis also devised by Smithies (1959) has been employed by Baron and Bell (1962) for the separation of the isoenzymes of human serum and tissue isocitrate dehydrogenase. Some aspects of the quantitation of starch-gel electrophoresis have been discussed by Pert and Pinteric (1964).

Polyacrylamide-gel electrophoresis
Since the introduction of polyacrylamide-gel as a medium for electrophoresis (Raymond and Weintraub, 1959; Ornstein and Davis, 1959), it has already found a considerable number of applications in the study of the molecular heterogeneity of enzymes. The gel is prepared by treating a solution of acrylamide containing 5% methylene-*bis*-acrylamide (Eastman Kodak Co.; American Cyanamid Co.)* in an aqueous buffer solution with small amounts of the catalysts, β-dimethylaminopropionitrile and ammonium persulphate (Raymond and Wang, 1960; Raymond and Nakamichi, 1962). Immediately after mixing, the solution is transferred to a vertical electrophoresis apparatus (Raymond, 1962) with the gel compartment supported horizontally. When the gel has set the apparatus is returned to its normal position and filled with buffer solution.

Ornstein and Davis (1959) carried out acrylamide-gel electrophoresis in cylindrical glass or rigid plastic tubes, a process to which they have given the name 'disc electrophoresis'. In this technique a thin 'spacer gel', prepared with a glycine buffer of pH 8·3, is applied above the small-pore gel (pH 9·3) in which electrophoresis is carried out. It is thus a discontinuous system, the main advantage of which is sharper resolution of the resulting protein bands. Details of the method and a discussion of the underlying theory is given by Ornstein (1964).

When 0·1M-tris-(hydroxymethyl)aminomethane buffer at pH 9·2 is used preparations of animal proteins (up to 100 μl. per 2-cm. slot) may be applied at the upper edge of the gel in the Raymond apparatus. When enzyme solutions are used a refrigerant solution at about 0° is passed through the cooling plates on each side of the gel supports and a

---

* According to the manufacturers, acrylamide and its derivatives are potentially neurotoxic. Several cases of poisoning through careless handling have been reported (Garland and Patterson. 1967). but with ordinary laboratory precautions there should be no health hazard from the use of this medium. The polymer. however. is non-toxic.

potential of about 25–30 v./cm. is applied across the gel. Under these conditions the proteins of blood serum are separated over a distance of 10–12 cm. in 1$\frac{1}{2}$–2 hr. The position occupied by albumin can be followed, as with other electrophoretic techniques, by incorporating a trace of bromophenol blue into the sample applied to the gel. The procedure is discussed in greater detail by Raymond (1964).

Polyacrylamide has advantages over other gels in that it is optically clear, and suitably stained bands can be determined by scanning procedures. It can be used over a wide pH range, though the procedure outlined above would need to be modified according to the isoelectric points of the proteins to be studied. Complications due to electro-osmotic flow do not normally occur with this medium.

Polyacrylamide gel has high resolving power and, like starch gel, behaves as a molecular sieve, enabling proteins to be separated according to their molecular sizes as well as to their charges. This feature has been further developed by Raymond and Aurell (1962), who introduced two-dimensional electrophoresis with this medium. After electrophoresis in a 5% acrylide gel, the strip containing the separated proteins may be incorporated into a second gel containing 8% acrylamide. Further electrophoresis with the same buffer solution leads to the production of a pattern of proteins, similar to the patterns obtained with simpler substances during two-dimensional paper chromatography. This procedure appears to offer an interesting new means for the investigation of the properties of proteins, since the only difference between the conditions employed in the two electrophoretic runs resides in the pore size of the gels used. Certain generalizations can thus be made: for example, proteins of equal molecular size lie on a straight line passing through the origin, whereas a series of polymeric proteins fall on a smooth curve asymptotic to the direction of electrophoresis in the 5% gel.

Wieme (1963) devised a horizontal unidimensional system, similar to that used for agar gel (pp. 14–16), in which thin layers of polyacrylamide gel are supported on Perspex bridges between the electrode compartments of a standard horizontal apparatus. As in his agar-gel system, cooling is effected by immersing the whole of the gel in light petroleum (b.p. 30–50°). A similar technique has also been reported by Allred and Keutel (1968).

Unidimensional acrylamide-gel electrophoresis has been applied to the separation of a wide variety of isoenzymes. Among the enzymes so treated are carbonic anhydrase (Jacey, Tappan and Boyden, 1963); catalase (Baumgarten, 1963); lactate dehydrogenase (Goldberg, 1963;

Fritz, 1963; Fritz and Jacobson, 1963; Wieme, 1963; Hemmingsen and Skov, 1968), malate dehydrogenase (Goldberg, 1963; Moore and Villee, 1963; Villee and Moore, 1963; Wieme, 1963), phosphatases (J. M. Allen, 1963a, 1963b; Allen and Gockerman, 1964; Goldberg, Takakura and Rosenthal, 1966; Kaschnitz, 1967), alcohol dehydrogenase (Grell, Jacobson and Murphy, 1965), glucose 6-phosphate dehydrogenase (Boivin, Piguet and Galand, 1966; Criss and McKerns, 1968) and phosphorylases (Fredrick, 1962; Davis, Schliselfeld, Wolf, Leavitt and Krebs, 1967). The lactate dehydrogenase isoenzymes have also been separated by the two-dimensional technique (Raymond and Aurell, 1962), and this procedure has now been applied to the study of normal and pathological human tissues (Johnston *et al.*, 1966; Fujimoto, Nazarian and Wilkinson, 1968).

Column electrophoresis

Electrophoresis in vertical columns packed with cellulose or starch (Flodin and Porath, 1954; Porath, 1954; Flodin and Kupke, 1956) has recently been applied to isoenzyme separation. The technique has been used for the quantitative fractionation of various tissue lactate dehydrogenases (Markert and Appella, 1961) and the separation of the isoenzymes of tissue and body fluid cholinesterases and other esterases (Augustinsson, 1961), the isoenzymes of aspartate aminotransferase (Jungner, 1957; Boyd, 1961), Torula yeast fumarases (Hayman and Alberty, 1961) and ribonucleases (Flodin and Kupke, 1956). The technique is capable of being applied on a relatively large scale, and thus appears to have potentialities in the enzyme field.

The packed column is surrounded by a cylindrical tube, and the annular space between them serves as one electrode compartment. It is filled with buffer, and electrical connection is made by suitably arranged stopcocks which enable the column to be removed when necessary (Porath, 1957). The top of the column is connected by a side arm with the second electrode compartment. The buffer solution is adjusted to the same level in both vessels so as to prevent disturbance of the material in the column.

After electrophoresis the various protein components may be recovered by simple elution using a fraction collector. There is, however, a broadening of the bands as they travel down the column so that the bands at the top tend to be more diffuse than those at the bottom. Consequently, there is some loss of resolution and the top fractions are spread over a greater number of tubes in the fraction collector than the lower fractions. Another disadvantage is that some degree of 'tailing' occurs during elution, and as often happens during protein fractionation, it is difficult to distinguish visually the positions occupied by the various zones (Kunkel and Trautman, 1959). On the other hand, the presence of the buffer solution in the electrode compartment surrounding the column exerts a considerable cooling effect and is a distinct advantage in the separation of isoenzymes.

A technique intermediate between column electrophoresis and the continuous systems described on pp. 13 and 14 is the large-scale continuous flow electrophoresis procedure of Winsten, Friedman and Schwartz (1963), in which siliconized glass beads are used as support phase. This method has now been applied for the separation of serum enzymes (Winsten, Jackson and Wolf, 1966). Almost quantitative recoveries of lactate dehydrogenase, leucine aminopeptidase and phosphohexose isomerase were obtained, but with aldolase and isocitrate dehydrogenase artifactually high recoveries, possibly due in part to a pH effect, were observed.

Isoelectric focusing

Isoelectric focusing is a form of electrophoresis in which a current is applied to a mixture of ampholytes, e.g. enzymes, proteins, dissolved in a suitable medium between electrolytes at different pH values. The various components of the mixture separate according to their isoelectric points. The carrier electrolytes usually consist of a series of amino acids (Svensson, 1961), now commercially available as 'Ampholines' from LKB-Producter AB, Bromma, Stockholm, Sweden. To prevent mixing after focusing, sucrose-gradient columns may be used (Kolin, 1954), but more recently Dale and Latner (1968) have used columns stabilized with acrylamide gel.

This potentially useful technique has been applied to the separation of serum and other proteins, but its use in the study of isoenzymes is so far rather limited (Ahlgren, Ericsson and Vesterberg, 1967; Carlström and Vesterberg, 1967). Dale and Latner (1968), however, reported the presence of several more than the five lactate dehydrogenase isoenzymes usually found in human heart and liver extracts by conventional starch-gel or acrylamide-gel electrophoresis.

Chromatographic techniques

Ion-exchange celluloses
Column-chromatographic procedures have recently been developed for protein fractionation, and as these enable the various fractions to be

separated on a preparative scale, such techniques would appear to have application to the separation of isoenzymes. The introduction of the ion-exchange celluloses, carboxymethylcellulose (CM-cellulose) and diethylaminoethylcellulose (DEAE-cellulose) (Sober and Peterson, 1956, 1958) for protein fractionation has been followed by the use of DEAE-cellulose for the separation of the lactate dehydrogenase isoenzymes (Hess, 1958; Hess and Walter, 1960, 1961*a*, 1961*b*) and the differentiation of the serum leucine aminopeptidases in liver diseases (Dioguardi, Agostoni, Fiorelli, Tittobello, Cirla and Schweizer, 1961).

DEAE-cellulose is a basic ether of cellulose in which the functional groups are $-OCH_2 \cdot CH_2 \cdot N(C_2H_5)_2$. It behaves as a monoacidic base whose apparent pK value is about 8·0–9·0, but the pH of half neutralization varies according to the nature of the electrolyte employed. It acts as an effective exchanger at pH values in the range 6·0–9·0, and adsorbed protein may be eluted by gradually reducing the pH and increasing the ionic strength of the medium.

CM-cellulose, on the other hand, is an acidic adsorbent containing $-OCH_2 \cdot COOH$ groups. Since introduction of this group leads to a progressive increase in aqueous solubility, the practical limit of substitution is less than 0·2 when the material is required for use in chromatography (Jakubovic and Knight, 1960). Although it has been used quite extensively for protein fractionation, it has so far found comparatively little application in enzymology.

The general arrangements for setting up the DEAE- and CM-cellulose columns and for gradient elution have been described by Tombs *et al.* (1961), and a detailed description would not be appropriate here. In applying DEAE-cellulose column chromatography to the separation of human serum and tissue lactate dehydrogenase isoenzymes, Hess and Walter (1960) eluted with a series of phosphate buffers of gradually increasing ionic strength, and a similar procedure was used by Dioguardi *et al.* (1961) for their fractionation of leucine aminopeptidase.

A simplified 'batch' procedure for the differential determination of serum lactate dehydrogenase isoenzymes has been devised by Hess and Walter (1961*a*, 1961*b*), who treat dialysed serum with a DEAE-cellulose suspension in phosphate buffer at pH 6·0. After centrifugation the lactate dehydrogenase remaining in the supernatant is compared with that of similarly diluted untreated serum. The more acidic (anodic) isoenzymes are preferentially adsorbed, while the more basic components remain in the supernatant. The opposite effects are observed when similar experiments are performed with CM-cellulose, for this exchanger preferentially adsorbs the more basic (cathodic) isoenzymes while the more acidic fractions remain in the supernatant (Withycombe and Wilkinson, 1963).

Gradient-elution chromatography on DEAE-cellulose columns has also been used for the separation of the multimolecular forms of rat-tissue phosphatases, glucose 6-phosphate dehydrogenase and 6-phosphogluconate dehydrogenase (Moore and Angeletti, 1961), mouse skeletal muscle creatine kinase (Hooton and Watts, 1966), yeast α-glucosidase (Yau and Lindegren, 1967), rat submandibular peptidase (Ekfors, Riekkinen, Malmiharju and Hopsu-Havu, 1967) and bacterial hyaluronate lyase (Greiling, Stuhlsatz and Eberhard, 1965).

Paper chromatography
DEAE-paper chromatography has been used for the separation of rat-tissue alkaline phosphatase isoenzymes (Hosoyama, Fukuda and Shimadate, 1967).

Hydroxylapatite
Multiple forms of human serum caeruloplasmin have been separated by gradient-elution chromatography on hydroxylapatite columns (Broman, 1958; Richterich, Gautier, Stillhart and Rossi, 1960; Curzon and Vallet, 1960; Hirschman, Morell and Scheinberg, 1960). Human- and pig-heart aspartate transaminases have also been purified by this procedure (Jenkins, Yphantis and Sizer, 1959; Bodansky, Schwartz and Nisselbaum, 1966).

Gel filtration
The technique of gel filtration (Porath and Flodin, 1959) has developed considerably since the introduction of the Sephadex series of modified dextrans (Pharmacia, Uppsala, Sweden). These swell considerably in water to produce three-dimensional matrices which can act as molecular sieves. Small molecules are capable of entering the interstices of the gel, and their passage down a Sephadex column is thereby retarded. Since substances of high molecular weight cannot enter the gel, they remain in the liquid medium surrounding the gel aggregates. A high degree of selectivity may be achieved by choosing a grade of Sephadex appropriate to the molecular size of the materials to be studied. For example, such columns may be used for protein fractionation, including the purification of enzymes (Kisliuk, 1960; Bjork, 1961).

Gel filtration appears to be an extremely promising technique for use in the study of isoenzymes, and several examples of its successful application appear in the literature. For instance, Wilding (1963) used it in the investigation of the apparent heterogeneity of human serum and urinary amylase. Moss (1963) found that human tissue alkaline phosphatase isoenzyme preparations exhibited similar chromatographic properties on a Sephadex column, and from this he concluded that the various isoenzymes had similar molecular weights. More recently, Abramson (1967) has separated partially purified staphylococcal hyaluronate lyase into four enzymically active fractions by means of gel filtration on Sephadex G-100, and a similar technique was used by DiPietro and Zengerle (1967) for the preparation of three distinct acid phosphatases from human placenta.

Sephadex is also available as an anion exchanger, DEAE-Sephadex, and as two cation exchangers, CM-Sephadex and SE-Sephadex. DEAE-Sephadex has been used for the chromatographic separation of lactate dehydrogenase isoenzymes (Wieland and Determann, 1962; Richterich *et al.*, 1962; Richterich and Burger, 1963; Richterich, Schafroth and Aebi, 1963; Wachsmuth and Pfleiderer, 1963), and for the preliminary isolation of an acid phosphatase from a cell-free extract of disintegrated *Escherichia coli* (von Hofsten and Porath, 1962). The phosphatase was found to consist of three electrophoretically distinct components which could be further purified by gel filtration.

DEAE-Sephadex has also been used for the purification and isolation of creatine kinase from ox brain. Enzyme activity was obtained in three peaks, two of which appeared to be interconvertible forms of the pure enzyme, while in the third the enzyme was associated with other proteins (Wood, 1963).

CM-Sephadex has recently found application for the purification of the Negro type variant A^+ of glucose 6-phosphate dehydrogenase (Yoshida, 1967).

REFERENCES

ABRAMSON, C. (1967), *Arch. Biochem. Biophys.* **121**, 103.

AHLGREN, E., ERICSSON, K. E. and VESTERBERG, O. (1967), *Acta Chem. Scand.* **21**, 937.

AHMED, Z., ABUL-FADL, M. A. M. and KING, E. J. (1959), *Biochim. biophys. Acta* **36**, 228.

ALLEN, J. M. (1961), *Ann. N.Y. Acad. Sci.* **94**, 937.

ALLEN, J. M. (1963a), *J. Histochem. Cytochem.* **11**, 167.

ALLEN, J. M. (1963b), *J. Histochem. Cytochem.* **11**, 542.

ALLEN, J. M. and GOCKERMAN, J. (1964), *Ann. N.Y. Acad. Sci.* **121**, 616.

ALLEN, S. L. (1961), *Ann. N.Y. Acad. Sci.* **94**, 753.

ALLEN, S. L., ALLEN, J. M. and LICHT, B. M. (1965), *J. Histochem. Cytochem.* **13**, 434.
ALLRED, R. J. and KEUTEL, H. J. (1968), *J. Lab. clin. Med.* **71**, 179.
AUGUSTINSSON, K. B. (1961), *Ann. N.Y. Acad. Sci.* **94**, 844.
BAKER, R. W. R. and PELLEGRINO, C. (1954), *Scand. J. clin. lab. Invest.* **6**, 94.
BARNETT, H. (1962), *Biochem. J.* **84**, 83P.
BARNETT, H. (1964), *J. clin. Path.* **17**, 567.
BARON, D. N. and BELL, J. L. (1962), *Proc. Ass. clin. Biochem.* **2**, 8.
BAUMGARTEN, A. (1963), *Blood* **22**, 466.
BECKMAN, L. and BECKMAN, G. (1967), *Biochem. Genet.* **1**, 145.
BELL, J. L. and BARON, D. N. (1962), *Biochem. J.* **82**, 5P.
BEUTLER, E. (1967), *Science*, **156**, 1516.
BEUTLER, E. and COLLINS, Z. (1966), *Experientia* **22**, 827.
BIER, M. (1959), *Electrophoresis*, New York: Academic Press.
BJORK, W. (1961), *Biochim. biophys. Acta* **49**, 195.
BLANCHAER, M. C. (1961a), *Clin. chim. Acta* **6**, 272.
BLANCHAER, M. C. (1961b), *Pure & appl. Chem.* **3**, 403.
BODANSKY, O., SCHWARTZ, M. K. and NISSELBAUM, J. S. (1966), *Advances in Enzyme Regulation*, ed. WEBER, G., Oxford: Pergamon Press, p. 299.
BODMAN, J. (1960), in *Chromatographic and Electrophoretic Techniques*. Vol. II. *Zone Electrophoresis*, ed. SMITH, I., London: Heinemann, p. 91.
BOIVIN, P., PIGUET, H. and GALAND, C. (1966), *Nouv. Rev. franc. Hematol.* **6**, 769.
BONAVITA, V. and GUARNERI, R. (1962), *Biochim. biophys. Acta* **59**, 634.
BONAVITA, V., PONTE, F. and AMORE, G. (1962), *Nature, Lond.* **196**, 576.
BOWMAN, J. E., CARSON, P. E., FRISCHER, H. and GARAY, A. L. DE (1966), *Nature, Lond.* **210**, 811.
BOYD, J. W. (1961), *Biochem. J.* **81**, 434.
BOYD, J. W. (1962), *Clin. chim. Acta* **7**, 424.
BOYDE, T. R. C. and LATNER, A. L. (1961), *Biochem. J.* **82**, 51P.
BOYER, S. H. (1961), *Science*, **134**, 1002.
BOYER, S. H., PORTER, I. H. and WEILBACHER, R. G. (1962), *Proc. Natl. Acad. Sci., Wash.* **48**, 1868.
BRODY, I. A. (1964), *Nature, Lond.* **204**, 685.
BROMAN, L. (1958), *Nature, Lond.* **182**, 1655.
BROWNLOW, E. K. and GAMMACK, D. B. (1967), *Biochem. J.* **103**, 47P.
BURGER, A., RICHTERICH, R. and AEBI, H. (1964), *Biochem. Z.* **339**, 305.
CAMPBELL, D. M. and MOSS, D. W. (1962), *Proc. Ass. clin. Biochem.* **2**, 10.
CARLSTRÖM, A. and VESTERBERG, O. (1967), *Acta Chem. Scand.* **21**, 271.
CARSON, P. E., AJMAR, F., HASHIMOTO, F. and BOWMAN, J. E. (1966), *Nature, Lond.* **210**, 813.
CARTER, N. D. and PARR, C. W. (1967), *Nature, Lond.* **216**, 511.
CHIANDUSSI, L., GREENE, S. F. and SHERLOCK, S. (1962), *Clin. Sci.* **22**, 425.
CHU, Y. E. (1967), *Jap. J. Genet.* **42**, 233.
CONKLIN, J. L. and NEBEL, E. J. (1965), *J. Histochem. Cytochem.* **13**, 510.
CRISS, W. E. and MCKERNS, K. W. (1968), *Biochemistry* **7**, 125.
CUATRECASAS, P. and SEGAL, S. (1967), *Science*, **156**, 1518.
CURZON, G. and VALLET, L. (1960), *Biochem. J.* **74**, 279.
DALE, G. and LATNER, A. L. (1968), *Lancet* i, 847.
DAVIDSON, R. G. and CORTNER, J. A. (1967), *Science*, **157**, 1569.

DAVIS, C. H., SCHLISELFELD, L. H., WOLF, D. P., LEAVITT, C. A. and KREBS, E. G. (1967), *J. biol. Chem.* **242**, 4824.

DAWSON, D. M., EPPENBERGER, H. M. and KAPLAN, N. O. (1967), *J. biol. Chem.* **242**, 210.

DEUL, D. H. and VAN BREEMEN, J. F. L. (1964), *Clin. chim. Acta* **10**, 276.

DEWEY, M. M. and CONKLIN, J. L. (1960), *Proc. Soc. exp. Biol., N.Y.* **105**, 492.

DIOGUARDI, N., AGOSTINI, A. and FIORELLI, G. (1964), *Protides of the Biological Fluids*, ed. PEETERS, H., Amsterdam: Elsevier. vol. 12, p. 278.

DIOGUARDI, N., AGOSTONI, A., FIORELLI, G., TITTOBELLO, A., CIRLA, E. and SCHWEIZER, M. (1961), *Enzymol. biol. clin.* **1**, 204.

DIPIETRO, D. L. and ZENGERLE, F. S. (1967), *J. biol. Chem.* **242**, 3391.

DURRUM, E. L. (1958), in *A Manual of Paper Chromatography and Paper Electrophoresis*, by BLOCK, R. J., DURRUM, E. L. and ZWEIG, G., New York: Academic Press, p. 487.

EATON, G. M., BREWER, G. J. and TASHIAN, R. E. (1966), *Nature, Lond.* **212**, 944.

EKFORS, T. O., RIEKKINEN, P. J., MALMIHARJU, T. and HOPSU-HAVU, V. K. (1967), *Z. physiol. Chem.* **348**, 111.

ENDO, T. (1967), *Radiat. Botany* **7**, 35.

ESTBORN, B. (1959), *Nature, Lond.* **184**, 1636.

ESTBORN, B. and SWEDIN, B. (1959), *Scand. J. clin. lab. Invest.* **11**, 235.

FILDES, R. A. and PARR, C. W. (1963), *Nature, Lond.* **200**, 890.

FISHMAN, W. H., INGLIS, N. I. and GHOSH, N. K. (1968), *Clin. chim. Acta* **19**, 71.

FISHMAN, W. H., INGLIS, N. I., STOLBACH, L. L. and KRANT, M. J. (1968), *Cancer Res.* **28**, 150.

FLEISHER, G. A., POTTER, C. S. and WAKIM, K. G. (1960), *Proc. Soc. exp. Biol. N.Y.* **103**, 229.

FLODIN, P. and KUPKE, D. W. (1956), *Biochim. biophys. Acta* **21**, 368.

FLODIN, P. and PORATH, J. (1954), *Biochim. biophys. Acta* **13**, 175.

FRANZINI, C. (1965), *J. clin. Path.* **18**, 664.

FREDRICK, J. F. (1962), *Phytochem.* **1**, 153.

FRITZ, P. J. (1963), *Federation Proc.* **22**, 485.

FRITZ, P. J. and JACOBSON, K. B. (1963), *Science*, **140**, 64.

FRÖHLICH, C. (1965), *Z. klin. Chem.* **3**, 137.

FUJIMOTO, Y., NAZARIAN, I. and WILKINSON, J. H. (1968), *Enzymol. biol. Clin.* **9**, 124.

GALSTON, A. W. and MCCUNE, D. C. (1961), Proc. 4th Internat. Conf. Growth Regulation, Yonkers, N.Y., ed. KLEIN, R. M., Ames, Iowa: Iowa State Univ. Press, p. 611.

GARAY, A. S. and FARAGÓ, M. (1959), *Phyton* **13**, 55.

GARLAND, T. O. and PATTERSON, M. W. H. (1967), *Br. med. J.* **4**, 134.

GERHARDT, W., CLAUSEN, J., CHRISTENSEN, E. and RIISHEDE, J. (1963), *Acta neurol. Scand.* **39**, 85.

GIBLETT, E. R. and SCOTT, N. M. (1965), *Am. J. Hum. Genet.* **17**, 425.

GOLDBERG, A. F., TAKAKURA, K. and ROSENTHAL, R. L. (1966), *Nature, Lond.* **211**, 41.

GOLDBERG, E. (1963), *Science*, **139**, 602.

GORDON, A. H., KEIL, B. and SEBESTA, K. (1949), *Nature, Lond.* **164**, 498.

GORDON, H., KERAAN, M. M. and VOOIJS, M. (1967), *Nature, Lond.* **214**, 466.

GRASSMANN, W. and HANNIG, K. (1950), *Naturwiss.* **37**, 397.

GREILING, H., STUHLSATZ, H. W. and EBERHARD, T. (1965), *Z. physiol. Chem.* **340**, 243.

GRELL, E. H., JACOBSON, K. B. and MURPHY, J. B. (1965), *Science*, **149**, 80.

GRUNDIG, E., CZITOBER, H. and SCHOBEL, B. (1965), *Clin. chim. Acta* **12**, 157.

HAIJE, W. G. and DE JONG, M. (1963), *Clin. chim. Acta* **8**, 620.

HANNIG, K. (1961), *Z. anal. Chem.* **181**, 244.

HARRIS, H., HOPKINSON, D. A. and ROBSON, E. B. (1962), *Nature, Lond.* **196**, 1296.

HAUGAARD, G. and KRONER, T. D. (1948), *J. Amer. chem. Soc.* **70**, 2135.

HAWKINS, D. F. and WHYLEY, G. A. (1962), *Lancet* i, 1126.

HAYMAN, S. and ALBERTY, R. A. (1961), *Ann. N.Y. Acad. Sci.* **94**, 812.

HAYWOOD, B. J., STARKWEATHER, W. H., SPENCER, H. H. and ZARAFONETIS, C. J. (1968), *J. Lab. clin. Med.* **71**, 324.

HEMMINGSEN, L. and SKOV, F. (1968), *Clin. Chim. Acta* **19**, 81.

HESS, A. R., ANGEL, R. W., BARRON, K. D. and BERNSOHN, J. (1963), *Clin. chim. Acta* **8**, 656.

HESS, B. (1958), *Ann. N.Y. Acad. Sc* 292.

HESS, B. and WALTER, S. I. (1960), *Klin. Wschr.* **38**, 1080.

HESS, B. and WALTER, S. I. (1961*a*), *Klin. Wschr.* **39**, 213.

HESS, B. and WALTER, S. I. (1961*b*), *Ann. N.Y. Acad. Sci.* **94**, 890.

HILL, B. R. (1958), *Ann. N.Y. Acad. Sci.* **75**, 292.

HILL, B. R. and MEACHAM, E. J. (1961), *Ann. N.Y. Acad. Sci.* **94**, 868.

HIRSCHMAN, S. Z., MORELL, A. G. and SCHEINBERG, I. H. (1961), *Ann. N.Y. Acad. Sci.* **94**, 960.

HOBART, M. H. and ROSE, C. F. M. (1955), *J. clin. Path.* **8**, 338.

HODSON, A. W., LATNER, A. L. and RAINE, L. (1962), *Clin. chim. Acta* **7**, 255.

HOOTON, B. T. and WATTS, D. C. (1966), *Biochem. J.* **99**, 53P.

HOSOYAMA, Y., FUKUDA, Y. and SHIMADATE, T. (1967), *Clin. chim. Acta* **18**, 141.

HUNTER, R. L. and MARKERT, C. L. (1957), *Science*, **125**, 1294.

HUNTER, R. L. and STRACHAN, D. S. (1961), *Ann. N.Y. Acad. Sci.* **94**, 861.

JACEY, M. J., TAPPAN, D. V. and BOYDEN, H. M. (1963), *Federation Proc.* **22**, 2845.

JACOB, S. H., HELDT, H. W. and KLINGENBURG, M. (1964), *Biochem. biophys. Res. Commun.* **16**, 516.

JAKUBOVIC, A. O. and KNIGHT, C. (1960), in *Chromatographic and Electrophoretic Techniques.* Vol. I. *Chromatography*, ed. SMITH, I., London: Heinemann, p. 557.

JENKINS, W. T., YPHANTIS, D. A. and SIZER, I. W. (1959), *J. biol. Chem.* **234**, 51.

JERMYN, M. A. and THOMAS, R. (1954), *Biochem. J.* **56**, 631.

JOHNSTON, H. A., WILKINSON, J. H., WITHYCOMBE, W. A. and RAYMOND, S. (1966), *J. clin. Path.* **19**, 250.

JUNGNER, G. (1957), *Scand. J. clin. lab. Invest.* **10**, Suppl. 31. 280.

KAJI, A., TRAYSER, K. A. and COLOWICK, S. P. (1961), *Ann. N.Y. Acad. Sci.* **94**, 798.

KAPLAN, N. O. and CAHN, R. D. (1962), *Proc. Natl. Acad. Sci., Wash.* **48**, 2123.

KAR, N. C. and PEARSON, C. M. (1965), *Am. J. clin. Path.* **43**, 207.

KECK, K. (1961), *Ann. N.Y. Acad. Sci.* **94**, 741.

KEIDING, N. R. (1959), *Scand. J. clin. lab. Invest.* **11**, 106.

KELLY, S., BELORIT, A. and COPELAND, W. (1967), *Clin. chim. Acta* **18**, 483.

KISLIUK, R. L. (1960), *Biochim. biophys. Acta* **40**, 531.

KITCHENER, P. N., NEALE, F. C., POSEN, S. and BRUDENELL-WOODS, J. (1965), *Am. J. clin. Path.* **44**, 654.

KOEN, A. L. (1967), *Biochim. biophys. Acta* **12**, 80.

KOHN, J. (1957a), Biochem. J. 65, 9.
KOHN, J. (1957b), Clin. chim. Acta 2, 297.
KOHN, J. (1958a), Clin. chim. Acta 3, 450.
KOHN, J. (1958b), Nature, Lond. 181, 839.
KOHN, J. (1960), in Chromatographic and Electrophoretic Techniques. Vol. II. Zone Electrophoresis, ed. SMITH, I., London: Heinemann, p. 56.
KOLIN, A. (1954), J. chem. Phys. 22, 1628.
KOWLESSAR, O. D., HAEFFNER, L. J. and RILEY, E. M. (1961), Ann. N.Y. Acad. Sci. 94, 836.
KOWLESSAR, O. D., HAEFFNER, L. J. and SLEISENGER, M. H. (1960), J. clin. Invest. 39, 671.
KREUTZER, H. H. and EGGELS, P. H. (1965), Clin. chim. Acta 12, 80.
KREUTZER, H. H. and FENNIS, W. H. S. (1964), Clin. chim. Acta 9, 64.
KREUTZER, H. H. and JACOBS, P. (1965), Clin. chim. Acta 11, 184.
KRIEG, A. F., ROSENBLUM, L. J. and HENRY, J. B. (1967), Clin. Chem. 13, 196.
KUNKEL, H. G. and SLATER, R. J. (1952), Proc. Soc. exp. Biol. N.Y. 80, 42.
KUNKEL, H. G. and TISELIUS, A. (1951), J. gen. Physiol. 35, 89.
KUNKEL, H. G. and TRAUTMAN, R. (1959), in Electrophoresis, ed. BIER, M., New York: Academic Press, p. 255.
LAI, L. Y. C. and KWA, S. B. (1968), Acta Genet., Basel 18, 45.
LATNER, A. L. (1961), Proc. 9th Colloq. Protides of Biol. Fluids, ed. PEETERS, H., Amsterdam: Elsevier.
LATNER, A. L. and SKILLEN, A. W. (1961), Lancet ii, 1286.
LATNER, A. L. and SKILLEN, A. W. (1962), Proc. Ass. clin. Biochem. 2, 3.
LAUFER, H. (1960), Ann. N.Y. Acad. Sci. 89, 490.
LAUFER, H. (1961), Ann. N.Y. Acad. Sci. 94, 825.
LAURSEN, T. (1962), Scand. J. clin. lab. Invest. 14, 152.
LAWRENCE, S. H., MELNICK, P. J. and WEIMER, H. E. (1960), Proc. Soc. exp. Biol. N.Y. 105, 572.
LEBHERZ, H. G. and RUTTER, W. J. (1967), Science, 157, 1198.
LINDER, D. and GARTLER, S. M. (1965), Science, 150, 67.
LOWENTHAL, A., VAN SANDE, M. and KARCHER, D. (1961), Ann. N.Y. Acad. Sci. 94, 988.
MANWELL, C. (1966), Comp. Biochem. Physiol. 17, 805.
MARKERT, C. L. and APPELLA, E. (1961), Ann. N.Y. Acad. Sci. 94, 678.
MARKERT, C. L. and MØLLER, F. (1959), Proc. Natl. Acad. Sci., Wash. 45, 753.
MEADE, B. W. and ROSALKI, S. B. (1963), J. Obst. Gyn. Brit. Cwlth. 70, 862.
MEADE, B. W. and ROSALKI, S. B. (1964), J. clin. Path. 17, 61.
MOORE, B. W. and ANGELETTI, P. U. (1961), Ann. N.Y. Acad. Sci. 94, 659.
MOORE, R. O. and VILLEE, C. A. (1963), Science, 142, 389.
MORELAND, B. and WATTS, D. C. (1967) Nature, Lond. 215, 1092.
MORELL, A. G. and SCHEINBERG, I. H. (1960), Science, 131, 930.
MOSS, D. W. (1962a), Nature, Lond. 193, 981.
MOSS, D. W. (1962b), Proc. Ass. clin. Biochem. 2, 5.
MOSS, D. W. (1963), Nature, Lond. 200, 1206.
MOSS, D. W., CAMPBELL, D. M., ANAGNOSTOU-KAKARAS, E. and KING, E. J. (1961a), Biochem. J. 81, 441.
MOSS, D. W., CAMPBELL, D. M., ANAGNOSTOU-KAKARAS, E. and KING, E. J. (1961b), Pure & appl. Chem. 3, 397.

MOUSTAFA, E. (1963), *Nature, Lond.* **199**, 1189.

MURPHEY, W. H., BARNABY. C., LIN, F. J. and KAPLAN, N. O. (1967), *J. biol. Chem.* **242**, 1548.

NAKAGAWA, Y. and NOLTMAN, E. A. (1967), *J. biol. Chem.* **242**, 4782.

OORT, J. and WILLIGHAGEN, R. G. J. (1961), *Nature, Lond.* **190**, 642.

ORNSTEIN, L. (1964), *Ann. N.Y. Acad. Sci.* **121**, 321.

ORNSTEIN, L. and DAVIS, B. J. (1959), *Disc Electrophoresis*, Distillation Products Industries (Division of Eastman Kodak Co.).

PAPADOPOULOS, N. M. and KINTZIOS, J. A. (1967), *Am. J. clin. Path.* **47**, 96.

PAPENBERG, J., VON WARTBURG, J. P. and AEBI, H. (1965), *Biochem. Z.* **342**, 95.

PARR, C. W. (1966), *Nature, Lond.* **210**, 487.

PARR, C. W. and FITCH, L. I. (1967), *Am. hum. Genet.* **30**, 339.

PAUL, J. and FOTTRELL, P. F. (1961), *Ann. N.Y. Acad. Sci.* **94**, 668.

PERT, J. H. and PINTERIC, L. (1964), *Ann. N.Y. Acad. Sci.* **121**, 310.

PFLEIDERER, G. and WACHSMUTH, E. D. (1961), *Biochem. Z.* **334**, 185.

PLA, G. W., PAPADOPOULOS, N. M. and ROSEN, S. (1968), *Enzymologia* **34**, 40.

PLAGEMANN, P. W. G., GREGORY, K. F. and WRÓBLEWSKI, F. (1960), *J. biol. Chem.* **235**, 2282.

PLUMMER, D. T., ELLIOTT, B. A., COOKE, K. B. and WILKINSON, J. H. (1963), *Biochem. J.* **87**, 416.

PORATH, J. (1954), *Acta chem. Scand.* **8**, 1813.

PORATH, J. (1957), *Arkiv. Kemi* **11**, 161.

PORATH, J. and FLODIN, P. (1959), *Nature, Lond.* **183**, 1657.

PORTER, I. H. (1968), *Ann. int. Med.* **68**, 251.

POSEN, S., NEALE, F. C., BIRKETT, D. J. and BRUDENELL-WOODS, J. (1967), *Am. J. clin. Path.* **48**, 81.

POULIK, M. D. (1957), *Nature, Lond.* **180**, 1477.

POULIK, M. D. and BEARN, A. G. (1962), *Clin. chim. Acta* **7**, 374.

POZNANSKA, H. and GORE, M. (1965), *Bull. Acad. Polon. Sci.* **13**, 629.

PRESTON, J. A., BRIERE, R. O. and BATSAKIS, J. G. (1965), *Am. J. clin. Path.* **43**, 256.

PRYSE-DAVIES, J. and WILKINSON, J. H. (1958), *Lancet* **i**, 1249.

RATTAZZI, M. C., BERNINI, L. F., FIORELLI, G. and MANNUCCI, P. M. (1967), *Nature, Lond.* **213**, 79.

RAYMOND, S. (1962), *Clin. Chem.* **8**, 455.

RAYMOND, S. (1964), *Ann. N.Y. Acad. Sci.* **121**, 350.

RAYMOND, S. and AURELL, B. (1962), *Science*, **138**, 152.

RAYMOND, S. and NAKAMICHI, M. (1962), *Anal. Biochem.* **3**, 23.

RAYMOND, S. and WANG, Y.-J. (1960), *Anal. Biochem.* **1**, 391.

RAYMOND, S. and WEINTRAUB, L. (1959), *Science*, **130**, 711.

RESSLER, N. and JOSEPH, R. R. (1962), *J. lab. clin. Med.* **60**, 349.

RESSLER, N. and MOY, T. (1959), *Clin. chim. Acta* **4**, 901.

RESSLER, N., SCHULZ, J. and JOSEPH, R. R. (1962), *Nature, Lond.* **197**, 872.

RESSLER, N., SCHULZ, J. and JOSEPH, R. R. (1963), *J. lab. clin. Med.* **62**, 571.

RICHTERICH, R. and BURGER, A. (1963), *Helv. physiol. Acta* **21**, 59.

RICHTERICH, R., GAUTIER, E., STILLHART, H. and ROSSI, E. (1960), *Helv. Paed. Acta* **5**, 424.

RICHTERICH, R., LOCKER, J., ZUPPINGER, K. and ROSSI, E. (1962), *Schweiz. med. Wschr.* **92**, 919.

RICHTERICH, R., SCHAFROTH, P. and AEBI, H. (1963), *Clin. chim. Acta* **8**, 178.

38 · Isoenzymes

ROSALKI, S. B. (1965), *Nature, Lond.* **207**, 414.

ROSALKI, S. B. and MONTGOMERY, A. (1967), *Clin. chim. Acta* **16**, 440.

ROSALKI, S. B. and WILKINSON, J. H. (1960), *Nature, Lond.* **188**, 1110.

ROSENBERG, I. N. (1959), *J. clin. Invest.* **38**, 630.

SAYRE, F. W. and HILL, B. R. (1957), *Proc. Soc. exp. Biol. N.Y.* **96**, 695.

SCANDALIOS, J. G. (1967), *Biochem. Genet.* **1**, 1.

SCHNEIDERMAN, H. (1967), *Nature, Lond.* **216**, 604.

SEARCY, R. L., UJIHIRA, I., HAYASHI, S. and BERK, J. E. (1964), *Clin. chim. Acta* **9**, 505.

SECCHI, G. C., PETRELLA, A., LOMANTO, B. and GERVASINI, N. (1967), *Enzymol. biol. clin.* **8**, 33.

SHAW, C. R. and KOEN, A. L. (1965), *J. Histochem. Cytochem.* **13**, 431.

SHAW, C. R. and KOEN, A. L. (1967), *Science*, **156**, 1517.

SHOWS, T. B. (1967), *Biochem. Genet.* **1**, 171.

SHREFFLER, D. C., BREWER, G. J., GALL, J. C. and HONEYMAN, M. S. (1967), *Biochem. Genet.* **1**, 101.

SJÖVALL, K. and VOIGT, A. (1964), *Nature, Lond.* **202**, 701.

SMITH, E. E. and RUTENBURG, A. M. (1963), *Nature, Lond.* **197**, 800.

SMITH, I. (1968), *Chromatographic and Electrophoretic Techniques.* Vol. II. *Zone Electrophoresis,* 2nd edit. London: Heinemann.

SMITHIES, O. (1955), *Biochem. J.* **61**, 629.

SMITHIES, O. (1959), *Biochem. J.* **71**, 585.

SOBER, H. A. and PETERSON, E. A. (1956), *J. Amer. chem. Soc.* **78**, 751.

SOBER, H. A. and PETERSON, E. A. (1958), *Federation Proc.* **17**, 1116.

STARKWEATHER, W. H., SPENCER, H. H., SCHWARZ, E. L. and SCHOCH, H. K. (1966), *J. Lab. clin. Med.* **67**, 329.

STAMATOYANNOPOULOS, G., YOSHIDA, A., BACAPOULOS, C. and MOTULSKY, A. G. (1967), *Science*, **157**, 831.

STERN, K. G. (1956), in *Physical Techniques in Biological Research,* ed. OSTER, G. and POLLISTER, A. W., New York: Academic Press, p. 243.

SUR, B. K., MOSS, D. W. and KING, E. J. (1962), *Proc. Ass. clin. Biochem.* **2**, 11.

SVENSSON, H. (1961), *Acta Chem. Scand.* **15**, 325.

THORNE, C. J. R., GROSSMAN, L. I. and KAPLAN, N. O. (1963), *Biochim. biophys. Acta* **73**, 193.

TOMBS, M. P., COOKE, K. B., BURSTON, D. and MACLAGAN, N. F. (1961), *Biochem. J.* **80**, 284.

TRUJILLO, J. M., WALDEN, B., O'NEIL, P. and ANSTALL, H. B. (1965), *Science*, **148**, 1630.

TSAO, M. U. (1960), *Arch. Biochem.* **90**, 234.

VAN DER HELM, H. J. (1961), *Lancet* **ii**, 108.

VAN DER HELM, H. J. (1962), *Clin. chim. Acta* **7**, 125.

VAN DER HELM, H. J., ZONDAG, H. A., HARTOG, H. A. P. and VAN DER KOOI, M. W. (1962), *Clin. chim. Acta* **7**, 540.

VESELL, E. S. (1962), *Nature, Lond.* **195**, 497.

VESELL, E. S. and BEARN, A. G. (1957), *Proc. Soc. exp. Biol. N.Y.* **94**, 96.

VESELL, E. S. and BEARN, A. G. (1962), *J. gen. Physiol.* **45**, 553.

VILLEE, C. A. and MOORE, R. O. (1963), *Biol. Bull.* **125**, 395.

VON HOFSTEN, B. and PORATH, J. (1962), *Biochem. biophys. Acta* **64**, 1.

VUYLSTEEK, K. and WIEME, R. J. (1958), *Belg. Tijds. Geneesk.* **14**, 750.

WACHSMUTH, E. D. and PFLEIDERER, G. (1963), *Biochem. Z.* **336**, 556.

WARBURTON, F. G. and WADDECAR, J. (1966), *J. clin. Path.* **19**, 517.

WATTS, D. C., DONNINGER, C. and WHITEHEAD, E. P. (1961), *Biochem. J.* **81**, 4P.

WIELAND, T. and DETERMANN, H. (1962), *Experientia* **18**, 431.

WIELAND, T. and PFLEIDERER, G. (1957), *Biochem. Z.* **329**, 112.

WIELAND, T., PFLEIDERER, G. and ORTANDERL, F. (1959), *Biochem. Z.* **331**, 103.

WIEME, R. J. (1959a), *Studies on Agar-gel Electrophoresis*, Brussels: Arscia.

WIEME, R. J. (1959b), *Clin. chim. Acta* **4**, 317.

WIEME, R. J. (1963), in *Protides of the Biological Fluids*, Vol. 10, ed. PEETERS, H., Amsterdam: Elsevier, p. 309.

WIEME, R. J. (1964), *Ann. N.Y. Acad. Sci.* **121**, 365.

WIEME, R. J. (1965), *Agar Gel Electrophoresis*, Amsterdam: Elsevier.

WIEME, R. J. and DEMEULENAERE, L. (1959), *Acta gastro-enterol. belg.* **22**, 69.

WIEME, R. J. and VAN MAERCKE, Y. (1961), *Ann. N.Y. Acad. Sci.* **94**, 898.

WILDING, P. (1963), *Clin. chim. Acta* **8**, 918.

WILKINSON, J. H., COOKE, K. B., ELLIOTT, B. A. and PLUMMER, D. T. (1961), *Biochem. J.* **80**, 298.

WINSTEN, S., FRIEDMAN, H. and SCHWARTZ, E. E. (1963), *Anal. Biochem.* **6**, 404.

WINSTEN, S., JACKSON, J. and WOLF, P. (1966), *Clin. Chem.* **12**, 497.

WITHYCOMBE, W. A. and WILKINSON, J. H. (1963), *Biochem. J.* **89**, 48P.

WOOD, T. (1963), *Biochem. J.* **87**, 453.

WRIGHT, E. J., CAWLEY, L. P. and EBERHARDT, L. (1966), *Am. J. clin. Path.* **45**, 737.

WRÓBLEWSKI, F. and GREGORY, K. F. (1961), *Ann. N.Y. Acad. Sci.* **94**, 912.

YAU, T. M. and LINDEGREN, C. C. (1967), *Biochem. Biophys. Res. Commun.* **27**, 305.

YOSHIDA, A. (1967), *Biochem. Genet.* **1**, 81.

ZONDAG, H. A. (1963), *Science*, **142**, 965.

CHAPTER 3

Detection and Determination of Isoenzymes

The choice of a method for the detection and determination of an isoenzyme is governed very largely by the technique used for its separation. Conventional assay methods can be used when the various fractions are eluted from a column or from segments of a starch block, and when it is required to determine activities with the highest degree of precision a method involving elution is undoubtedly to be preferred. Elution of a large number of fractions, however, is a time-consuming business, and when detection or roughly quantitative assessment of the activities of the isoenzymes will suffice one of a number of highly elegant staining techniques may be used. These were originally developed for the cytochemical localization of enzymes (reviewed by Pearse, 1960), but have since found many applications for the detection of isoenzymes separated by micro-electrophoresis on starch or agar gel or on cellulose acetate.

It is obvious that if the activities of the relatively minor components are to be determined with reasonable accuracy methods with high sensitivity must be used, and in some cases new techniques have been specially devised for use with isoenzymes. An example of this is the fluorimetric procedure for alkaline phosphatase using β-naphthyl phosphate as substrate, which enabled Michaelis constants to be determined on the minute fractions separated by starch-gel electrophoresis (Moss, Anagnostou-Kakaras, Campbell and King, 1961).

Elution techniques

When enzymic activity is to be determined on fractions eluted from paper, starch blocks, ion-exchange cellulose columns or Sephadex columns certain precautions must be taken to avoid loss of activity. For example, the temperature should be kept in the range $0-4°$, and excessive dilution should be avoided, since this might lead to adsorption on glass surfaces of disproportionate amounts of enzyme protein as well as to inactivation. Practical details of the methods of determination are given in such standard texts as *Methods in Enzymology* (Colowick and Kaplan, 1955–70) *Methods in Enzymatic Analysis* (Bergmeyer, 1970),

Practical Clinical Enzymology (King, 1965) and in the references cited. In this chapter therefore it is not necessary to give more than an outline of the procedures used.

NAD*- and NADP*-dependent dehydrogenases
Since the multimolecular forms of lactate dehydrogenase have been studied to a greater extent than those of other enzymes, determination of its activity serves as a good example of the procedures used for NAD- and NADP-dependent dehydrogenases. This enzyme catalyses the reversible oxidation of lactate to pyruvate in the presence of NAD:

$$CH_3 \cdot CHOH \cdot COO^- + NAD \rightleftharpoons CH_3 \cdot CO \cdot COO^- + NADH_2$$

Its activity may be determined spectrophotometrically with either lactate or pyruvate as substrate, provided that the appropriate oxidized or reduced form of the coenzyme is incorporated into the reaction mixture. The most convenient procedure is that of Kubowitz and Ott (1943) modified by Wróblewski and LaDue (1955), in which pyruvate serves as substrate and $NADH_2$ as coenzyme in 0·067M-phosphate buffer at pH 7·4. The rate of oxidation of $NADH_2$ at 25° is followed by measuring the fall in absorbance at 340 nm at regular time intervals of 15–60 sec. A straight-line relationship indicates that the reaction is of zero order, and from the slope may be calculated the enzyme activity. This is usually expressed in International units, i.e. μmoles substrate reacted per minute (*Enzyme Nomenclature Recommendations 1964 of the International Union of Biochemistry* (1965) Amsterdam: Elsevier).

Certain precautions should be taken in carrying out this procedure; thus, before starting the reaction by the addition of substrate it is necessary to pre-incubate the enzyme solution with the coenzyme in order to eliminate any impurities which might utilize $NADH_2$. The reaction is inhibited by excess pyruvate, and it is important not to exceed greatly the optimal substrate concentration. Moreover, preparations containing different isoenzymes may exhibit different temperature coefficients, and it is therefore desirable to avoid temperature corrections by working within a narrow temperature range (e.g. $\pm 0·5°$) (Plummer and Wilkinson, 1963). Until recently it was standard practice to measure enzyme activities at 25°, but it is now officially recommended that whenever possible activities should be determined at 30°.

---

* NAD = nicotinamide adenine dinucleotide formerly known as diphosphopyridine nucleotide (DPN). NADP = nicotinamide-adenine dinucleotide phosphate, formerly known as triphosphopyridine nucleotide (TPN).

The method outlined has been criticized because solutions of $NADH_2$ develop an inhibitor on storage (Fawcett, Ciotti and Kaplan, 1961; Dalziel, 1961). Although this effect is most marked with alcohol dehydrogenase, lactate dehydrogenase and possibly other oxidoreductases are also inhibited, and it is therefore recommended that solutions of $NADH_2$ should be freshly prepared daily (Plummer, Elliott, Cooke and Wilkinson, 1963).

The alternative technique involves the oxidation of lactate with NAD as coenzyme. It is carried out in buffer at pH 9–10, and the increase in the extinction at 340 nm is measured (Von Euler, Adler and Hellström, 1936; Wacker, Ulmer and Vallee, 1956). The reaction is powerfully inhibited by pyruvate, and the rate ceases to be of zero order in a very short time, but the inhibitory effect is less marked when a large excess of the coenzyme is used (Wacker and Dorfman, 1962). The increase in absorbance with this procedure is appreciably slower than the fall in absorbance when pyruvate is used as substrate. Consequently, the pyruvate technique is more sensitive, and would therefore appear to be more suitable for the determination of the activities of minor isoenzymes.

Colorimetric methods involving the formation of pyruvate 2,4-dinitrophenylhydrazone are generally not sufficiently sensitive for determination of the minor lactate dehydrogenase isoenzymes, but it is possible that a colorimetric procedure based upon the use of tetrazolium salts might be suitable for these (Nachlas, Margulies, Goldberg and Seligman, 1960; Babson and Phillips, 1965).

Spectrophotometric techniques similar to those described for lactate dehydrogenase may be employed with 2-oxobutyrate or 2-hydroxybutyrate as substrate in order to measure '2-hydroxybutyrate dehydrogenase' activity (Rosalki and Wilkinson, 1960; Elliott and Wilkinson, 1961). The use of the keto acid is strongly recommended, as a concentration of the hydroxy acid of about 4M is needed to obtain maximum reaction rates, and even then the formation of $NADH_2$ is inconveniently slow (Plummer *et al.*, 1963).

Other NAD- and NADP-dependent isoenzymes may be determined by similar procedures, and some of the more important are listed in Table 3.

Normally, isoenzymes can be eluted from non-adsorptive materials, such as starch grain, by displacement, but a modification which appears to be of practical value when separation on a preparative scale is required was introduced by Wieland, Pfleiderer and Rettig (1958). These authors made a paper print of the starch-block separation of lactate dehydrogenase isoenzymes in order to ascertain their positions in the block. Subsequently, the enzyme proteins were transferred by further electrophoresis to glass powder, from which they were readily eluted. Lactate dehydrogenase isoenzymes have also been eluted from acrylamide gel by the recently introduced technique of elution-convection electrophoresis (Raymond, 1964). Satisfactory recoveries of the fast-migrating isoenzymes (LD_1 and LD_2) have been obtained in the writer's laboratory, but the relatively immobile LD_5 is difficult to remove from the gel by this technique.

TABLE 3 *NAD- and NADP-dependent isoenzymes determined spectrophotometrically*

Enzyme	Substrate	Coenzyme Activator	Products	References
Malate dehydro genase	Malate	NAD	Oxaloacetate	Siegel and Bing (1956)
6-Phosphogluconate dehydrogenase	6-Phosphogluconate	NADP	Ribulose 5-phosphate + CO_2	Glock and McLean (1953) Wolfson and Williams-Ashman (1957)
Glucose 6-phosphate dehydrogenase	Glucose 6-phosphate	NADP Mg^{2+}	6-Phosphogluconate	Zinkham and Lenhard (1959)
Isocitrate dehydrogenase	Isocitrate	NADP Mn^{2+}	2-Oxoglutarate + CO_2	Glock and McLean (1953) Wolfson and Williams-Ashman (1957)
α-Glycerophosphate dehydrogenase	Dihydroxyacetone phosphate	$NADH_2$	L-α-glycerophosphate	Baranowski (1949)
Alcohol dehydrogenase	Ethanol	NAD	Acetaldehyde	Theorell and Bonnichsen (1951)
Galactose dehydrogenase	Galactose	NAD	Galactonate	Cuatrecasas and Segal (1966a)
Glutamate dehydrogenase	Glutamate	NAD	2-Oxoglutarate + NH_3	Strecker (1955)

Although it is not strictly an elution procedure, it is convenient to refer at this point to the ingenious technique of Plagemann, Gregory and Wróblewski (1960), who, after separating rabbit and human tissue lactate dehydrogenase isoenzymes by starch-gel electrophoresis, dissolved the individual fractions of the gel by treating them with α-amylase. Enzyme activity was then determined in the resulting solution by the method of Wróblewski and LaDue (1955) outlined above. In applying this procedure, however, there is a risk that some samples of amylase may contain traces of trypsin which could hydrolyse the enzyme protein. The method would, of course, not be suitable for the study of phosphatases or any other enzyme whose kinetics might be affected by the relatively high concentrations of maltose produced by the hydrolysis of starch.

The activities of certain other enzymes may be determined by coupling their reactions with secondary reactions catalysed by NAD-dependent dehydrogenases. An example of such an enzyme is aspartate aminotransferase (glutamate-oxaloacetate transaminase), which catalyses the reversible transfer of the amino group from aspartate to 2-oxoglutarate with the formation of glutamate and oxaloacetate (see Chapter 6). In the presence of $NADH_2$ excess of malate dehydrogenase converts oxaloacetate into malate as fast as it is formed. Consequently, the rate of consumption of $NADH_2$, measured spectrophotometrically, provides a convenient index of the aminotransferase activity (Karmen, 1955).

Creatine kinase may also be determined with the aid of an auxiliary enzyme, hexokinase and a coupled indicator reaction (Oliver, 1955; Rosalki, 1965). Creatine kinase catalyses the reaction of creatine phosphate with ADP to form creatine and ATP. The ATP formed then reacts with glucose in the presence of hexokinase to form glucose 6-phosphate. This is the substrate for the indicator enzyme, glucose 6-phosphate dehydrogenase, which is NAD-dependent.

Fluorimetric methods have also been applied in the determination of NAD-dependent dehydrogenases (Laursen and Hansen, 1958) and have now been used in isoenzyme studies, since their sensitivities are appreciably greater than those of spectrophotometric techniques (Laursen, 1962).

Esterases

Elution techniques have found comparatively little application in the study of the isoenzymes of esterases generally, and of the reports that have appeared most relate to phosphatases separated by paper or starch-block electrophoresis (Baker and Pellegrino, 1954; Eisfeld and Koch, 1954; Taleisnik, Paglini and Zeitune, 1955; Wolfson, 1957; Fahey, McCoy and Goulian, 1958). In their studies of the distribution of the serum alkaline phosphatase in bone and liver diseases Rosenberg (1959) and Keiding (1959) eluted the enzyme from sections of a starch block and determined its activity in each separate fraction using *p*-nitrophenyl phosphate and phenyl phosphate as substrate respectively. Cooke and Zilva (1961) developed this elution technique further by determining enzyme activity in an 'AutoAnalyzer'.

Similar enzyme assay processes may also be used for the isoenzymes of acid phosphatase. The activities of the fractions separated by gradient-elution chromatography have been determined using *p*-

nitrophenyl phosphate as substrate in citrate buffer at pH 5·2 (Moore and Angeletti, 1961). This procedure has the merit of simplicity, since the addition of alkali after incubation produces a yellow colour, the intensity of which gives a direct measure of the enzyme activity.

Perhaps the most elegant of all the elution techniques for the separation and determination of alkaline phosphatase isoenzymes so far reported is that of Moss *et al.* (1961). After separating the isoenzymes by starch-gel electrophoresis, the positions of the principal bands are located by taking a print on paper moistened with α-naphthyl phosphate (p. 53). This substrate is preferred to the β-isomer used subsequently, since the fluorescence of α-naphthol is more easily seen than is that of β-naphthol. The enzyme from a narrow strip cut across the gel is then eluted by macerating with bicarbonate–carbonate buffer at pH 10·0 and centrifuging. This procedure is preferred to disrupting the gel by freezing and thawing. The recovery of enzyme activity is about 30%. Determination of phosphatase activity is then made by the spectrofluorimetric technique of Moss (1960*a*, 1960*b*), using β-naphthyl phosphate as substrate. Since the excitation and emission spectra of the liberated β-naphthol and its ester differ considerably in alkaline solution, it is possible to determine traces of β-naphthol in the presence of a large excess of the phosphate using an excitation wavelength of 350 nm and measuring the fluorescence intensity at 420 nm. The authors report a standard deviation of 3% for nine determinations of 0·2 nmole.

Alkaline phosphatase, 5′-nucleotidase and leucine aminopeptidase in starch-gel electropherograms may also be determined simply by cutting the gel into 0·5-cm. sections and placing each into tubes containing the appropriate buffered substrate (Kowlessar, Haeffner and Riley, 1961). For alkaline phosphatase and 5′-nucleotidase these workers used β-glycerophosphate and adenosine 5′-phosphate respectively (Dixon and Purdom, 1954), and for leucine aminopeptidase the substrate was L-leucyl-β-naphthylamide (Goldbarg and Rutenburg, 1958). This technique was used for the determination of human serum leucine aminopeptidase fractionated by paper electrophoresis (Smith, Pineda and Rutenburg, 1962). Strips (0·5 cm.) are cut from the electropherograms and incubated with the substrate, which undergoes hydrolysis to β-naphthylamine. This is diazotized and coupled with N-(1-naphthyl)-ethylenediamine to give a blue azo dye (Bratton and Marshall, 1939).

Elution techniques have also been applied to the determination of esterase activity in various fractions of human serum. After column electrophoresis Augustinsson (1959, 1961) assayed the eluate fractions for activity against a number of esters, and his characterization of the various components was further refined by the use of specific cholin-esterase inhibitors such as di-isopropyl phosphorofluoridate (DFP) and physostigmine (eserine). These procedures have also been used by Wilde and Kekwick (1964), who separated the serum arylesterases by column electrophoresis and by chromatography on DEAE-cellulose. They determined enzyme activity with phenyl acetate as substrate at pH 7·4 by the colorimetric technique of Aldridge (1954). Fractions were incubated for 30 min. at 37° with buffered substrate, and the liberated phenol was determined by coupling with 4-aminoantipyrine. Alternatively, for their kinetic studies, these investigators maintained the unbuffered substrate and esterase fraction at pH 7·4 by continuous titration with 0·2N-NaOH from a micrometer syringe.

Peroxidases
The multiple peroxidases of corn, separated by starch-block electrophoresis, have been eluted and determined spectrophotometrically in phosphate buffer at pH 6·1, with guaiacol as substrate in the presence of hydrogen peroxide. The rate of formation of tetraguaiacol is followed by measuring the extinction at 470 nm (McCune, 1961).

Detection of isoenzymes in electrophoretic media
Several methods have been devised for the detection of isoenzymes separated by electrophoresis without recourse to their elution. Among these is the technique of 'enzymoelectrophoresis' devised by Wieme and his colleagues for the detection of lactate dehydrogenase isoenzymes in agar gel, and as this appears to be capable of being applied to other NAD- and NADP-dependent isoenzymes and also to other gel media, a brief description is included here.

'Enzymoelectrophoresis'
This term has been used by Wieme (1959*a*, 1959*b*) to describe his technique for the detection and determination of lactate dehydrogenase isoenzymes *in situ* after electrophoresis in agar gel. A second agar gel containing the substrate, pyruvate and $NADH_2$ in barbitone buffer of the same pH (8·4) and ionic strength as that used for electrophoresis is prepared on a glass slide identical in size with that used to support the electrophoretic gel. The latter is carefully superimposed on the substrate gel, and the paired slides are immediately transferred to a spectrophotometer modified so that readings of the absorbance at 340 nm can be made at 0·25-mm. intervals. As the enzyme reaction proceeds $NADH_2$ is oxidized and the optical density decreased, hence by repeated scan-

ning of the paired gels the kinetics of the reactions catalysed by the individual isoenzymes can be followed. The differences between the initial scans and those obtained after a given incubation period are proportional to the isoenzyme activities (Wieme and Van Maercke, 1961; Kamarýt and Zázvorka, 1963).

This technique has a number of advantages over the histochemical procedures described below, since spreading of the isoenzyme bands is kept to a minimum and there is no risk of enzyme action being affected by extraneous chemical substances required in staining procedures. Moreover, enzymoelectrophoresis provides quantitative information which is likely to be more reliable than that obtained by scanning a stained gel. On the other hand, this technique requires a specially modified spectrophotometer which may not be readily available in many laboratories.

Tetrazolium-staining techniques
Tetrazolium salts, particularly triphenyltetrazolium chloride, have been used during the past two decades for the colorimetric detection of lactate dehydrogenase (Kun and Abood, 1949), owing to the ease with which they undergo reduction to sparingly soluble, but intensely coloured, formazans:

$$C_6H_5 \cdot C \underset{N-N \cdot C_6H_5}{\overset{\overset{+}{N}=N \cdot C_6H_5}{\Big|}} |Cl|^- + NADH_2 \underset{transferring\ agent}{\overset{Hydrogen}{\rightleftharpoons}} NAD + C_6H_5 \cdot C \underset{N-NH \cdot C_6H_5}{\overset{N=N \cdot C_6H_5}{}} + HCl$$

2,3,5-Triphenyltetrazolium   Triphenylformazan
chloride

More sensitive derivatives have been synthesized, and of these, 2,5-diphenyl-3-(4,5-dimethylthiazolyl-2)-tetrazolium bromide (MTT) has been widely used for the cytochemical localization of dehydrogenases (Nineham, 1955; Pearse, 1960; Seligman, 1963).

$$\begin{array}{c} CH_3 \cdot C - N \\ \parallel \qquad \qquad C - N - N \\ CH_3 \cdot C - S \qquad \qquad C \cdot C_6H_5 \ |Br|^- \\ \qquad C_6H_5 \cdot \overset{+}{N} = N \end{array}$$

2,5-Diphenyl-3-(4,5-dimethylthiazolyl-2)-tetrazolium bromide (MTT)

This compound has the advantage over earlier compounds of this type, since the corresponding formazans are more intensely coloured, and less soluble in lipids, so that they do not diffuse away from their sites of formation so readily. These histochemical techniques have now been successfully applied to the direct detection of the isoenzymes of NAD- and NADP-dependent dehydrogenases in such media as starch gel, agar gel, cellulose acetate and acrylamide gel.

The first successful application of this procedure was that of Markert and Møller (1959), who overlaid a starch-gel lactate dehydrogenase electropherogram with an agar gel containing phosphate buffer, lactate, NAD, hydrazine, diaphorase, methylene blue and a tetrazolium salt.

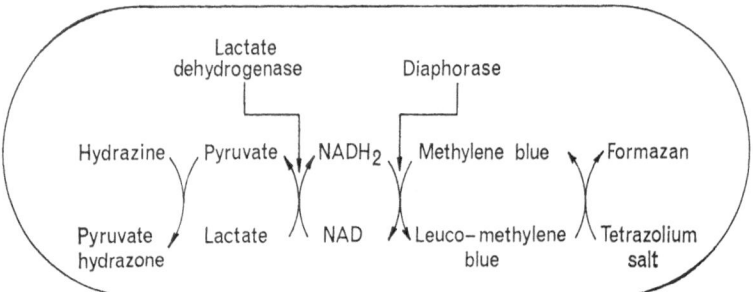

Fig. 4. Reactions involved in the detection of lactate dehydrogenase isoenzymes by the technique of Markert and Møller (1959).

The reactions involved are quite complex (Fig. 4). The pyruvate formed as a result of the enzyme action combines with the hydrazine, and so is removed from the scene, while the $NADH_2$ formed at the same time is oxidized back to NAD by the diaphorase, the hydrogen being transferred by the methylene blue to the tetrazolium salt, which as a result is reduced to the formazan.

$NADH_2$ does not react directly with tetrazolium salts, and some hydrogen-transferring agent is required as an intermediate. A major advance followed the use by Dewey and Conklin (1960), Van der Helm

$$[CH_3 \cdot SO_4]^-$$

Methylphenazonium methosulphate

(1961) and Latner and Skillen (1961) of methylphenazonium methosulphate, which had originally been used by Dickens and McIlwain (1938) as an intermediate electron carrier. This compound catalyses the reaction between $NADH_2$ and the tetrazolium salt, thereby rendering unnecessary the use of diaphorase and methylene blue. Moreover, when methylphenazonium methosulphate is used gels stain satisfactorily when treated with an aqueous solution of the reagents. Methylphenazonium methosulphate, however, is light sensitive and should be stored in the dark and added last to solutions of the remaining reagents.

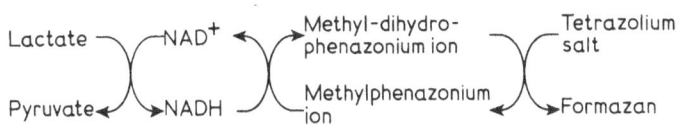

Fig. 5. Reactions involved in the simplified staining procedure for lactate dehydrogenase isoenzymes.

Many investigators have used cyanide instead of hydrazine with the object of trapping pyruvate as fast as it is formed, but this is not necessary, and in its absence the staining solution remains clear for a much longer period.

The reactions involved in the simplified staining procedure are summarized in Fig. 5.

An even more sensitive reagent than MTT is now available; this is the ditetrazolium salt, 2,2′-di-*p*-nitrophenyl-5,5′-diphenyl-3,3′-(3,3′-dimethoxy-4,4′-diphenylene)ditetrazolium chloride, also known as Nitro Blue Tetrazolium (Tsou, Cheng, Nachlas and Seligman, 1956). The corresponding diformazan remains more compactly at the site of its formation than does that of MTT, and in the writer's experience gives

$$N = N \cdot C_6H_4 \cdot NO_2 \qquad NO_2 \cdot C_6H_4 \cdot N = N$$

$$C_6H_5 \cdot C \qquad \qquad C \cdot C_6H_5$$

$$N - N \qquad \qquad N - N$$

$$2|Cl|$$

$$CH_3O \qquad \qquad OCH_3$$

2,2′-Di-*p*-nitrophenyl-5,5′-diphenyl-3,3′-(3,3′-dimethoxy-4,4′-diphenylene)di-tetrazolium chloride (Nitro Blue Tetrazolium) sharper definition of the isoenzyme bands separated by gel electrophoresis. It has been used for this purpose by Markert and Møller (1959), Vesell (1961), Van der Helm (1962), Wieme, Van Sande, Karcher, Löwenthal and Van der Helm (1962), and by Gerhardt, Clausen, Christensen and Riishede (1963). Nitro Blue Tetrazolium is rather troublesome to dissolve, however, but it is effective at much lower concentrations than is MTT.

A suitable mixture which gives satisfactory staining of lactate dehydrogenase isoenzymes in a variety of gels or on cellulose acetate foil is the following: NAD (50 mg.), MTT (15 mg.) or Nitro Blue Tetrazolium (5 mg.) and 0·5M-sodium lactate (20 ml.) in sufficient 0·1M-tris buffer (pH 9·2) to produce 100 ml. This solution is stable for a week or more if kept in a refrigerator. Immediately before use, methylphenazonium methosulphate (approx. 0·5 mg.) should be dissolved in each 20 ml. Staining is best carried out at a temperature about 25–35° away from strong light. A similar solution may be used for other isoenzymes, e.g. isocitrate dehydrogenase, by replacing lactate with the appropriate substrate and, if necessary, the NAD by NADP. Furthermore, by coupling the dehydrogenase reaction with another system this technique can be extended to the study of other isoenzymes such as aspartate aminotransferase (Boyde and Latner, 1961) and creatine kinase (Rosalki, 1965).

By the employment of a suitable scanning technique a roughly quantitative assessment of the relative activities of the various isoenzymes can be obtained (Laycock, Thurman and Boulter, 1965; Preston, Briere and Batsakis, 1965; Papadopoulos and Kintzios, 1967), but it should be borne in mind that variations in the conditions might lead to serious inaccuracies. For example, excessively prolonged incubation might lead to complete utilization of substrate in the vicinity of the major bands of activity, while the reaction catalysed by the minor bands would still continue. Thus, serious over-estimation of the minor isoenzymes might result. Diffusion of the dye from intensely stained zones might produce the same effect. Errors might also be introduced by the use of a fixed substrate concentration, since isoenzymes may differ considerably in their substrate affinities.

'Nothing dehydrogenase' effect

A curious phenomenon frequently observed when tetrazolium-staining techniques are applied to the detection (especially on cellulose acetate foil, but also in various gels) of lactate dehydrogenase isoenzymes is the 'nothing dehydrogenase' effect (Pearse, 1961; Barnett, 1962, personal communication). 'Blank' preparations from which substrate has been omitted sometimes exhibit faintly stained replicas of the patterns observed with the normal mixture. This effect cannot be due to free lactate, which would be removed during electrophoresis, but may possibly be attributed to substrate bound to the enzyme protein.

Other suggestions, however, have recently been put forward, and there is strong evidence that the 'nothing dehydrogenase' effect is

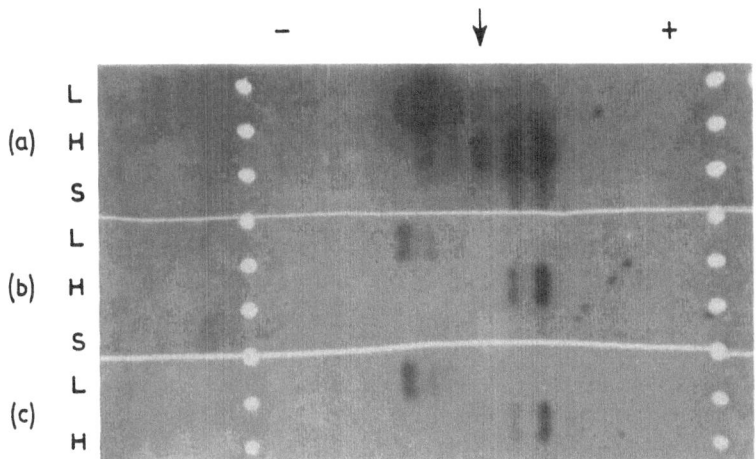

Fig. 6. Cellulose acetate electrophoresis of human liver (L) and heart (H) extracts and human serum (S) stained to show the 'nothing dehydrogenase' effect. The Beckman 'Microzone' cell was used and the unstained foil was cut into three parts: (*a*) stained with the complete tetrazolium system with lactate as substrate; (*b*) as (*a*), except that substrate was omitted; (*c*) as (*a*), except that ethanol replaced lactate as substrate. The arrow indicates the point of application of the samples. Note the congruity of LD_1 and LD_2 of the heart extracts with the corresponding 'nothing dehydrogenase' and alcohol dehydrogenase bands, and also that the liver extract exhibits an extra 'nothing dehydrogenase' band migrating farther towards the cathode than LD_5. This band is coincident with the most cationic alcohol dehydrogenase component.

largely due to alcohol dehydrogenase (Shaw and Koen, 1965). Like alcohol dehydrogenase, 'nothing dehydrogenase' is inhibited by *p*-hydroxymercuribenzoate. Another possibility is galactose dehydrogenase suggested by Cuatrecasas and Segal (1966*b*, 1967), but further evidence has been reported in support of the view that alcohol dehydrogenase is responsible (Beutler, 1967; Shaw and Koen, 1967). It has been established that several of the materials used, even when highly purified, contain traces of alcohol. This applies not only to reagents such as phenazine methosulphate but also to galactose and even to the starch used for the preparation of the gels. It has been shown that drying the media and other materials at 80° eliminates the 'nothing dehydrogenase' reaction (Beutler, 1967).

Nevertheless, it is rather striking that when human heart and serum extracts are subjected to electrophoresis the mobilities of the 'nothing dehydrogenase' bands are very similar to those of lactate dehydrogenase, though an extra cationic band of 'nothing dehydrogenase' can sometimes be found in liver extracts. Figure 6 shows the patterns observed when cellulose acetate electropherograms of human liver and heart extracts were subjected to the tetrazolium reaction: (a) with lactate as substrate; (b) without substrate; and (c) with ethanol as substrate. The similarity between the 'nothing dehydrogenase' in (b) and the alcohol dehydrogenase pattern in (c) is apparent, as is that between the electrophoretic mobilities in all three cases.

The observation by Scandalios (1967) that the same patterns were observed when maize endosperm was subjected to electrophoresis irrespective of whether ethanol, xanthine, lactate, malate, glucose, glucose 6-phosphate, isocitrate or galactose was used as substrate or even when no substrate was added, could be interpreted in the same way.

Whether or not 'nothing dehydrogenase' is finally proved to be alcohol dehydrogenase, it is clear that its existence must be taken into account when scanning techniques are employed for the quantitation of enzyme electropherograms.

Staining techniques for esterases

Most staining techniques in widespread use for the location of esterases in electrophoretic media are based upon the technique originally devised by Nachlas and Seligman (1949) for use in histochemistry, and since modified by Gomori (1952). A β-naphthyl ester is used as substrate, and the liberated β-naphthol is coupled with tetrazotized o-dianisidine (Brentamine Fast Blue B salt) to produce an intense purple dye. This is a very versatile technique which can readily be adapted for the detection of more specific groups of esterases, such as the phosphatases and cholinesterases. Certain other stable diazotized aromatic amines also serve as satisfactory coupling agents.

The nature of the electrophoretic medium is not critical, and satisfactory results have been obtained when paper, agar gel or starch gel has been used.

β-Naphthyl acetate is the substrate usually but by no means invariably used for the location of non-specific esterase activity (Hunter and Markert, 1957; Allen and Hunter, 1959; Oort and Willighagen, 1961;

Paul and Fottrell, 1961; Allen, 1961; Hunter and Strachan, 1961; Hess, Angel, Barron and Bernsohn, 1963), but α-naphthyl acetate, propionate and butyrate have also been employed, as well as certain long-chain β-naphthyl esters. α-Naphthyl phosphate has been widely used as a phosphatase substrate (Kind, 1958; Estborn, 1959; Moss, 1962; Chiandussi, Greene and Sherlock, 1962).

Some investigators have prepared paper strips for the detection of phosphatases by soaking filter paper in a solution of the buffered substrate and the coupling agent (Kind, 1958; Estborn, 1959). After drying, the test papers may be stored in the dark until required. A second filter paper moistened with the buffer is placed on the electropherogram for about 30 sec., after which it is placed in contact with the test paper. Purple bands appear on both papers in 0·5–2 min. according to the activities of the enzyme bands.

A similar printing procedure has been used by Moss *et al.* (1961) for the fluorimetric detection of alkaline phosphatase (see p. 45). The test papers, containing α-naphthyl phosphate in carbonate–bicarbonate buffer at pH 10, are placed in contact with the gel and incubated at 37° for 5–10 min., after which the paper is removed. The gel is then viewed under ultraviolet light, when the zones of enzyme activity appear as areas of pale-blue fluorescence.

An alternative staining technique suitable for use after agar-gel electrophoresis is that of Stevenson (1961), who incubates the gel with β-glycerophosphate at pH 9·4 in a barbitone buffer containing calcium and magnesium nitrates. The slide is then dried with filter paper and thoroughly washed with 0·0001N-ammonium hydroxide, after which it is immersed in 0·5% silver nitrate solution. Dark bands denote the zones of enzyme activity. This technique has been used by Haije and de Jong (1963) for the location of the alkaline phosphatases of human sera and tissue extracts after agar-gel electrophoresis.

The β-naphthyl acetate technique coupled with the use of specific inhibitors (p. 134) may be employed for the detection of cholinesterases (Harris, Hopkinson and Robson, 1962), but the histochemical method of Koelle and Friedenwald (1949), in which acetylthiocholine is used as substrate, has also been adapted for the location of human serum cholinesterase after starch-gel electrophoresis (Bernsohn, Barron and Hess, 1961). The gel is incubated with a solution of the substrate and copper glycinate saturated with copper thiocholine, and after appropriate washing it is immersed in a solution of ammonium sulphide. The enzyme releases thiocholine from the ester, and this forms a very sparingly soluble white copper complex. Subsequent treatment with ammonium sulphide converts this into the dark-brown or black copper sulphide.

Another histochemical technique suitable for the detection of esterases on electropherograms is that of Holt (1952), which depends upon the hydrolysis of indoxyl acetate and subsequent oxidation of the indoxyl to indigo with potassium ferricyanide (S. J. Holt and P. C. Barrow, personal communication). This procedure has been used for the location of esterases separated by starch-gel electrophoresis (Hunter and Burstone, 1960; Barron, Bernsohn and Hess, 1961). Indoxyl acetate may also be used in conjunction with a diazo-coupling reagent, and the writer is indebted to Mrs P. C. Barrow for details of her experiments in advance of publication. It was found that the nine esterase bands in rat-liver homogenates detectable with β-naphthyl acetate and Brentamine fast blue B salt also stained with the indoxyl substrate and the coupling agent, but the intensity of the bands was greater with β-naphthyl acetate than with the alternative substrate.

However, Holt and his colleagues have recently extended this technique to the study of esterases by electron microscopy. An indoxyl ester, e.g. the acetate, serves as substrate, and the liberated indoxyl is coupled with diazo reagent to produce a dye which has a unique ability to chelate osmium (Holt and Hicks, 1966; Hicks, 1966). It seems that this procedure might well be applicable in isoenzyme studies.

REFERENCES

ALDRIDGE, W. N. (1954), *Biochem. J.* **57**, 693.

ALLEN, J. M. and HUNTER, R. L. (1959), *J. Histochem. Cytochem.* **8**, 50.

ALLEN, S. L. (1961), *Ann. N.Y. Acad. Sci.* **94**, 753.

AUGUSTINSSON, K.-B. (1959), *Acta chem. Scand.* **13**, 571.

AUGUSTINSSON, K.-B. (1961), *Ann. N.Y. Acad. Sci.* **94**, 844.

BABSON, A. L. and PHILLIPS, G. E. (1965), *Clin. chim. Acta* **12**, 210.

BAKER, R. W. R. and PELLEGRINO, C. (1954), *Scand. J. clin. lab. Invest.* **6**, 94.

BARANOWSKI, T. (1949), *J. biol. Chem.* **180**, 535.

BARRON, K. D., BERNSOHN, J. and HESS, A. R. (1961), *J. Histochem. Cytochem.* **9**, 656.

BERGMEYER, H. U. (1970), *Methods in Enzymatic Analysis*, 2nd ed., in press.

BERNSOHN, J., BARRON, K. D. and HESS, A. R. (1961), *Proc. Soc. exp. Biol. N.Y.* **108**, 71.

BEUTLER, E. (1967), *Science*, **156**, 1516.

BOYDE, T. R. C. and LATNER, A. L. (1961), *Biochem. J.* **82**, 51P.

BRATTON, A. C. and MARSHALL, E. K. (1939), *J. biol. Chem.* **128**, 537.

CHIANDUSSI, L., GREENE, S. F. and SHERLOCK, S. (1962), *Clin. Sci.* **22**, 425.

COLOWICK, S. P. and KAPLAN, N. O. (1955–70), *Methods in Enzymology*, Vols. 1–14, New York: Academic Press.

COOKE, K. B. and ZILVA, J. F. (1961), *J. clin. Path.* **14**, 500.
CUATRECASAS, P. and SEGAL, S. (1966a), *J. biol. Chem.* **241**, 5904, 5910.
CUATRECASAS, P. and SEGAL, S. (1966b), *Science*, **154**, 133.
CUATRECASAS, P. and SEGAL, S. (1967), *Science*, **156**, 1518.
DALZIEL, K. (1961), *Nature, Lond.* **191**, 1098.
DEWEY, M. and CONKLIN, J. (1960), *Proc. Soc. exp. Biol. N.Y.* **105**, 492.
DICKENS, F. and MCILWAIN, H. (1938), *Biochem. J.* **32**, 1615.
DIXON, T. F. and PURDOM, M. (1954), *J. clin. Path.* **7**, 341.
EISFELD, G. and KOCH, E. (1954), *Z. ges. inner. Med.* **9**, 514.
ELLIOTT, B. A. and WILKINSON, J. H. (1961), *Lancet* **i**, 698.
ESTBORN, B. (1959), *Nature, Lond.* **184**, 1636.
FAHEY, J. L., MCCOY, P. F. and GOULIAN, M. (1958), *J. clin. Invest.* **37**, 272.
FAWCETT, C. P., CIOTTI, M. M. and KAPLAN, N. O. (1961), *Biochim. biophys. Acta* **54**, 210.
GERHARDT, W., CLAUSEN, J., CHRISTENSEN, E. and RIISHEDE, J. (1963), *Acta neurol. Scand.* **39**, 85.
GLOCK, G. E. and MCLEAN, P. (1953), *Biochem. J.* **55**, 400.
GOLDBARG, J. A. and RUTENBURG, A. M. (1958), *Cancer* **11**, 283.
GOMORI, G. (1952), *Microscopic Histochemistry*, Chicago Univ. Press, p. 211.
HAIJE, W. G. and DE JONG, M. (1963), *Clin. chim. Acta* **8**, 620.
HARRIS, H., HOPKINSON, D. A. and ROBSON, E. B. (1962), *Nature, Lond.* **196**, 1296.
HESS, A. R., ANGEL, R. W., BARRON, K. D. and BERNSOHN, J. (1963), *Clin. chim. Acta* **8**, 656.
HICKS, R. M. (1966), *Proc. Assoc. clin. Biochem.* **4**, 101.
HOLT, S. J. (1952), *Nature, Lond.* **161**, 171.
HOLT, S. J. and HICKS, R. M. (1966), *J. cell. Biol.* **29**, 214.
HUNTER, R. L. and MARKERT, C. L. (1957), *Science*, **125**, 1294.
HUNTER, R. L. and STRACHAN, D. S. (1961), *Ann. N.Y. Acad. Sci.* **94**, 861.
HUNTER, R. L. and BURSTONE, M. S. (1960), *J. Histochem. Cytochem.* **8**, 58.
KAMARÝT, J. and ZÁZVORKA, Z. (1963), *Science Tools*, LKB-Produkten, Stockholm. **10**, 21.
KARMEN, A. (1955), *J. clin. Invest.* **34**, 131.
KEIDING, N. R. (1959), *Scand. J. clin. lab. Invest.* **11**, 106.
KIND, S. S. (1958), *Nature, Lond.* **182**, 1372.
KING, J. (1965), *Practical Clinical Enzymology*, London: Van Nostrand.
KOELLE, G. B. and FRIEDENWALD, J. S. (1949), *Proc. Soc. exp. Biol., N.Y.* **70**, 617.
KOWLESSAR, O. D., HAEFFNER, L. J. and RILEY, E. M. (1961), *Ann. N.Y. Acad. Sci.* **94**, 836.
KUBOWITZ, F. and OTT, P. (1943), *Biochem. Z.* **314**, 94.
KUN, E. and ABOOD, L. G. (1949), *Science*, **109**, 144.
LATNER, A. L. and SKILLEN, A. W. (1961), *Lancet* **ii**, 1286.
LAURSEN, T. (1962), *Scand. J. clin. lab. Invest.* **14**, 152.
LAURSEN, T. and HANSEN, P. (1958), *Scand. J. clin. lab. Invest.* **10**, 53.
LAYCOCK, M. V., THURMAN, D. A. and BOULTER, D. (1965), *Clin. chim. Acta* **11**, 98.
MCCUNE, D. C. (1961), *Ann. N.Y. Acad. Sci.* **94**, 723.
MARKERT, C. L. and MØLLER, F. (1959), *Proc. Natl. Acad. Sci., Wash.* **45**, 753.
MOORE, B. W. and ANGELETTI, P. U. (1961), *Ann. N.Y. Acad. Sci.* **94**, 659.
MOSS, D. W. (1960a), *Biochem. J.* **76**, 32P.
MOSS, D. W. (1960b), *Clin. chim. Acta* **5**, 283.

MOSS, D. W. (1962), *Nature, Lond.* **193**, 981.

MOSS, D. W., ANAGNOSTOU-KAKARAS, E., CAMPBELL, D. M. and KING, E. J. (1961*a*), *Biochem. J.* **81**, 441.

NACHLAS, M. M., MARGULIES, S. I., GOLDBERG, J. D. and SELIGMAN, A. M. (1960), *Anal. Biochem.* **1**, 317.

NACHLAS, M. M. and SELIGMAN, A. M. (1949), *J. Natl. Cancer Inst.* **9**, 415.

NINEHAM, A. W. (1955), *Chem. Rev.* **55**, 355.

OLIVER, I. T. (1955), *Biochem. J.* **61**, 116.

OORT, J. and WILLIGHAGEN, R. G. J. (1961), *Nature, Lond.* **190**, 642.

PAPADOPOULOS, N. M. and KINTZIOS, J. A. (1967), *Am. J. clin. Path.* **47**, 96.

PAUL, J. and FOTTRELL, P. F. (1961), *Ann. N.Y. Acad. Sci.* **94**, 668.

PEARSE, A. G. E. (1960), *Histochemistry, Theoretical and Applied,* 2nd ed., London: Churchill.

PLAGEMANN, P. G. W., GREGORY, K. F. and WRÓBLEWSKI, F. (1960), *J. biol. Chem.* **235**, 2282.

PLUMMER, D. T., ELLIOTT, B. A., COOKE, K. B. and WILKINSON, J. H. (1963), *Biochem. J.* **87**, 416.

PLUMMER, D. T. and WILKINSON, J. H. (1963), *Biochem. J.* **87**, 423.

PRESTON, J. A., BRIERE, R. O. and BATSAKIS, J. G. (1965), *Am. J. clin. Path.* **43**, 256.

RAYMOND, S. (1964), *Science,* **146**, 406.

ROSALKI, S. B. (1965), *Nature, Lond.* **217**, 414.

ROSALKI, S. B. and WILKINSON, J. H. (1960), *Nature, Lond.* **188**, 1110.

ROSENBERG, I. N. (1959), *J. clin. Invest.* **38**, 630.

SCANDALIOS, J. G. (1967), *Biochem. Genet.* **1**, 1.

SELIGMAN, A. M. (1963), *Methods in Enzymology,* ed. COLOWICK, S. P. and KAPLAN, N. O., Vol. 6, New York: Academic Press, p. 889.

SHAW, C. R. and KOEN, A. L. (1965), *J. Histochem. Cytochem.* **13**, 431.

SHAW, C. R. and KOEN, A. L. (1967), *Science,* **156**, 1517.

SIEGEL, A. and BING, R. J. (1956), *Proc. Soc. exp. Biol. N.Y.* **91**, 604.

SMITH, E. E., PINEDA, E. P. and RUTENBURG, A. M. (1962), *Proc. Soc. exp. Biol. N.Y.* **110**, 683.

STEVENSON, D. E. (1961), *Clin. chim. Acta* **6**, 142.

STRECKER, H. J. (1955), *Methods in Enzymology,* ed. COLOWICK, S. P. and KAPLAN, N. O., Vol. 2, New York: Academic Press, p. 220.

TALEISNIK, A., PAGLINI, S. and ZEITUNE, V. (1955), *C. R. Soc. Biol., Paris* **149**, 1790.

THEORELL, H. and BONNICHSEN, R. K. (1951), *Acta chem. Scand.* **5**, 1105.

TSOU, K. C., CHÉNG, C. S., NACHLAS, M. M. and SELIGMAN, A. M. (1956), *J. Am. chem. Soc.* **78**, 6139.

VAN DER HELM, H. J. (1961), *Lancet* **ii**, 108.

VAN DER HELM, H. J. (1962), *Clin. chim. Acta* **7**, 124.

VESELL, E. S. (1961), *Ann. N.Y. Acad. Sci.* **94**, 877.

VON EULER, H., ADLER, E. and HELLSTRÖM, H. (1936), *Z. physiol. Chem.* **241**, 239.

WACKER, W. E. C. and DORFMAN, L. E. (1962), *J. Amer. med. Assoc.* **181**, 972.

WACKER, W. E. C., ULMER, D. D. and VALLEE, B. L. (1956), *New Engl. J. Med.* **255**, 449.

WIELAND, T., PFLEIDERER, G. and RETTIG, H. L. (1958), *Angew. Chem.* **70**, 341.

WIEME, R. J. (1959*a*), *Studies on Agar-Gel Electrophoresis,* Brussels: Arscia.

WIEME, R. J. (1959*b*), *Clin. chim. Acta* **4**, 46.

WIEME, R. J., VAN SANDE, M., KARCHER, D., LÖWENTHAL, A. and VAN DER HELM, H. J. (1962), *Clin. chim. Acta* **7**, 750.

WIEME, R. J. and VAN MAERCKE, Y. (1961), *Ann. N.Y. Acad. Sci.* **94**, 898.

WILDE, C. E. and KEKWICK, R. G. O. (1964), *Biochem. J.* **91**, 297.

WOLFSON, S. K. and WILLIAMS-ASHMAN, H. G. (1957), *Proc. Soc. exp. Biol. N.Y.* **96**, 231.

WOLFSON, W. Q. (1957), *Nature, Lond.* **180**, 550.

WRÓBLEWSKI, F. and LADUE, J. S. (1955), *Proc. Soc. exp. Biol. N.Y.* **90**, 210.

ZINKHAM, W. H. and LENHARD, R. E. (1959), *J. Pediat.* **55**, 319.

CHAPTER 4

The Chemical Nature of Isoenzymes

Several different structural concepts have been proposed to explain the occurrence of enzyme heterogeneity. Most of the chemical variations observed in various non-catalytic proteins have been reported to occur in enzymes. Thus the multiple forms of lactate dehydrogenase, creatine kinase and certain other enzymes have been attributed to the combination of different types of sub-unit. The various alkaline phosphatases appear to contain different amounts of neuraminic (sialic) acid, and hence differ in their electric charges and electrophoretic mobilities. The polymorphism of certain cholinesterases seems to be due to polymerization, while conformational differences and association with different carrier proteins have been suggested to account for other groups of isoenzymes.

The molecular basis of enzyme heterogeneity has recently been reviewed by Markert and Whitt (1968).

Sub-unit structure
At the present time a considerable amount of knowledge has accrued concerning isoenzymes resulting from the combination of sub-units of more than one type. Most of this has been obtained from the study of the structure of lactate dehydrogenase, and this enzyme will therefore be considered first.

Lactate dehydrogenase
Reference has already been made to the presence of lactate dehydrogenase in most human and animal tissues in five electrophoretically distinct forms. In a given species all of these isoenzymes have approximately the same molecular weight, about 135,000, as measured by light scattering and ultracentrifugation. A simple demonstration of this is provided by Raymond and Aurell (1962), who showed that after two-dimensional electrophoresis in acrylamide gels of different pore sizes the isoenzymes were obtained in a straight line passing through the origin. Another striking feature is the regular manner in which they are spaced on a unidimensional electropherogram (Fig. 9).

This regular difference in net electrical charge suggests that the amino-acid composition of LD_1 differs from that of LD_2 by the same increment as that between other pairs of neighbouring isoenzymes, a view which is supported by the experimental studies of Wieland and Pfleiderer and their colleagues. The peptides resulting from the tryptic digestion of highly purified samples of the five individual isoenzymes gave 'fingerprint' patterns which differed according to the origin of the isoenzyme. Though a surprising similarity was observed between the peptide patterns of ox-heart and rat-heart isoenzymes, more pronounced species differences were found between those of the principal rat- and rabbit-skeletal muscle isoenzymes (Wieland, Pfleiderer and Rajewsky, 1960).

More drastic hydrolysis with 6N-hydrochloric acid followed by quantitative amino-acid analysis has provided evidence for the differences between various isoenzyme proteins (Wieland and Pfleiderer, 1961). Rat-heart LD_1, for example, contains appreciably more of the acidic amino acids, e.g. aspartic acid (138 moles per mole of enzyme), than does rat-skeletal muscle LD_5 (118 moles of aspartic acid per mole of enzyme), but the latter contains more of the basic amino acid, lysine (100 moles per mole of enzyme) than the former (89 moles per mole of

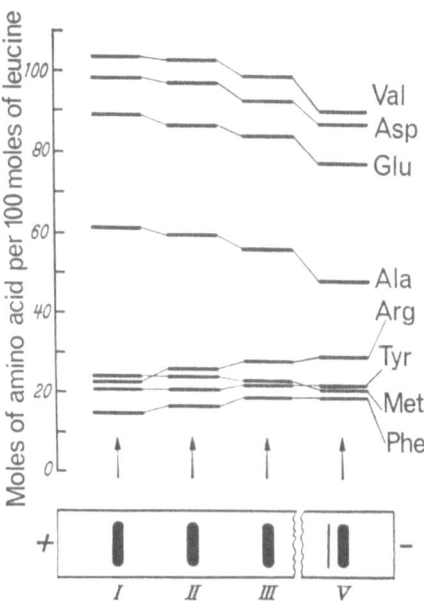

Fig. 7. Differences in the numbers of molecules of certain amino acids per molecule of enzyme in four human lactate dehydrogenase isoenzymes. (Wachsmuth *et al.*, 1964.)

enzyme). The striking regular differences in the amino-acid compositions of the four human isoenzymes, LD_1, LD_2, LD_3 and LD_5, are
shown in Fig. 7 (Wachsmuth, Pfleiderer and Wieland, 1964). The
gradual increase in the content of lysine and arginine and the regular fall
in aspartic and glutamic acid moieties from LD_1 to LD_5 explain the
changes in electrophoretic mobilities.

Little is known at present of the amino-acid sequence in the enzyme
molecule, but the demonstration that each isoenzyme is a tetramer
which may be split into four monomers on treatment with 12M-urea or
5M-guanidine hydrochloride has had dramatic consequences which
extend far beyond structural considerations (Appella and Markert,
1961; Cahn, Kaplan, Levine and Zwilling, 1962; Markert and Appella,
1963). It is known that the enzyme combines with four molecules of
coenzyme per molecule, and it has been demonstrated that the four
identical B(H)* sub-units produced by the dissociation of each molecule
of LD_1 differ from the A(M)* sub-units, four of which are obtainable
from each molecule of LD_5. The molecular weights of both H and M
sub-units are about 35,000, and each of these contains only one-quarter
of the number of lysine and arginine residues found in the whole enzyme
molecule (Cahn *et al.*, 1962; Appella and Markert, 1962). It has therefore been suggested that each of the isoenzymes of intermediate mobility
is a hybrid composed of mixed sub-units according to the scheme shown
in Fig. 8.

Considerable evidence has now accumulated in support of this
theory, which provides a convincing explanation of most of the
observed facts, including the regular change in the electric charge on the
isoenzyme molecules. Markert (1963*b*) has confirmed the findings of
Wieland and Pfleiderer (1961) by showing that the amino-acid composition of LD_3 is almost exactly midway between those of LD_1 and LD_5. A
tryptic digest of LD_2 from pig-heart and pig-skeletal muscle has been
shown to give a peptide ('fingerprint') pattern on two-dimensional thin-
layer chromatography almost identical with that produced by a 3 : 1
mixture of LD_1 and LD_5 (Wieland, Georgopoulos, Kampe and
Wachsmuth, 1964). Further evidence has also been obtained from
immunochemical studies (p. 64).

It is not possible to restore enzymic activity after urea treatment
either by removing the urea by dialysis or by diluting it to an ineffective

---

* Appella and Markert (1961) describe the electrophoretically slowest component as
LD_1 and the fastest as LD_5, and the corresponding sub-units as 'A' and 'B' respectively.
In this monograph the more descriptive 'H' and 'M' notation of Cahn *et al.* (1962) is
preferred (H = heart, M = muscle).

concentration. It seems that the breakdown of hydrogen bonds leads to irreversible changes in the secondary and tertiary structure of the monomers which prevent them from recombining to produce an active enzyme. However, Markert (1963a) devised a much milder procedure for causing dissociation. When the tetramers are frozen in molar sodium chloride and thawed after several hours dissociation occurs, and this is followed by recombination of the monomers to form active enzymes. Convincing proof of the validity of the theory was obtained by

Electrophoretically separated isoenzymes M.W. 130,000

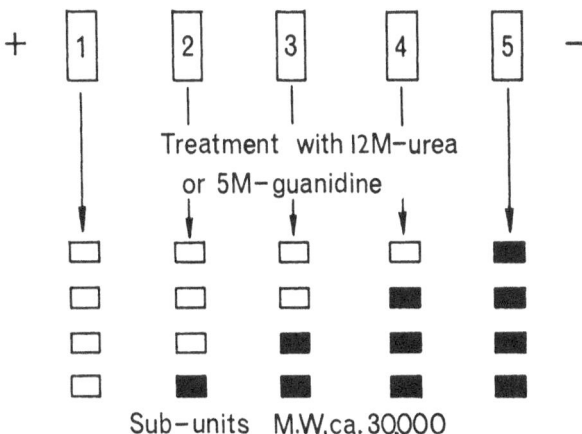

Sub-units M.W. ca. 30,000

Fig. 8. The tetrameric nature of the five lactate dehydrogenase isoenzymes indicating their formation from two different types of monomer, M (black) and H (white).

applying this treatment to a mixture of equal amounts of the carefully purified ox-heart isoenzymes LD_1 and LD_5. Subsequently the presence was observed of all five isoenzymes in approximately the calculated proportions of 1 : 4 : 6 : 4 : 1, which would be expected if random re-association occurs, while electrophoresis after similar separate treatment of LD_1 or LD_5 showed the presence of the single isoenzyme only (Fig. 10) (Markert, 1963a).

Markert also reports that during these experiments there was little loss of enzyme activity and suggests that the monomers might possess activity, an idea which finds support from the observation that a lactate dehydrogenase of molecular weight 72,000 has been found to be active in dilute solution (Millar, 1962). This would appear to be a dimer.

However, Markert and Massaro (1968) have recently re-investigated the effect of dilution, and have concluded that the tetramer is the only enzymatically active component. They demonstrated that on dilution lactate dehydrogenase not only dissociates into monomers but also simultaneously undergoes changes in tertiary structure which inactivate the enzyme. Such dissociation is strongly inhibited by the presence of homologous proteins.

The relative amounts of H and M polypeptide sub-units in cells of a given type may vary considerably, thus it is to be expected that an excess of H sub-units would give rise to a preponderance of the fast-moving isoenzymes, LD_1 and LD_2, and vice versa. It also follows from this theory that all tissues of a given species are likely to contain all five possible isoenzymes, though in some cases one of the extreme forms (LD_1 or LD_5) might not be present in detectable quantity.

The synthesis of the two different polypeptide sub-units, H and M, is regarded as being under the control of separate genes, and the iso-enzyme patterns found in different tissues, which depend upon the relative amounts of H and M polypeptides, consequently are controlled by the relative genetic activities (Markert and Ursprung, 1962; Cahn et al., 1962; Markert, 1963a). Thus, it can be seen how only two separate genes can control the synthesis of five distinct isoenzymes, a mechanism which provides an extension of the one gene–one enzyme hypothesis.

The mechanism by which dissociation and recombination of lactate dehydrogenase monomers takes place during freezing and thawing in vitro has been investigated by Chilson, Costello and Kaplan (1965). Hybridization appears to involve conformational changes which occur as a result of the increased salt and protein concentrations and possibly decreased pH near the eutectic point. The presence of NADH appears to be essential for the recovery of maximal activity after dissociation and recombination (Clausen and Hustrulid, 1968).

Electron microscopy of the crystalline rabbit-muscle enzyme has given results consistent with the tetrameric structure. Four units, almost square in cross-section, 95 Å wide, appear to be combined in fibrils, rectangular in cross-section and 200 Å wide (Hruban, Slesers and Orlando, 1967).

It has been widely assumed that the isoenzyme molecules are produced by random association of sub-units into tetramers, but Goldberg and Wuntch (1967) have suggested that their finding of a relative deficiency of LD_2 and LD_4 isoenzymes in the frog may be due to the combination of both types of sub-unit into homodimers which subsequently undergo random association into tetramers.

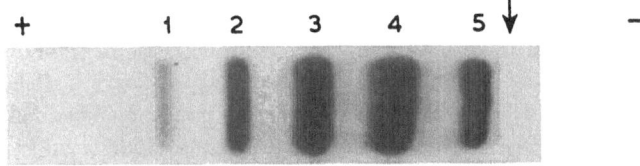

Fig. 9. Uni-dimensional acrylamide-gel electrophoresis showing the regular spacing of the lactate dehydrogenase isoenzymes of a human skeletal muscle extract (gastrocnemius).

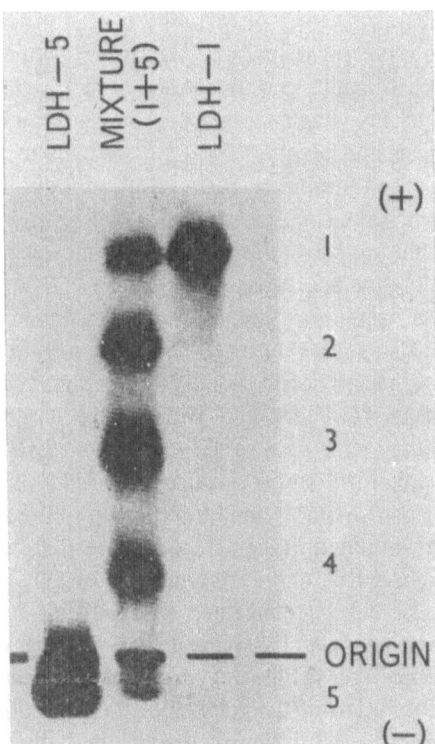

Fig. 10. Starch-gel electrophoresis of purified ox-heart LD₁ and LD₅ showing single bands only. The centre run is that of a mixture of LD₁ and LD₅ which had been subjected to dissociation and reassociation. (Markert, 1963a.)

Though examples of restricted association may occur, the structural requirements for the reassociation of monomers are not very critical, since interspecific *in vitro* hybridization from such diverse combinations as chicken LD_1 + haddock LD_1 and beef LD_1 + frog LD_5 has been demonstrated. In each case five-membered patterns including three artificial hybrid isoenzymes were observed (Salthe, Chilson and Kaplan, 1965).

Further evidence in support of the sub-unit structure of lactate dehydrogenase has been obtained from immunochemical studies. For many years specific antisera have been used to detect the presence of particular protein components (antigens) in complex mixtures, and recently these have been applied to the study of lactate dehydrogenase. Differences in the structures of the rabbit-heart (principally LD_1) and rabbit-skeletal muscle (principally LD_5) lactate dehydrogenases were established by Nisselbaum and Bodansky (1959), who injected these enzymes into chickens which produced antisera. A high degree of specificity was shown, and sera containing antibodies to the rabbit-heart enzyme had a marked inhibitory effect on its activity, but showed little action against the rabbit-muscle enzyme. Conversely, the latter was inhibited by the homologous antiserum, which, however, displayed little cross-action against the heart enzyme.

Similar results were also reported by Plagemann *et al.* (1960*a*, 1960*b*), who used an anti-rabbit-skeletal muscle lactate dehydrogenase, produced in chickens and partly purified by fractional precipitation with ammonium sulphate (Gregory and Wróblewski, 1958). This showed increasing inhibitory effects on rabbit isoenzymes LD_2, LD_3, LD_4 and LD_5, but relatively little action against LD_1. The anti-rabbit-muscle component rather surprisingly inhibited human-tissue isoenzymes, but whereas LD_1 was inhibited to the extent of 40%, there was 75% inhibition of LD_3, LD_4 and LD_5 and 55% inhibition of LD_2.

These results have been confirmed with purified human-heart and -liver lactate dehydrogenases (Nisselbaum and Bodansky, 1960, 1961*a*, 1961*b*). Anti-human-liver lactate dehydrogenase, prepared in the rabbit, produced 75% inhibition of the activity of the crystalline human-liver enzyme, but only 14% inhibition of that of the human-heart enzyme. Anti-human-heart lactate dehydrogenase was prepared in the chicken and again specific antigen–antibody reactions were observed. Crystalline human-heart lactate dehydrogenase was inhibited to the extent of 75% by the homologous antiserum, which, however, only inhibited the liver enzyme by 21%. The degrees of inhibition observed by these investigators show a close parallel with the isoenzymic compositions of the various tissue enzymes (Table 4). This suggests that in a given species all the individual isoenzymes have identical protein structures irrespective of the tissue from which they are derived.

T A B L E 4 *Relation between the isoenzyme composition of human-tissue lactate dehydrogenases and the degrees of inhibition with antisera to the human heart and liver enzymes*

(Based upon the results of Nisselbaum and Bodansky, 1961*b*)

| | | Percentage inhibition with antisera | |
Tissue	Principal isoenzymes	Anti-human-liver lactate dehydro-genase	Anti-human-heart lactate dehydro-genase
Heart	LD_1 and LD_2	2	60
Heart (crystal enzyme)	LD_1	14	75
Erythrocytes	LD_1 and LD_2	0	58
Kidney	LD_1 and LD_2	15	55
Skeletal muscle	LD_4 and LD_5	51	16
Liver	LD_4 and LD_5	60	11
Liver (crystal enzyme)	LD_5	75	21

The results obtained with the human-heart and -liver enzymes are paralleled by those found in inhibition studies with anti-chicken-heart and -muscle lactate dehydrogenases produced in the rabbit (Cahn *et al.*, 1962).

This conclusion has been confirmed and extended by Markert and Appella (1963), who used a variety of different techniques with antisera produced in rabbits to the following antigens: ox-muscle LD_5, crystalline ox-skeletal-muscle lactate dehydrogenase containing all five isoenzymes (LD_{1-5}) and crystalline ox-heart lactate dehydrogenase containing isoenzymes, LD_1, LD_2 and LD_3.

In immunodiffusion analysis a well-marked line of precipitation was formed between anti-LD_5 and LD_5, but no cross-reaction with LD_1 was observed. LD_1, however, was precipitated by antisera to $LD_{1\,5}$ or to LD_{1-3}. Anti-$LD_{1\,3}$ also precipitated LD_5. Moreover, LD_5 from different tissues of the same animal was precipitated by the same antisera. These results are consistent with the view (p. 136) that, with the exception of LD_x, all LD isoenzymes in a given species are produced by combinations of the same H and M polypeptide sub-units, and that enzymes containing H or M sub-units possess different antigenic properties. The hybrid isoenzymes, LD_2, LD_3 and LD_4, however, behave as antigens provoking the production of antibodies which precipitate all five isoenzymes.

Markert and Appella (1963) confirmed these conclusions by quantitative inhibition studies: they found that, whereas anti-LD_5 has no effect on the activity of LD_1, the homologous enzyme was about 80% inhibited and LD_3 was inhibited by about 45%. Inhibition of LD_3, however, was increased to about 80% by the inclusion of anti-LD_{1-3}, a result which suggests that half the active centres of LD_3 are affected by anti-LD_5, while the remaining centres are inhibited by anti-LD_1. In all cases about 20% of the original activity remained after antibody treatment, and this was attributed to about 20% of the antibodies being non-neutralizing. These non-neutralizing antibodies might combine with the enzyme in such a way as to protect the active centres from inactivation while still allowing substrate molecules free access (cf. Cinader and Lafferty, 1963).

Absorption of anti-LD_{1-5} with increasing amounts of LD_5 not only caused the preparation to lose its power to inactivate LD_5 but also produced the rather surprising loss of about 50% of its inhibitory action against LD_1. This result is explained by Markert and Appella (1963) on the assumption that a fourth antibody capable of reacting with both LD_1 and LD_5 is produced in the rabbit, but for which neither LD_1 nor LD_5 is the provoking antigen.

Markert and Appella conclude their remarkable paper by suggesting that the antigenic specificity of the enzyme resides in its tertiary and quaternary structure, since when the various isoenzymes were dissociated into their polypeptide sub-units the products were completely devoid of antigenic activity. Furthermore, the sub-units did not absorb antibodies from antisera, nor did they have any effect upon the enzyme-inhibitory properties of the antisera prepared against the whole enzyme. It seems, therefore, that more than one polypeptide chain is required to provoke an antigenic reaction and, if $\alpha\alpha$ antibodies are produced in response to the LD_5 antigen which contains only MM chains and if $\beta\beta$ antibodies are solely produced by the HH chains of LD_1, it is conceivable that LD_{1-5} might provoke the production of $\alpha\beta$ antibodies which would be capable of inhibiting both LD_1 and LD_5 isoenzymes.

Although the tetrameric structure of lactate dehydrogenase is supported by an impressive volume of evidence, a number of reports indicating the presence of more than five components in certain tissues is not easily reconciled with this concept.

Wieland and Pfleiderer (1961) refer to their occasional observations of extra bands of activity in rat tissues separated by paper or cellulose acetate electrophoresis. Allen (1961) found a total of seven anodic fractions and two cathodic fractions in various mouse tissues, and

Blanchaer (1962) observed that in various human-tissue homogenates and pathological human sera as many as eight fractions could be detected on starch-gel electrophoresis. Several investigators have found one or more additional bands in the sera of patients with malignant disease (Beautyman, 1962; Latner, 1964; Soetens, Karcher, Van Sande and Lowenthal, 1965; Fujimoto, Nazarian and Wilkinson, 1968).

A possible explanation for some of these anomalous findings is provided by the observation in the writer's laboratory that a portion of liver tumour contained a set of five isoenzymes regularly spaced on acrylamide-gel electrophoresis in which the slowest component migrated appreciably faster than the normal LD_5, obtained from the same patient (Fig. 43, p. 183). Each of the hybrid isoenzymes also migrated faster than the corresponding normal isoenzymes. These results are explained by assuming that an abnormal M sub-unit was elaborated by the tumour and that the extra bands were due to hybridization with the normal H sub-unit (Fujimoto *et al.*, 1968). This observation is discussed in greater detail in Chapter 6 (p. 183).

Another observation which cannot easily be explained on the basis of the sub-unit hypothesis is that of Fritz and Jacobson (1963), who found that the five major bands of enzymatic activity in extracts of mouse tissues could be split into a total of fifteen bands during electrophoresis on polyacrylamide gels when $0.005\text{M-}\beta$-mercaptoethanol was previously incorporated into the gel. Similar observations were made with the crystalline rabbit-muscle enzyme (Fritz, 1963) and in the writer's laboratory with both rabbit- and human-muscle preparations. Since a similar effect is found during starch-gel electrophoresis, the possibility of a polyacrylamide artifact can be discounted. Fritz and Jacobson postulated that the multiplicity of bands might be due to partial dissociation of the sub-units from bound coenzyme molecules, but this theory has not been confirmed (Fritz and Jacobson, 1965; Fujimoto and Wilkinson, unpublished observations). Examples of the multiple sub-bands of lactate dehydrogenase are illustrated in Fig. 11.

The occurrence of multiple sub-bands in a variety of mouse tissues has been confirmed (Koen and Shaw, 1965; Houssais, 1966; Koen, 1967*a*, 1967*b*). Vesell and Brody (1964) found evidence that human-muscle extracts produce split isoenzyme bands on storage. Koen (1967*a*) observed that the sub-band patterns of mouse-muscle and -liver extracts also change on ageing, but in this species there is a change in their relative concentrations rather than their number. There is shift towards the anodic sub-bands during the ageing process, an effect which is accelerated by the presence of pyruvate but not of lactate. The shift is

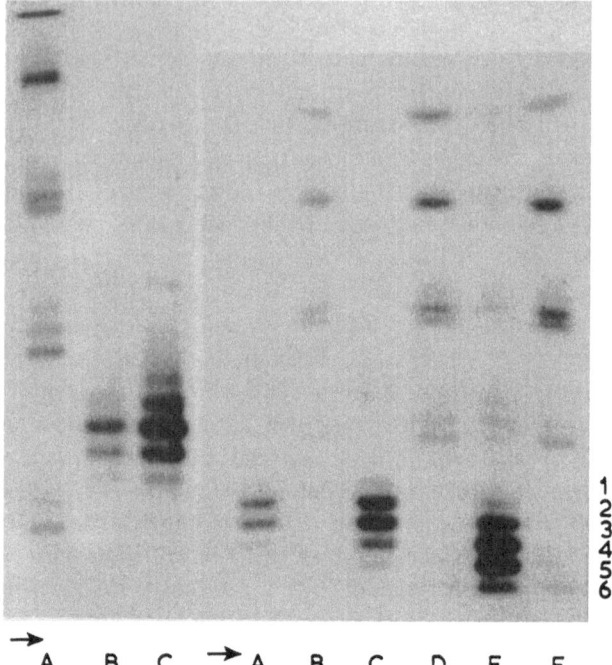

Fig. 11. Sub-bands of lactate dehydrogenase isoenzymes of mouse-tissue extracts separated by starch-gel electrophoresis. Left: (*a*) kidney; (*b*) muscle; (*c*) liver. Right: (*a*) 8-day-old muscle; (*b*) 8-day-old kidney; (*c*) 4-day-old muscle; (*d*) 4-day-old kidney; (*e*) 1-day-old muscle; (*f*) 1-day-old kidney. The arrow indicates the origin, and the anode is at the top. (Koen, 1967*a*.)

prevented by the addition of mouse-kidney extracts. Although Koen (1967*a*) reports the presence of sub-bands in tissues from several other species, including pig, rabbit, horse and man, she detected the anodal shift only in the mouse (Fig. 11). She suggests that it is due to conformational rearrangements, controlled to some extent by the presence or absence of a macromolecular component in tissue extracts.

Koen (1967*a*) also found that the sub-bands of mouse liver LD_5 separately eluted from starch gel and again subjected to electrophoresis gave in addition to the original sub-band a new Z band which migrated almost as far as LD_4 (Fig. 12). Five different Z bands can be produced from the five separate LD_5 sub-bands. The formation of the Z band appears to be reversible, since on re-electrophoresis a small amount of the original sub-band is formed. This phenomenon appears to be unrelated to the anodal shift discussed above.

Complex bands have also been found in pre-pubertal testis by Blanco and Zinkham (1963), but the mobilities of the main fractions corresponded with those of the five bands found in other tissues, and the authors attribute the presence of the minor and double bands to interaction with the electrophoretic medium. A sixth isoenzyme of a rather different type, however, occurs in post-pubertal human testis (Blanco and Zinkham, 1963; Goldberg, 1963). This has a mobility intermediate between those of LD_3 and LD_4 and has been provisionally designated LD_x. LD_x appears to be related to spermatogenesis, and the spermatozoal lactate dehydrogenase consists very largely of this isoenzyme.

LD_x also occurs in the testes of other species, but its electrophoretic distribution differs from that found in the human being (Zinkham, Blanco and Kupchyk, 1963). This aspect is discussed in Chapter 6.

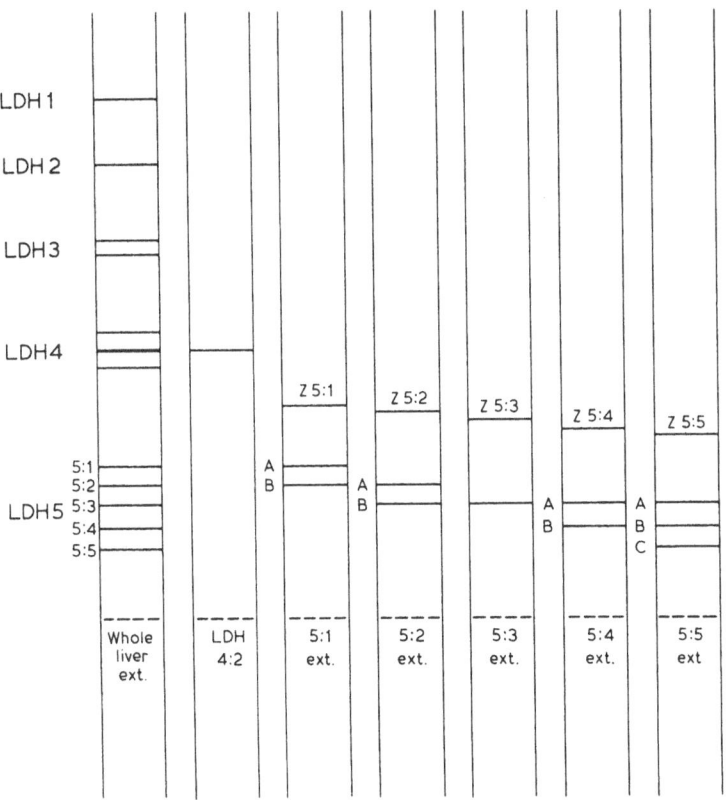

Fig. 12. Diagram of the primary, secondary and Z bands of mouse LD_5, after re-electrophoresis on starch-gel. (Koen, 1967*b*.)

The existence of the sixth isoenzyme (LD_x) in human testis has also been explained on the sub-unit hypothesis (Fig. 13). Zinkham *et al.* (1963) have demonstrated that when a testis homogenate containing the six isoenzymes LD_{1-5} and LD_x is subjected to dissociation and recombination the amounts of all except LD_3 are decreased, but new bands appear between LD_2 and LD_3, between LD_3 and LD_x and between LD_4 and LD_5. No new bands are found when tissues other than the testis are similarly treated. The authors therefore suggest that LD_x is a tetramer consisting of four identical X sub-units which differ from the H and M sub-units found in other isoenzymes. On dissociation LD_{1-5} would give

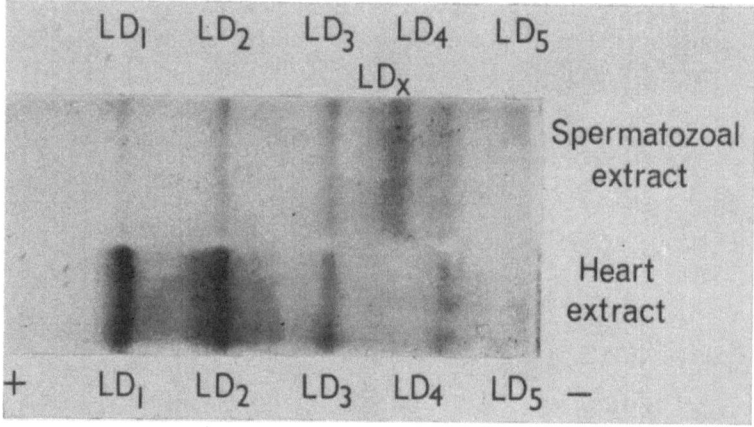

Fig. 13. Human-tissue lactate dehydrogenase isoenzymes separated by acrylamide-gel electrophoresis in tris buffer at pH 9·2.

rise to H and M sub-units, which would be free to combine with the X sub-units derived from LD_x. Thus X polypeptides would be present in relatively high concentration and could form hybrid isoenzymes such as M_3X on reassociation. This new isoenzyme would be expected to have an electrophoretic mobility between that of LD_4 and LD_5, and might be one of the new forms found experimentally. Similarly, the new component migrating between LD_2 and LD_3 might be composed of HM_2X, while that with a mobility between LD_3 and LD_x might consist of H_2MX, MX_3 or HMX_2 or a mixture of these.

Evidence in support of this conclusion has been obtained from dissociation and recombination studies with electrophoretically separated human LD_1 and LD_x when a new series of hybrid isoenzymes was obtained (Wilkinson and Withycombe, 1965).

Similar observations have also been made with the LD_x isoenzyme from rabbit testes. It is suggested that the existence of multiple LD_x bands in the testes of some animals, such as the bull, may be due to the presence of more than one new polypeptide rather than to the combination of X sub-units with H or M polypeptides (Zinkham *et al.*, 1963). The absence of HX and MX combinations in homogenates of rabbit or human testes is possibly due to the fact that the gene controlling the synthesis of X sub-units becomes active only at puberty, i.e. after the H and M combinations have become established. Since LD_x is the major isoenzyme in spermatozoa, X sub-units will predominate to such an extent that there will be little opportunity for significant quantities of mixed recombinants to be formed *in vivo*.

The properties of LD_x have been investigated in considerable detail in a variety of species, and the results are consistent with a chemical structure consisting of four X sub-units (Zinkham, Blanco and Kupchyk, 1964; Zinkham, Kupchyk, Blanco and Isensee, 1965; Goldberg, 1965; Clausen and Øvlisen, 1965; Wilkinson and Withycombe, 1965; Zinkham, 1968; Hawtrey and Goldberg, 1958; Clausen, 1969). The genetic aspects and the properties of this isoenzyme are discussed in Chapter 6.

An alternative theory to account for the anomalous properties of LD_x and the sub-bands described above has recently been proposed by Stambaugh and Buckley (1967). These investigators found rabbit heart, kidney and the ampulla of Vater to contain small amounts of LD_x, and that the various sub-bands of LD_1 do not all have the same kinetic properties. On the basis of the observation by Steginc and Vestling (1966) that the rabbit H and M sub-units each contain two polypeptide chains, Stambaugh and Buckley suggest that each H sub-unit consists of an enzymatically active chain (h) combined with a non-enzymatically active chain (A). The latter would contribute to the electric charge on the isoenzyme molecule, but not to its kinetic properties. Similarly the M sub-unit might contain an active chain (m) coupled with an inactive chain (C). It is suggested that cross-over combinations might occur.

The various LD_1 sub-bands would thus have the composition:

$$H_4 - 1 = (hA)_4$$
$$H_4 - 2 = (hA)_3(mA)$$
$$H_4 - 3 = (hA)_2(mA)_2$$
$$H_4 - 4 = (hA)(mA)_3$$
$$H_4 - 5 = (mA)_4$$

and the LD_5 sub-bands would have the following structures:

$$M_4 - 1 = (mC)_4$$
$$M_4 - 2 = (mC)_3(hC)$$
$$M_4 - 3 = (mC)_2(hC)_2$$
$$M_4 - 4 = (mC)(hC)_3$$
$$M_4 - 5 = (hC)_4$$

In the rabbit, the LD_x band has the same electrophoretic mobility as the $M_4 - 5$ sub-band and has similar kinetic properties to LD_1, and the proposed structure therefore provides a plausible explanation (Fig. 14).

If it is further assumed that the A and C chains can also undergo independent crossing-over, a very large number of electrophoretic variants is possible and it is conceivable that the existence of human LD_x with a mobility between LD_3 and LD_4 could be explained on a structural rather than a genetic basis. It might have a structure similar to $(hA)_2(hC)_2$, but proof of the existence of such a combination must await further investigation.

Some evidence in support of this proposal is provided by the observation of Appella (1964) that lactate dehydrogenase contains eight N-terminal valine groups. Later work (Appella and Zito, 1968) failed to confirm this, however, though the finding of seven to eight moles of acetate per mole of enzyme suggests that the technique used in the earlier experiments may have caused the acetyl groups to be hydrolysed from the terminal valine groups.

Creatine kinase

Mammalian creatine kinase also owes its molecular heterogeneity to the occurrence of more than one type of sub-unit. The polymorphism of this enzyme was first demonstrated by Deul and van Breeman (1964*a*, 1964*b*), who found human-brain and human-skeletal-muscle creatine kinase to differ considerably in their electrophoretic mobilities. The brain enzyme migrates towards the anode, while the muscle component remains near the point of application. The human heart contains both components, together with an extra zone of intermediate mobility.

The chemical nature of the various isoenzymes has been established mainly with four purified creatine kinases from rabbit muscle and brain, and chicken muscle and heart (Eppenberger, Dawson and Kaplan, 1967). These four components differ considerably in their amino-acid composition and their properties, though all four have similar molecular weights, about 82,000. In the presence of dilute hydrochloric acid and

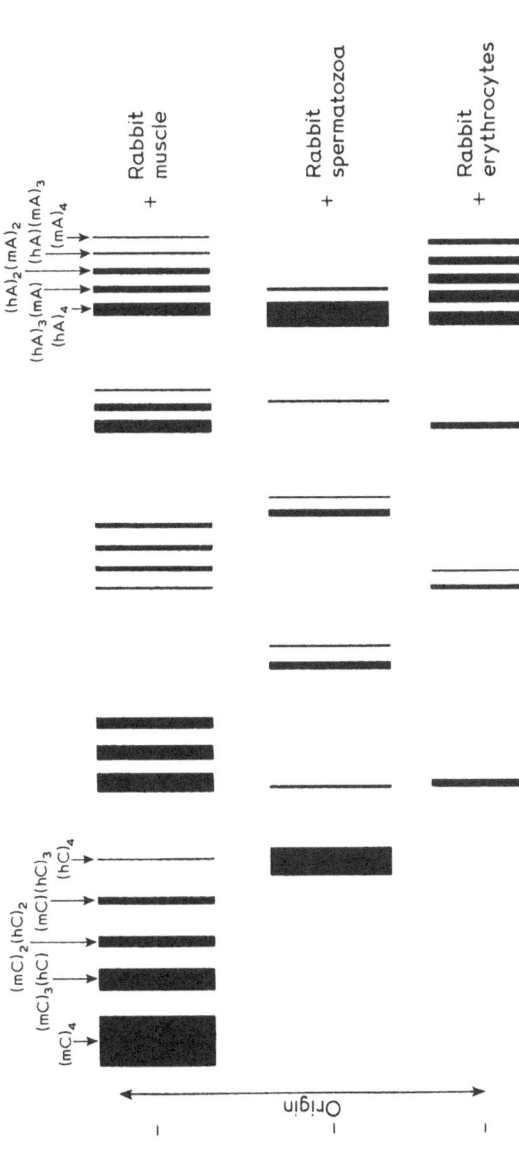

Fig. 14. Scheme proposed to account for the multiple sub-bands of lactate dehydrogenase and to explain the occurrence of spermatozoal components in the rabbit. (Stambaugh and Buckley. 1967.)

dithiothreitol, however, ultracentrifuge studies showed the chicken-muscle enzyme to be homogeneous and to have a molecular weight of 42,000 (Dawson, Eppenberger and Kaplan, 1967).

On starch-gel electrophoresis, Dawson *et al.* (1967) showed each of the purified enzymes to migrate as a single band, though occasionally the chicken-heart enzyme appeared as a double band. Treatment with 8M-urea dissociated each enzyme into its monomers, which recombined on subsequent dialysis. When mixtures of two enzymes were treated in this manner a third hybrid component with an intermediate electrophoretic mobility was formed in each case. Typical results are shown in Fig. 15.

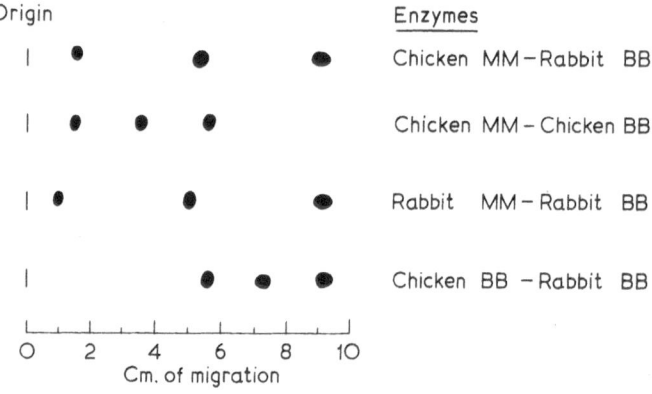

Fig. 15. Formation of hybrid creatine kinases from chicken and rabbit muscle and brain demonstrated by starch-gel electrophoresis. (Dawson, Eppenberger and Kaplan, 1967.)

Additional evidence in support of the dimeric structure of creatine kinase is provided by the finding that carboxyalkylation of two sulphhydryl groups per molecule of enzyme with iodoacetate abolishes its activity (Watts, Rabin and Crook, 1961; Dawson *et al.*, 1967).

Strong support is also provided by the immunochemical investigations of Eppenberger *et al.* (1967), who prepared antibodies against the chicken-muscle and chicken-brain enzymes. The homologous antigen, in each case, reacted with the antibody at high dilution while the hybrid enzyme reacted at higher concentration, while the other enzyme did not react at all.

This and other evidence is summarized in a recent report by Dawson, Eppenberger and Eppenberger (1968), but as is the case with lactate dehydrogenase, there is evidence that the structure of creatine kinase may be more complicated. Differences in the electrophoretic mobilities of the mitochondrial and soluble creatine kinases of rat-skeletal muscle, heart and brain (Jacobs, Heldt and Klingenberg, 1964), and of the pig-heart enzyme (Keto and Doherty, 1968) have been reported. Moreover, Swanson (1967) described differences in the kinetic properties of the soluble and mitochondrial enzymes of guinea-pig brain.

Keto and Doherty (1968) have recently prepared the pig-heart mitochondrial enzyme in crystalline form and have found it to have a sedimentation coefficient of 11·7, which indicates a molecular weight much higher than that of the purified creatine kinases discussed above. This component differs from the corresponding soluble form in requiring a protective thiol. Keto and Doherty found that in its electrophoretic mobility it consisted largely of two cationic forms. It is not yet known whether the mitochondrial enzyme is capable of hybridizing with the soluble enzyme.

Another anomalous form of creatine kinase has been found in dystrophic mice by Hooton and Watts (1966). The dystrophic enzyme differs from the normal muscle form in having only half the specific activity and in containing only one reactive thiol group. The authors obtained evidence from fingerprint patterns that the dystrophic enzyme may contain glutamic acid in place of cysteine at the catalytic site, a substitution which might explain its similarity to the normal enzyme in its molecular weight and electrophoretic mobility. In this connection it is interesting to note that Rosalki (1965) found the serum enzyme in human muscular dystrophy to have the same electrophoretic mobility as the normal muscle enzyme.

Aldolase
Although there is an impressive volume of evidence suggesting that the aldolase molecule consists of three sub-units (Rutter, 1962), the formation of five-membered sets of artificial hybrids indicates a tetrameric structure (Foxwell, Cran and Baron, 1966; Penhoet, Kochman, Valentine and Rutter, 1967). This interpretation, however, has been challenged by Chan, Morse and Horecker (1967), but recent reports suggest that the tetrameric structure is probably correct. The evidence is discussed more fully in Chapter 5 (pp. 117–122).

Glucose 6-phosphatase dehydrogenase
Recent work by Yoshida (1966) suggests that normal human erythrocytic (B+) glucose 6-phosphate dehydrogenase may be a hexamer with a molecular weight of about 240,000 and composed of six identical sub-

units. In high dilution the enzyme appears to dissociate into a less active trimer (Nevuldine and Levy, 1965; Yoshida, 1966).

The common Negro variant (A+) has been shown to resemble the B+ form very closely in its sub-unit structure, but studies of its amino-acid composition show that an asparagine residue of the B+ form is replaced by aspartic acid in the A+ isoenzyme (Yoshida, 1967, 1968).

The structure of glucose 6-phosphate dehydrogenase is considered in greater detail in Chapter 5 (pp. 101–112).

Tryptophan synthetase

The isoenzymes of *Escherichia coli* tryptophan synthetase also appear to owe their existence to variations in their sub-unit composition (Yanofsky, 1960; Wilson and Crawford, 1965; Creighton and Yanofsky, 1966). The enzyme molecule consists of four polypeptide sub-units combined in $\alpha_2\beta_2$ aggregates, which dissociate during starch-gel electrophoresis. The α and β components have been isolated, and both have been shown to possess limited enzyme activity. Neither is capable of converting indoleglycerol phosphate into L-tryptophan, the reaction catalysed by the intact enzyme, but the α-sub-unit catalyses the splitting of indoleglycerol phosphate into indole and glycerol phosphate, while the β-fraction couples indole with L-serine to form L-tryptophan (Crawford and Yanofsky, 1958; Crawford, Ito and Hatanaka, 1968).

Catalase

The five electrophoretically distinct forms of *Zea mays* catalase have recently been shown to resemble those of lactate dehydrogenase in being the products of two different types of sub-unit. Certain newly discovered variants, however, exhibit only two hybrid forms together with the parental isoenzymes. It has therefore been suggested that active trimers of maize catalase may occur (Scandalios, 1968).

Caeruloplasmin

Caeruloplasmin also appears to have a sub-unit structure which accounts for its electrophoretic heterogeneity. Poulik (1968) has suggested that the enzyme is apparently an octomer comprising four non-identical sub-units (α, β, γ, δ). The δ chain appears to have a higher molecular weight than the α, β and γ-polypeptides, and Poulik postulates an alternative structure consisting of a pair of short-chain (α, β, γ) and a pair of long-chain (δ) sub-units.

Presence of charged groups in isoenzyme molecules

Alkaline phosphatase

Though the existence of a two-sub-unit system seems to be responsible for the occurrence of the alkaline phosphatase isoenzymes of *Escherichia coli* (Schlesinger and Levinthal, 1963; Schlesinger, 1965, 1967*a*, 1967*b*; Lazdunski and Lazdunski, 1968; Schlesinger and Andersen, 1968), an entirely different mechanism has been suggested to account for the heterogeneity of mammalian alkaline phosphatases. Present knowledge of the structure of this enzyme will therefore be briefly reviewed, though this is necessarily incomplete.

It has been clearly demonstrated in the case of several purified mammalian and bacterial alkaline phosphatases that the active centre of the enzyme molecule contains a serine residue. The amino-acid sequence containing this reactive group has been partially resolved (Schwartz, Crestfield and Lipmann, 1963; Engström, 1964; Milstein, 1964; Zwaig and Milstein, 1964; Lazdunski and Lazdunski, 1968).

Present evidence suggests that the carbohydrate content of alkaline phosphatase may be responsible for the variation in electric charge on which the electrophoretic mobility depends. Ahmed and King (1960) found no more than traces of carbohydrate in the human placental enzyme, but significant amounts have been detected in intestinal alkaline phosphatase (Portmann, Rossier and Chardonnens, 1960; Engström, 1961). A development highly germane to the present discussion was the observation of Robinson and Pierce (1964) that treatment with neuraminidase reduces the anodic migration on starch-gel electrophoresis of three of the four human serum alkaline phosphatase isoenzymes. This result is interpreted as indicating that in three of the isoenzymes there is a terminal neuraminic (sialic) acid residue which is absent in the fourth. Its removal will reduce the magnitude of the negative charge on the enzyme molecule (Fig. 16).

These results have been confirmed by other investigators who found that the mobility of the intestinal enzyme was unaffected by treatment with neuraminidase (Moss, Eaton, Smith and Whitby, 1966; Butterworth and Moss, 1966; Ghosh and Fishman, 1967; Fishman, Inglis and Ghosh, 1968). Fishman and his co-workers identified sialic acid among the products of hydrolysis of highly purified placental alkaline phosphatase, and showed that after treatment with thiobarbituric acid the absorption spectra of the products of hydrolysis and of pure sialic acid were comparable (Ghosh and Fishman, 1967; Ghosh, Goldman and Fishman, 1967). Fishman and Ghosh (1967) have therefore proposed that all human alkaline phosphatases are sialoproteins, and suggest that the insensitivity of the intestinal enzyme to neuraminidase might be due to the sialic acid residue either not being terminal or being occluded in the three-dimensional structure of the enzyme.

It seems, however, that more than one sialic acid residue is combined in each molecule of alkaline phosphatase, for Robson and Harris (1966) have obtained as many as eight bands of activity when the human placental enzyme is treated with graded concentrations of neuraminidase prior to electrophoresis. This conclusion was confirmed by Saraswathi and Bachhawat (1968), who treated the two alkaline phosphatases of sheep brain with varying amounts of neuraminidase.

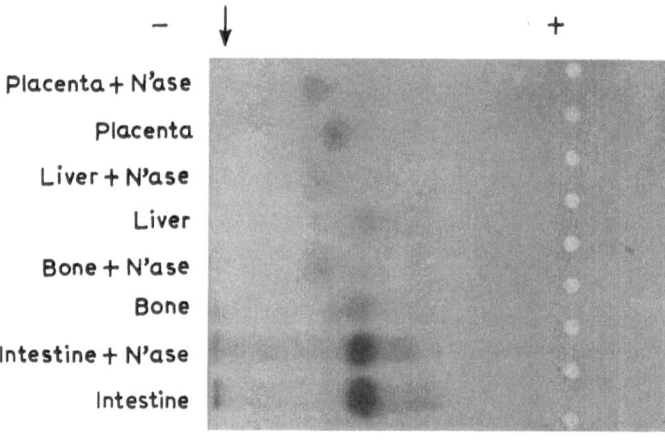

Fig. 16. Effect of incubation with neuraminidase on the electrophoretic mobilities of human-tissue alkaline phosphatases on Cellogel at pH 9·0.

Enzyme I was found to contain only about one-third of the amount of N-acetylneuraminic acid in enzyme II. The chromatographic properties of enzyme I were unaffected by treatment with neuraminidase, while similar treatment caused those of enzyme II to approach those of enzyme I.

Treatment with neuraminidase has little or no effect upon the kinetic properties of alkaline phosphatases, and hence it appears that the differences in structure affecting electrophoretic mobility and chromatographic properties do not involve the active centre of the enzyme (Butterworth, 1968; Saraswathi and Bachhawat, 1968). This confirms the earlier conclusions of Moss (1964), who reported that the two principal intestinal alkaline phosphatase bands separated by DEAE-Sephadex column chromatography, and which differ in their electrophoretic mobilities and thermal properties, have similar kinetic properties and are similarly inhibited by L-phenylalanine (p. 255).

Full understanding of the heterogeneity of alkaline phosphatase must await further study of the enzyme structure, and it is interesting to note that certain crystalline alkaline phosphatases have now been prepared (Malamy and Horecker, 1964; Ghosh and Fishman, 1968). Though other factors are concerned, differences in sialic acid content are undoubtedly responsible in part for the chromatographic and electrophoretic variation. It may well be that other enzymes also owe their heterogeneity to differences in the proportions of charged groups in their molecules.

Indeed, Svensmark (1961) and Gaffney (1968) have shown that hydrolysis of the terminal sialic acid residues of cholinesterase by neuraminidase reduces the electrophoretic mobility of the cholinesterase. However, there is strong evidence, discussed below, for other mechanisms for the heterogeneity of the cholinesterases, and it is unlikely that variations in sialic acid content can be a major factor.

Polymerization as the source of enzyme heterogeneity
Although the combination of polypeptide sub-units to form a molecule of enzyme protein might be regarded as polymerization, we are now concerned with groups of isoenzymes which appear to consist of separable aggregates in different states of polymerization. Such a mechanism has been proposed to account for the multiplicity of the human serum cholinesterases.

Cholinesterases
On starch-gel electrophoresis, Bernsohn, Barron and Hess (1961) found human serum cholinesterase to separate into seven zones of activity, all of which were sensitive to inhibition by eserine. Such inhibition was complete in all cases except the fastest band, which was only partially inhibited. These studies have been extended to other species, including the rat, monkey and cat, which also show multiple bands of eserine-sensitive esterase activity (Hess, Angel, Barron and Bernsohn, 1963).

Bernsohn *et al.* (1961) suggested that combination with inert protein carriers might account for the heterogeneity of the cholinesterases, but Harris, Hopkinson and Robson (1962) found that the single band of human serum cholinesterase obtained after paper electrophoresis is resolved into at least four fractions when subjected to starch-gel electrophoresis (Fig. 61, p. 288). As discussed in Chapter 2 (p. 10), electrophoresis on paper separates proteins solely according to their electric charges, whereas starch gel exerts a molecular-sieving effect superimposed upon electrophoretic migration. It appears therefore that the four bands obtained by Harris *et al.* have similar electric charges, but differ in their molecular dimensions.

These results have been confirmed by combined chromatographic and electrophoretic techniques (Harris and Robson, 1963; Goedde, Gehring and Hofmann, 1965), but the most convincing evidence so far

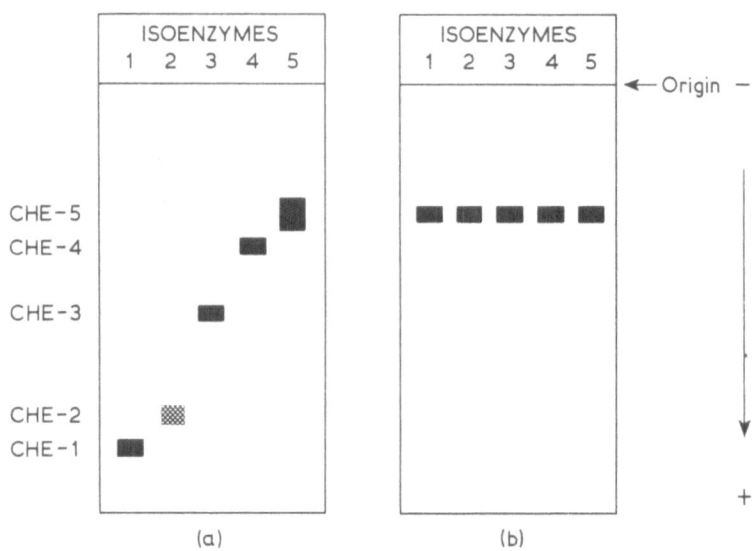

Fig. 17. (*a*) Diagram of individual cholinesterase isoenzymes separated by starch-gel electrophoresis. On repeated electrophoresis the isoenzymes migrate to the same positions. (*b*) Diagram of isoenzyme pattern obtained after eluting the various isoenzymes from (*a*) and concentrating them approximately twofold by treatment with ammonium sulphate before the second electrophoresis. Each isoenzyme behaves like CHE-5. (LaMotta, McComb and Wetstone, 1965.)

has come from a demonstration of the interconvertibility of the cholinesterase isoenzymes.

LaMotta, McComb and Wetstone (1965) showed that a purified concentrate of human pool serum cholinesterase separates into five bands on starch-gel electrophoresis (CHE-1–5). Most activity is associated with the slowest band, CHE-5. Similar patterns are obtained when the electrophoresis is repeated under the same conditions, but when each isoenzyme is separated and concentrated by precipitation with ammonium sulphate a single band only is observed. This corresponds in mobility to CHE-5 (Fig. 17). An increase in this component occurs when the eluates are concentrated by dialysis. Elution without concentration causes partial conversion of CHE-5 into CHE-1, CHE-2, CHE-3 and CHE-4. These results are interpreted as indicating stepwise polymerization leading towards an increase in CHE-5 during concentration, while depolymerization occurs on dilution, leading to the progressive breakdown of CHE-5 into four different smaller molecules.

Subsequently LaMotta's group observed two additional components in human serum. These are cathodic to CHE-5 and are described as CHE-6 and CHE-7 (LaMotta, McComb, Noll, Wetstone and Reinfrank, 1968). The new isoenzymes are completely converted into CHE-5 on concentration and partly converted on storage. The authors suggest that CHE-6 and CHE-7 represent molecular heterogeneity of a different type, since they are converted into CHE-5 under the same conditions as the smaller aggregates. It is possible that the new forms represent conformational isomers (conformers) (*v. inf.*), but as yet there is no evidence to support this proposal.

Amylase
Amylase also appears to exist in polymeric forms. The recently recognized condition of macroamylasaemia (Berk, Kizu, Wilding and Searcy, 1967) has been attributed to the occurrence in affected individuals of a polymer of the enzyme, the molecular weight of which is too great to permit filtration by the kidney.

Conformational isomerism
It has recently been suggested that relatively stable variations in the three-dimensional structure of enzymes might account for the existence of certain groups of isoenzymes. Such a mechanism was proposed by Sundaram and Fincham (1964) to explain the heterogeneity of glutamate dehydrogenase, and by Kitto, Wassarman and Kaplan (1966) for the malate dehydrogenases, which appear to have the same amino-acid sequence and the same molecular weight but differ in the way in which the polypeptide chains are folded. The various isomeric forms of haemoglobin H had previously been attributed to such a mechanism by Benesch, Benesch, Ranney and Jacobs (1962).

Several other enzyme systems have since been shown to form a series of isoenzymes for which three-dimensional isomerism of a single amino-acid chain may be responsible. Among these are pig-heart aspartate aminotransferase (Martinez-Carrion, Turano, Chiancone, Bossa, Giartosio, Riva and Fasella, 1967), the L-amino-acid oxidases of snake venom (Wellner and Hayes, 1968) and the bacterial penicillinases (Pollock, 1968).

Such isomeric forms have been described as 'conformational isoenzymes' or 'conformers' (Kitto *et al.*, 1966).

The thermodynamic basis has recently been discussed by Epstein and Schechter (1968), who suggest that in solution peptide chains take up configurations such that the free energy of the system is minimal. Stable conformational variants are those in which change from the metastable form to the usual form of the enzyme is prevented by a relatively large energy barrier. In the case of readily interchangeable conformers, on the other hand, the energy barrier is relatively small.

Epstein and Schechter differentiate between true conformational variants and ligand-induced variants, in which changes in configuration are brought about by combination with a small molecule. While small molecules have long been known to act as inhibitors or activators by combining with the active centre of an enzyme molecule, such interaction may also occur at secondary allosteric sites some distance from the catalytically active centre (Monod, Wyman and Changeux, 1965). Such combination may affect the catalytic activity of the enzyme (allosteric activation or inhibition), but will almost certainly influence the quaternary structure sufficiently to alter certain physical properties.

Malate dehydrogenase
The mitochondrial and supernatant mammalian malate dehydrogenases differ from each other in many respects, and must be regarded as separate enzymes which differ in their primary structure (Wieland, Pfleiderer, Haupt and Wörner, 1959; Delbrück, Zebe and Bücher, 1959; Thorne, 1960; Grimm and Doherty, 1961). Each form, however, is heterogeneous, and on electrophoresis gives rise to several (usually five or six) sub-bands. The sub-bands, moreover, appear to have identical catalytic properties (Thorne, Grossman and Kaplan, 1963).

Since there is evidence that the enzyme molecule consists of two identical sub-units, a mechanism different from that concerned in the heterogeneity of the lactate dehydrogenases must be responsible. Kitto *et al.* (1966) postulated that the electrophoretically distinct sub-bands are conformational isoenzymes on the basis of studies on various avian mitochondrial malate dehydrogenases. These investigators were impressed by the close similarity between the electrophoretic patterns of different species, though their absolute mobilities differed considerably. Optical rotatory dispersion measurements on the various sub-bands indicated that they have significantly different structures.

Further evidence has been obtained by the partial iodination of the chicken mitochondrial enzyme, both in its unresolved and separated forms, and by reversible denaturation (Kaplan, 1968). On electrophoresis, each iodinated band was found to produce new sub-bands which migrated more rapidly towards the anode (Kitto, Wassarman, Michejda and Kaplan, 1966).

This interpretation, however, has recently been challenged. Schechter and Epstein (1968) found, with a series of partially resolved pools of chicken-heart mitochondrial malate dehydrogenases, that reversible denaturation with 7·6M-guanidine hydrochloride did not affect the relative distribution of the various isoenzymes on subsequent electrophoresis. They point out that maintenance of the pattern differences under such conditions cannot easily be reconciled with a primary conformational heterogeneity.

Mann and Vestling (1968) have recently obtained evidence by means of reversible acid denaturation that the three principal components of rat-liver mitochondrial malate dehydrogenase are sub-unit isoenzymes. They suggest that two different sub-units combine to form three isoenzymes in much the same manner as the creatine kinase sub-units discussed on p. 72. This theory accounts for the presence of three of the bands observed on electrophoresis, but does not explain the presence of the remaining bands. Nevertheless, the occurrence of the genetic variant described by Davidson and Cortner (1967) is more easily explained on the basis of the sub-unit hypothesis (see Chapter 7).

At the present time therefore it is uncertain whether or not conformational isomerism plays an important part in establishing the heterogeneity of the malate dehydrogenases.

Differences in amino-acid sequence

The simplest possible model for the occurrence of isoenzymes involves variation in the primary structure of the enzyme protein. Such a mechanism has recently been established to account for the existence of the isoenzymes of carbonic anhydrase, three electrophoretically and chromatographically distinct forms of which occur in human erythrocytes (Nyman, 1961; Rickli and Edsall, 1962; Laurent, Charrel, Luccioni, Autran and Derrien, 1965).

The enzyme molecule consists of a single peptide chain having a molecular weight of about 30,000. The C-terminal amino-acid sequences of the three isoenzymes have been elucidated by Nyman, Strid and Westermark (1966). The three forms thus differ in their primary structure. This conclusion has been confirmed by Edsall and his co-workers, who observed differences in circular dichroism and optical rotatory dispersion (Rickli, Ghazanfar, Gibbons and Edsall, 1964; Armstrong, Myers, Verpoorte and Edsall, 1966; Beychok, Armstrong, Lindblow and Edsall, 1966; Edsall, 1968).

Combination with carrier proteins
It is attractive to propose that an enzyme might be associated with several non-enzymatic proteins with different chromatographic and electrophoretic properties. At the time of writing, however, the author is not aware of any enzyme whose heterogeneity can be attributed to differences in the nature of inert carrier proteins.

REFERENCES

AHMED, Z. and KING, E. J. (1960), *Biochim. biophys. Acta* **45**, 581.
ALLEN, J. M. (1961), *Ann. N.Y. Acad. Sci.* **94**, 937.
APPELLA, E. (1964), *Brookhaven Symp. Biol.* **17**, 131.
APPELLA, E. and MARKERT, C. L. (1961), *Biochem. biophys. Res. Commun.* **6**, 171.
APPELLA, E. and MARKERT, C. T. (1962), *Federation Proc.* **21**, 253.
APPELLA, E. and ZITO, R. (1968), *Ann. N.Y. Acad. Sci.* **151**, 568.
ARMSTRONG, J. M., MYERS, D. V., VERPOORTE, J. A. and EDSALL, J. T. (1966), *J. biol. Chem.* **241**, 5137.
BEAUTYMAN, W. (1962), *Lancet* **ii**, 305.
BENESCH, R., BENESCH, R. E., RANNEY, H. M. and JACOBS, A. S. (1962), *Nature, Lond.* **194**, 840.
BERK, J. E., KIZU, H., WILDING, P. and SEARCY, R. L. (1967), *New Engl. J. Med.* **277**, 941.
BERNSOHN, J., BARRON, K. D. and HESS, A. (1961), *Proc. Soc. exp. Biol., N.Y.* **108**, 71.
BEYCHOK, S., ARMSTRONG, J. M., LINDBLOW, C. and EDSALL, J. T. (1966), *J. biol. Chem.* **241**, 5150.
BLANCHAER, M. C. (1962), *Pure appl. Chem.* **3**, 403.
BLANCO, A. and ZINKHAM, W. H. (1963), *Science,* **139**, 601.
BUTTERWORTH, P. J. (1968), *Biochem. J.* **107**, 467.
BUTTERWORTH, P. J. and MOSS, D. W. (1966), *Nature, Lond.* **209**, 805.
CAHN, R. D., KAPLAN, N. O., LEVINE, L. and ZWILLING, E. (1962), *Science,* **136**, 962.
CHAN, W., MORSE, D. E. and HORECKER, B. L. (1967), *Proc. Natl. Acad. Sci., Wash.* **57**, 1013.
CHILSON, C. A., COSTELLO, L. A. and KAPLAN, N. O. (1965), *Biochemistry,* **4**, 271.
CINADER, B. and LAFFERTY, K. J. (1963), *Ann. N.Y. Acad. Sci.* **103**, 653.
CLAUSEN, J. (1969), *Biochem. J.* **111**, 207.
CLAUSEN, J. and HUSTRULID, R. (1968), *Biochem. biophys. Acta* **167**, 221.
CLAUSEN, J. and ØVLISEN, B. (1965), *Biochem. J.* **97**, 513.
CRAWFORD, I. P. and YANOFSKY, C. (1958), *Proc. Natl. Acad. Sci., Wash.* **44**, 1161.
CRAWFORD, I. P., ITO, J. and HATANAKA, M. (1968), *Ann. N.Y. Acad. Sci.* **151**, 171.
CREIGHTON, T. E. and YANOFSKY, C. (1966), *J. biol. Chem.* **241**, 980.

DAVIDSON, R. G. and CORTNER, J. A. (1967), *Nature, Lond.* **215**, 761.

DAWSON, D. M., EPPENBERGER, H. M. and EPPENBERGER, M. E. (1968), *Ann. N.Y. Acad. Sci.* **151**, 616.

DAWSON, D. M., EPPENBERGER, H. M. and KAPLAN, N. O. (1967), *J. biol. Chem.* **242**, 210.

DELBRÜCK, A., ZEBE, E. and BÜCHER, T. (1959), *Biochem. Z.* **331**, 273.

DEUL, D. H. and VAN BREEMAN, J. F. L. (1964a), Abstracts, 1st meeting, Federation of European Biochemical Societies, p. 52.

DEUL, D. H. and VAN BREEMAN, J. F. L. (1964b), *Clin. Chim. Acta* **10**, 276.

EDSALL, J. T. (1968), *Ann. N.Y. Acad. Sci.* **151**, 41.

ENGSTRÖM, L. (1961), *Biochim. biophys. Acta* **52**, 36.

ENGSTRÖM, L. (1964), *Biochim. biophys. Acta* **92**, 78.

EPPENBERGER, H. M., DAWSON, D. M. and KAPLAN, N. O. (1967), *J. biol. Chem.* **242**, 204.

EPSTEIN, C. J. and SCHECHTER, A. N. (1968), *Ann. N.Y. Acad. Sci.* **151**, 85.

FISHMAN, W. H. and GHOSH, N. K. (1967), *Advances in Clinical Chemistry*, ed. BODANSKY, O. and STEWART, C. P., Vol. 10, New York: Academic Press, p. 255.

FISHMAN, W. H., INGLIS, N. I. and GHOSH, N. K. (1968), *Clin. Chim. Acta* **19**, 71.

FOXWELL, C. J., CRAN, E. J. and BARON, D. N. (1966), *Biochem. J.* **100**, 44P.

FRITZ, P. J. (1963), *Federation Proc.* **22**, 241.

FRITZ, P. J. and JACOBSON, K. B. (1963), *Science,* **140**, 64.

FRITZ, P. J. and JACOBSON, K. B. (1965), *Biochemistry* **4**, 282.

FUJIMOTO, Y., NAZARIAN, I. and WILKINSON, J. H. (1968), *Enzymol. biol. clin.* **9**, 124.

GAFFNEY, P. J., JR (1968), *Biochem. J.* **110**, 12P.

GHOSH, N. K. and FISHMAN, W. H. (1967), *Federation Proc.* **26**, 558.

GHOSH, N. K. and FISHMAN, W. H. (1968), *Biochem. J.* **108**, 779.

GHOSH, N. K., GOLDMAN, S. S. and FISHMAN, W. H. (1967), *Enzymologia* **33**, 113.

GOEDDE, H. W., GEHRING, D. and HOFMANN, R. (1965), *Z. anal. Chem.* **212**, 238.

GOLDBERG, E. (1963), *Science,* **139**, 602.

GOLDBERG, E. (1965), *Arch. Biochem. Biophys.* **109**, 134.

GOLDBERG, E. and WUNTCH, T. (1967), *J. exp. Zool.* **165**, 101.

GREGORY, K. F. and WRÓBLEWSKI, F. (1958), *J. Immunol.* **81**, 359.

GRIMM, F. C. and DOHERTY, D. G. (1961), *J. biol. Chem.* **236**, 1980.

HARRIS, H., HOPKINSON, D. A. and ROBSON, E. B. (1962), *Nature, Lond.* **196**, 1296.

HARRIS, H. and ROBSON, E. B. (1963), *Biochim. biophys. Acta* **73**, 649.

HAWTREY, C. and GOLDBERG, E. (1968), *Ann. N.Y. Acad. Sci.* **151**, 611.

HESS, A., ANGEL, R. W., BARRON, K. D. and BERNSOHN, J. (1963), *Clin. Chim. Acta* **8**, 656.

HOOTON, B. T. and WATTS, D. C. (1966), *Biochem. J.* **99**, 53P.

HOUSSAIS, J. F. (1966), *Biochim. biophys. Acta* **128**, 239.

HRUBAN, Z., SLESERS, A. and ORLANDO, R. (1967), *Lab. Invest.* **16**, 550.

JACOBS, H., HELDT, H. W. and KLINGENBERG, M. (1964), *Biochem. biophys. Res. Commun.* **16**, 516.

KAPLAN, N. O. (1968), *Ann. N.Y. Acad. Sci.* **151**, 382.

KETO, A. I. and DOHERTY, M. D. (1968), *Biochim. biophys. Acta* **151**, 721.

KITTO, G. B., WASSARMAN, P. M. and KAPLAN, N. O. (1966), *Proc. Natl. Acad. Sci., Wash.* **56**, 578.

KITTO, G. B., WASSARMAN, P. M., MICHEJDA, J. and KAPLAN, N. O. (1966), *Biochem. biophys. Res. Commun.* **22**, 75.

KOEN, A. L. (1967a), *Biochim. biophys. Acta* **140**, 487.

KOEN, A. L. (1967b), *Biochim. biophys. Acta* **140**, 496.

KOEN, A. L. and SHAW, C. R. (1965), *Biochim. biophys. Acta* **96**, 231.

LA MOTTA, R. V., MCCOMB, R. B., NOLL, C. R., JR, WETSTONE, H. J. and REINFRANK, R. F. (1968), *Arch. Biochem. Biophys.* **124**, 299.

LA MOTTA, R. V., MCCOMB, R. B. and WETSTONE, H. J. (1965), *Can. J. Physiol. Pharmacol.* **43**, 313.

LATNER, A. L. (1964), *Proc. Ass. clin. Biochem.* **3**, 120.

LAURENT, G., CHARREL, M., LUCCIONI, F., AUTRAN, M. F. and DERRIEN, Y. (1965), *Bull. Soc. Chim. Biol.* **47**, 1101.

LAZDUNSKI, C. and LAZDUNSKI, M. (1967), *Biochim. biophys. Acta* **147**, 280.

MALAMY, M. H. and HORECKER, B. L. (1964), *Biochemistry* **3**, 1893.

MANN, K. G. and VESTLING, C. S. (1968), *Biochim. biophys. Acta* **159**, 567.

MARKERT, C. L. (1963a), *Science*, **140**, 1329.

MARKERT, C. L. (1963b), *Cytodifferential and Macromolecular Synthesis*, New York: Academic Press, p. 65.

MARKERT, C. L. and APPELLA, E. (1963), *Ann. N.Y. Acad. Sci.* **103**, 915.

MARKERT, C. L. and MASSARO, E. J. (1968), *Science*, **162**, 695.

MARKERT, C. L. and URSPRUNG, H. (1962), *Devel. Biol.* **5**, 363.

MARKERT, C. L. and WHITT, G. S. (1968), *Experientia* **24**, 977.

MARTINEZ-CARRION, M., TURANO, C., CHIANCONE, E., BOSSA, F., GIARTOSIO, A., RIVA, F. and FASELLA, P. (1967), *J. biol. Chem.* **242**, 2397.

MILLAR, D. B. S. (1962), *J. biol. Chem.* **237**, 2135.

MILSTEIN, C. (1964), *Biochem. J.* **92**, 410.

MONOD, J., WYMAN, J. and CHANGEUX, J.-P. (1965), *J. molec. Biol.* **12**, 88.

MOSS, D. W. (1964), *Biochem. J.* **92**, 16P.

MOSS, D. W., EATON, R. H., SMITH, J. K. and WHITBY, L. G. (1966), *Biochem. J.* **98**, 32C.

NEVULDINE, B. H. and LEVY, H. R. (1965), *Biochem. biophys. Res. Commun.* **21**, 28.

NISSELBAUM, J. S. and BODANSKY, O. (1959), *J. biol. Chem.* **234**, 3276.

NISSELBAUM, J. S. and BODANSKY, O. (1960), *Federation Proc.* **19**, 336.

NISSELBAUM, J. S. and BODANSKY, O. (1961a), *J. biol. Chem.* **236**, 323.

NISSELBAUM, J. S. and BODANSKY, O. (1961b), *J. biol. Chem.* **236**, 401.

NYMAN, P.-O. (1961), *Biochim. biophys. Acta* **52**, 1.

NYMAN, P.-O., STRID, L. and WESTERMARK, G. (1966), *Biochim. biophys. Acta* **122**, 554.

PENHOET, E., KOCHMAN, M., VALENTINE, R. and RUTTER, W. J. (1967), *Biochemistry* **6**, 2940.

PLAGEMANN, P. G. W., GREGORY, K. F. and WRÓBLEWSKI, F. (1960a), *J. biol. Chem.* **235**, 2282.

PLAGEMANN, P. G. W., GREGORY, K. F. and WRÓBLEWSKI, F. (1960b), *J. biol. Chem.* **235**, 2288.

POLLOCK, M. R. (1968), *Ann. N.Y. Acad. Sci.* **151**, 502.

PORTMANN, P., ROSSIER, R. and CHARDONNENS, H. (1960), *Helv. physiol. pharmacol. Acta* **18**, 414.

POULIK, M. D. (1968), *Ann. N.Y. Acad. Sci.* **151**, 476.

RAYMOND, S. and AURELL, B. (1962), *Science*, **138**, 152.

RICKLI, E. E. and EDSALL, J. T. (1962), *J. biol. Chem.* **237**, PC 258.

RICKLI, E. E., GHAZANFAR, S. A. S., GIBBONS, B. H. and EDSALL, J. T. (1964), *J. biol. Chem.* **239**, 1065.

ROBINSON, J. C. and PIERCE, J. E. (1964), *Nature, Lond.* **204**, 472.

ROBSON, E. B. and HARRIS, H. (1966), *Ann. hum. Genet.* **30**, 219.

ROSALKI, S. B. (1965), *Nature, Lond.* **207**, 414.

RUTTER, W. J. (1962), *The Enzymes*, ed. BOYER, P. D., LARDY, H. A. and MYRBÄCK, K., 2nd ed., Vol. 5, New York: Academic Press, p. 341.

SALTHE, S. N., CHILSON, O. P. and KAPLAN, N. O. (1965), *Nature, Lond.* **207**, 723.

SARASWATHI, S. and BACHHAWAT, B. K. (1968), *Biochem. J.* **107**, 185.

SCANDALIOS, J. G. (1968), *Ann. N.Y. Acad. Sci.* **151**, 274.

SCHECHTER, A. N. and EPSTEIN, C. J. (1968), *Science*, **159**, 997.

SCHLESINGER, M. J. (1965), *J. biol. Chem.* **240**, 4293.

SCHLESINGER, M. J. (1967*a*), *J. biol. Chem.* **242**, 1599.

SCHLESINGER, M. J. (1967*b*), *J. biol. Chem.* **242**, 1604.

SCHLESINGER, M. J. and ANDERSEN, L. (1968), *Ann. N.Y. Acad. Sci.* **151**, 159.

SCHLESINGER, M. J. and LEVINTHAL, C. (1963), *J. molec. Biol.* **7**, 1.

SCHWARTZ, J. H., CRESTFIELD, A. M. and LIPMANN, F. (1963), *Proc. Natl. Acad. Sci., Wash.* **49**, 722.

SOETENS, A., KARCHER, D., VAN SANDE, M. and LOWENTHAL, A. (1965), in *Enzymes in Clinical Chemistry*, ed. RUYSSEN, R. and VANDENDRIESSCHE, L., Amsterdam: Elsevier, p. 130.

STAMBAUGH, R. and BUCKLEY, J. (1967), *J. biol. Chem.* **242**, 4053.

STEGINK, L. D. and VESTLING, C. S. (1966), *J. biol. Chem.* **241**, 4923.

SUNDARAM, T. K. and FINCHAM, J. R. S. (1964), *J. molec. Biol.* **10**, 423.

SVENSMARK, O. (1961), *Danish med. Bull.* **8**, 28.

SWANSON, P. D. (1967), *J. Neurochem.* **14**, 343.

THORNE, C. J. R. (1960), *Biochim. biophys. Acta*, **42**, 175.

THORNE, C. J. R., GROSSMAN, L. I. and KAPLAN, N. O. (1963), *Biochim. biophys. Acta* **73**, 193.

VESELL, E. S. and BRODY, I. A. (1964), *Ann. N.Y. Acad. Sci.* **121**, 544.

WACHSMUTH, E. D., PFLEIDERER, G. and WIELAND, T. (1964), *Biochem. Z.* **340**, 80.

WATTS, D. C., RABIN, B. R. and CROOK, E. M. (1961), *Biochim. biophys. Acta* **48**, 380.

WELLNER, D. and HAYES, M. B. (1968), *Ann. N.Y. Acad. Sci.* **151**, 118.

WIELAND, T., GEORGOPOULOS, D., KAMPE, H. and WACHSMUTH, E. D. (1964), *Biochem. Z.* **340**, 483.

WIELAND, T. and PFLEIDERER, G. (1961), *Ann. N.Y. Acad. Sci.* **94**, 691.

WIELAND, T., PFLEIDERER, G., HAUPT, I. and WÖRNER, W. (1959), *Biochem. Z.* **331**, 103.

WIELAND, T., PFLEIDERER, G. and RAJEWSKY, K. (1960), *Naturforsch.* **15b**, 434.

WILKINSON, J. H. and WITHYCOMBE, W. A. (1965), *Biochem. J.* **97**, 663.

WILSON, D. A. and CRAWFORD, I. P. (1965), *J. biol. Chem.* **240**, 4801.

YANOFSKY, C. (1960), *Bacteriol. Rev.* **24**, 221.

YOSHIDA, A. (1966), *J. biol. Chem.* **241**, 4966.

YOSHIDA, A. (1967), *Proc. Natl. Acad. Sci., Wash.* **57**, 835.

YOSHIDA, A. (1968), *Ann. N.Y. Acad. Sci.* **151,** 145.

ZINKHAM, W. H. (1968), *Ann. N.Y. Acad. Sci.* **151,** 598.

ZINKHAM, W. H., BLANCO, A. and KUPCHYK, L. (1963), *Science,* **142,** 1303.

ZINKHAM, W. H., BLANCO, A. and KUPCHYK, L. (1964), *Science,* **144,** 1353.

ZINKHAM, W. H., KUPCHYK, L., BLANCO, A. and ISENSEE, H. (1965), *Nature, Lond.* **208,** 284.

ZWAIG, N. and MILSTEIN, C. (1964), *Biochem. J.* **92,** 421.

CHAPTER 5

Enzyme Multiplicity in the Glycolytic Pathway and the Pentose–Phosphate Cycle

During the past few years many of the enzymes catalysing the reactions of the Embden–Meyerhof glycolytic pathway and the pentose–phosphate cycle have been demonstrated to occur in multiple forms. The former are, of course, of major interest in connection with the investigation of the mechanism of action of insulin, but at the present time it must be admitted that many of our conclusions are hypothetical. The isoenzymes of the pentose–phosphate cycle are of great importance in the study of hereditary anaemias, both from the diagnostic and genetic points of view. The principal enzymes concerned are shown in the diagram of both pathways illustrated in Fig. 18.

The present chapter is therefore devoted to a consideration of the more important enzymes concerned with the intermediary metabolism of glucose, which are known to exist in multiple forms: hexokinase, phosphoglucomutase, glucose 6-phosphate dehydrogenase, 6-phosphogluconate dehydrogenase, aldolase, glyceraldehyde 3-phosphate dehydrogenase and pyruvate kinase. Lactate dehydrogenase should also be included in this group, but it is of such general interest that it merits separate discussion (Chapter 6).

Hexokinase

The first indication of the heterogeneity of hexokinase was obtained by Kunitz and McDonald (1946), who observed that the yeast enzyme exhibits a double boundary on ultracentrifugation. This observation has been confirmed by column chromatography and by electrophoresis (Lazarus, Ramel, Rustum and Barnard, 1966; Derechin, Ramel, Lazarus and Barnard, 1966; Schulze, Gazith and Gooding, 1967). The separation of six components on a DEAE-cellulose column (Trayser and Colowick, 1961; Kaji, Trayser and Colowick, 1961) has now been shown to be due to proteolytic degradation during preparation, and it is generally agreed that two forms only occur in yeast (Gazith, Schulze, Gooding, Womack and Colowick, 1968).

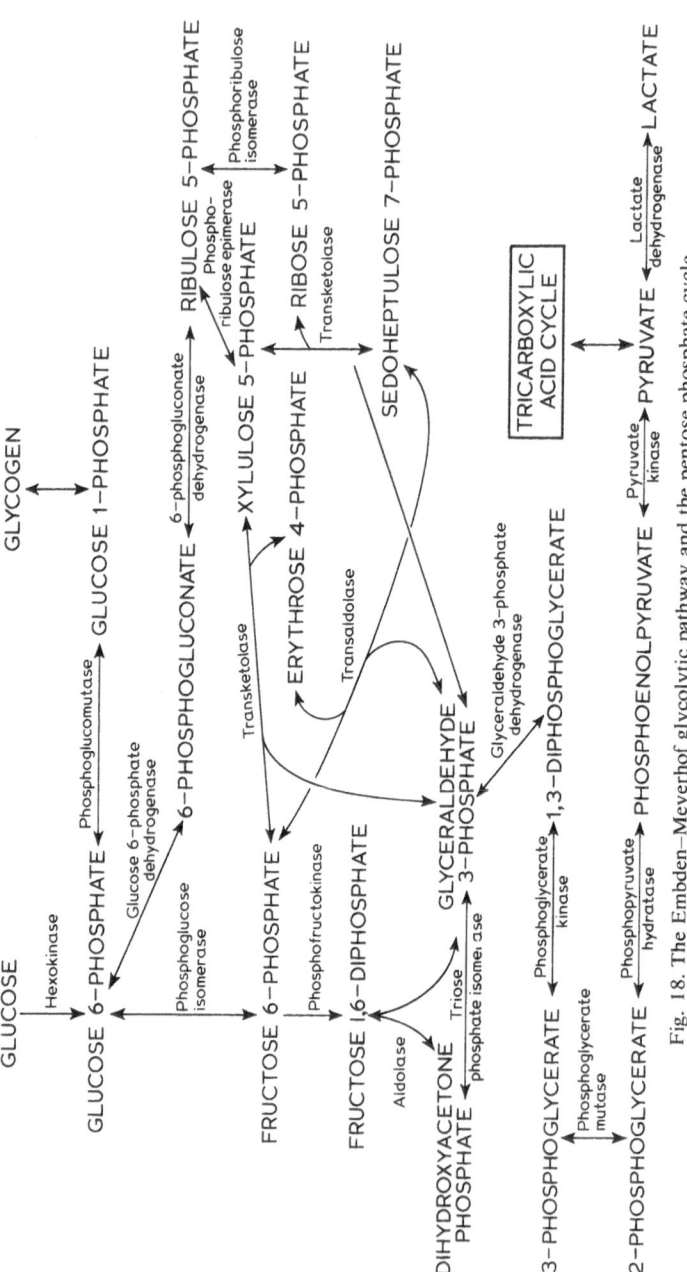

Fig. 18. The Embden–Meyerhof glycolytic pathway and the pentose phosphate cycle.

The two forms differ in amino-acid composition, substrate affinity and enzymatic activity. Each is converted without loss of activity into a more acidic form by mild treatment with a protease in the presence of glucose. The acidic forms have stronger substrate-binding capacities. The enzyme molecule appears to be a dimer consisting of two polypeptide sub-units, each with a molecular weight of about 24,000. Normal dimers show a greater tendency than the acidic dimers to polymerize into tetramers, which are reversibly dissociated into the corresponding dimeric forms by glucose. Gazith *et al.* (1968) conclude that glucose is preferentially bound to the dimeric isoenzymes.

In early investigations mammalian liver was shown to contain two hexokinase fractions, but most other tissues contain only a single component (Walker, 1963; Sharma, Manjeshwar and Weinhouse, 1963; Vinuela, Salas and Sols, 1963). The two liver components were characterized as 'low-K_m hexokinase' and 'high-K_m hexokinase', but subsequently evidence was obtained showing that the former occurs in multiple forms. Gonzales, Ureta, Sanchez and Niemeyer (1964) separated four hexokinase fractions from rat liver by DEAE-cellulose column chromatography. The presence of these fractions was confirmed by starch-gel electrophoresis by Katzen and Schimke (1965), who described them as I to IV in order of their increasing mobility towards the anode. This is contrary to the numbering system now officially recommended for isoenzymes, but as it has been followed by other investigators, it will be retained during this discussion.

Types I to III are 'low-K_m hexokinases', while type IV is a 'high-K_m hexokinase' which can be demonstrated only when a relatively high concentration of substrate (1M-glucose) is used in the staining reagent (Fig. 19). The presence of multiple hexokinases was detected in other rat tissues and in the tissues of other species, but the 'high-K_m glucokinase' appears to be confined to the liver and kidney (Katzen and Schimke, 1965; Pilkis and Hansen, 1968). An additional 'low-K_m hexokinase', which on starch-gel electrophoresis in veronal buffer at pH 7·4 migrates towards the cathode, has recently been detected in human and dog liver and adipose tissue (Brown, Miller, Holloway and Leve, 1967).

Space limitations preclude comprehensive discussion of the general properties of the hexokinases in this monograph, and the reader is referred to the excellent reviews by Walker (1966) and Katzen (1967).

Detection of hexokinase activity after zone electrophoresis
Hexokinase isoenzymes may be detected in gel electropherograms by a modification of the tetrazolium procedure discussed in Chapter 3. The

staining mixture contains glucose and ATP, which serve as substrates for the hexokinase reaction: the product, glucose 6-phosphate, is oxidized by glucose 6-phosphate dehydrogenase in the presence of Mg^{2+} and the coenzyme NADP. The reduced form of NADP is coupled to the tetrazolium–formazan system by means of methylphenazonium methosulphate (Katzen and Schimke, 1966; Eaton, Brewer and Tashian, 1966; Brown *et al.*, 1967).

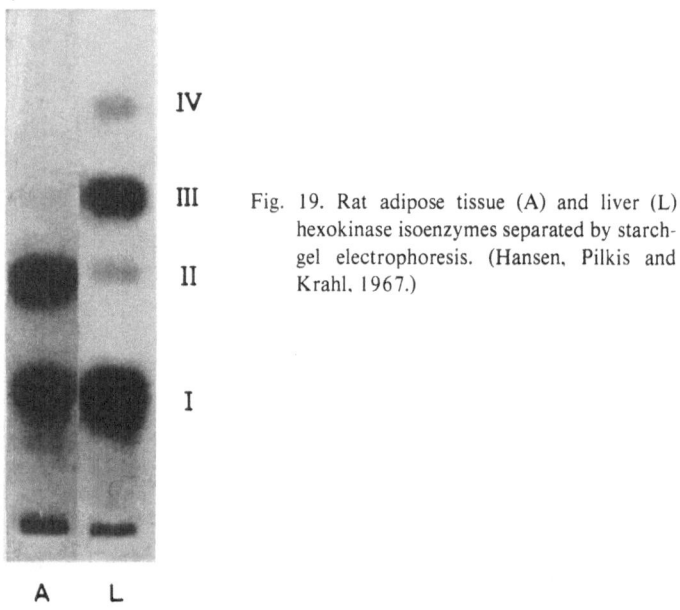

Fig. 19. Rat adipose tissue (A) and liver (L) hexokinase isoenzymes separated by starch-gel electrophoresis. (Hansen, Pilkis and Krahl, 1967.)

Earlier Trayser and Colowick (1961) had adapted the assay procedure of Darrow and Colowick (1961), in which the change in pH during the hexokinase reaction was detected in an agar overlay containing the reactants and a suitable indicator, cresol red.

A more recent development of particular interest in the study of erythrocytic hexokinase is the fluorescent technique devised by Kaplan and Beutler (1968). The hexokinase reaction is coupled with that of glucose 6-phosphate dehydrogenase as in the tetrazolium method, but the presence of reduced NADP is detected by ultra-violet photography. This method has the advantage of avoiding interference due to the presence of haemoglobin in the gel.

Hexokinase isoenzymes in the erythrocyte
Although the heterogeneity of erythrocytic hexokinase was first reported at the end of 1966, several groups have already described their

electrophoretic studies, and some of the findings have formed the subject of a lively controversy. Eaton, Brewer and Tashian (1966) found seven bands of activity (B 1–7) after the starch-gel electrophoresis of human haemolysates. The slowest of these bands (B 7) migrated slightly faster than haemoglobin at pH 8·0. Greatest activity was found in the B 5 band. Qualitatively, similar patterns were observed in American Caucasians, American Negroes and Israelis, the only variation being in the intensity of staining of the various bands.

A much faster (A) band was found in reticulocyte-rich haemolysates, an observation consistent with the finding that the total hexokinase activity of the erythrocyte decreases with ageing (Brewer and Powell, 1963; Brewer, Powell, Swanson and Alving, 1964; Valentine, Oski, Paglia, Baughan, Schneider and Naiman, 1967). Eaton *et al.* (1966) suggest that it is associated with the reticulum in the red cell, since the enzyme is known to be associated with particulate matter in other tissues (Crane and Sols, 1953).

Holmes, Malone, Winegrad and Oski (1967), however, found two bands of hexokinase activity which migrated on starch gel like bands I and III of rat liver. Both bands stained more intensely when 5×10^{-4} M-glucose rather than 0·1M-glucose was used as substrate, and hence were of the 'low-K_m' type. Band III was detected in all of 21 haemolysates, but band I was found in only 17. Type II hexokinase was demonstrated in each of 17 haemolysates from blood of the new-born, but in none of the adult samples. In 5 cases type II was the only hexokinase observed, but in the remaining new-born samples type I or III (or both) was also present. This is consistent with the observation of Katzen and Schimke (1965) that type II hexokinase occurs in the liver of new-born rats but can scarcely be demonstrated in adult rat liver.

Holmes *et al.* (1967) suggest a possible relationship between the presence of type II hexokinase and that of foetal haemoglobin, since they were able to demonstrate type II hexokinase in haemolysates from an individual known to be homozygous for the persistent foetal haemoglobin trait and in members of his family heterozygous for the trait. Furthermore, the red cells of a 1-day-old dysmature infant contained only haemoglobin A and type I and III hexokinase.

These authors suggested that the additional bands of activity reported by Eaton *et al.* (1966) might possibly result from technical differences in preparing the haemolysates and in performing the electrophoretic separation. Brewer and Knutsen (1968), however, obtained seven bands of hexokinase activity when using a technique similar to that of Holmes *et al.* (1968) and counter-suggested that the latter authors had failed to resolve all the components. Moreover, they were unable to confirm any association between any of their hexokinase bands and the presence of foetal haemoglobin, an observation which led to the further suggestion that problems of interpretation might arise as a result of the encroachment of adult haemoglobin into the hexokinase patterns obtained by Holmes *et al.* (1967).

The latter, however, are in accordance with the patterns obtained with other mammalian tissues (Katzen, 1967; Brown *et al.*, 1967). Holmes *et al.* (1968) refer to unpublished work by Oski and Rose in which hexokinase types I and II have been isolated from haemolysates by column chromatography on DEAE-cellulose. With the aid of a slightly modified electrophoretic technique they observed faint bands of type II in the haemolysates prepared from the red cells of a proportion of normal adults. They also report finding increased amounts of type II hexokinase in the red cells of adults homozygous for haemoglobin C or S, and suggest that the proportions of the three hexokinases may also be under genetic control.

Using the fluorometric procedure outlined above (p. 92), Kaplan and Beutler (1968) have also investigated human erythrocytic hexokinase, and have obtained results which might explain the presence of the additional bands found by Eaton *et al.* (1966). They observed that in normal adults a much weaker type III component was accompanied by a double type I band. In most individuals the two type I bands were of equal intensity. The slow band comparable with type I of other tissues was described as type I_A and the other factor as type I_F. Type I_F was found to be more prominent in cord-blood haemolysates. When electrophoresis was carried out at pH 8·6 they observed that haemoglobin A tends to obscure band I_F and haemoglobin F masks band I_A, so that type I_A alone can be seen in adult red cells and type I_F in foetal red cells. At pH 8·0 haemoglobin does not interfere with the patterns and both sub-fractions can be clearly seen. This was confirmed by examining the hexokinases of haemolysates from which haemoglobin had been removed with DEAE-cellulose.

Other investigators have reported double type II and type IV bands in rat-liver and other tissues (Katzen and Schimke, 1965; Hansen, Pilkis and Krahl, 1967a, 1967b) and Kaplan and Beutler (1968) suggest that such duplication might account for the extra bands found by Eaton *et al.* (1966). However, they were unable to confirm the presence of type II hexokinase in foetal red cells. At the present time therefore the status of the various erythrocytic hexokinases remains to be settled.

Hexokinases in other tissues

The occurrence of three 'low-K_m hexokinases' (glucose–ATP–6-phosphotransferase) and one 'high-K_m glucokinase' in a variety of rat tissues has been briefly discussed in the introduction to this section. Their relative distribution is summarized in Table 5, from which it will be seen that the type IV glucokinase is almost exclusively confined to the liver and kidney. The major hexokinase of most tissues is the type I isoenzyme, but in skeletal muscle and certain others type II predominates.

TABLE 5 *Relative distribution of hexokinase isoenzymes in rat tissues* |Based upon Katzen (1967) and Hansen *et. al.* (1967*a*, 1967*b*)|

Tissue	Insulin sensitivity	Hexokinase isoenzymes				
		Sperm type	I	II	III	IV
Brain	Insensitive	---	+ + +	+	tr.	0
Erythrocytes	Insensitive	---	+ + +	+	tr.	0
Testis	Insensitive	+ +	+ + +	+ + +	tr.	0
Kidney	Insensitive	---	+ + +	+	+ +	+
Liver	Insensitive	---	+ + +	+	+ + +	+ + +
Lung	Insensitive	--	+ + +	+ + +	+ + +	0
Diaphragm	Sensitive	---	+	+ + +	+	0
Sk. muscle	Sensitive	---	+	+ + +	+	0
Fat pad (young)	Sensitive	--	+	+ + +	+	0
Heart	Sensitive	---	+ +	+ + +	+	0

An extra band of activity, which remains near the point of application during starch-gel electrophoresis, occurs in extracts of rat testis and the epididymal fat pad. This cannot be detected in the fat pads of young, immature rats, and appears to be derived from the sperm (Katzen, 1967). The presence of a testis-specific hexokinase has also been reported in *Drosophila melanogaster*. It first appears during the pupal period (Murray and Ball, 1967).

The various rat-tissue hexokinases have been purified and their properties compared. They differ considerably in their affinities for glucose (Table 6). Type III has the lowest K_m value, and is sometimes difficult to detect on starch-gel electrophoresis owing to inhibition by the high substrate concentrations necessary to obtain adequate staining of the bands with lower affinities. Schimke and Grossbard (1968) report marked differences not only in their catalytic properties but also in their sensitivity to chemical inactivation and their stability at 45˙. In both respects type I is the most stable and type II the least stable of the hexokinases. Type IV glucokinase is also susceptible to heat inactivation. Two electrophoretically distinct components of type IV glucokinase have recently been isolated and shown to differ from each other in their salt solubilities and immunochemical properties, though their Michaelis constants for glucose are similar and they appear to have comparable molecular weights (Pilkis and Hansen, 1968).

TABLE 6 K_m values of rat-tissue hexokinase isoenzymes (Katzen, 1967)

Hexokinase type	Tissue	K_m glucose (M)
Sperm type	Testis	ca. 10^{-4}
I	Liver	$2 \cdot 5 \times 10^{-5}$
II	Muscle	$2 \cdot 0 \times 10^{-4}$
III	Liver	$5 \cdot 0 \times 10^{-6}$
IV	Liver	$1 \cdot 6 \times 10^{-2}$

Chromatography on Sephadex G-100 indicates that the type IV component has a molecular weight of about 48,000 (Pilkis and Krahl, 1966), whereas those of types I, II and III are about 96,000 (Grossbard and Schimke, 1966). This suggests the possibility of a sub-unit structure and a hybridization mechanism to account for the band duplication referred to on p. 94, and also for the occurrence of the sperm-type enzyme, but this remains to be verified. However, some recent immunochemical studies hardly support the idea of a structural relationship between type IV glucokinase and the hexokinases, for antisera to a highly purified rat glucokinase did not inhibit the 'low-K_m hexokinases' (Pilkis, Hansen and Krahl, 1968).

Before the heterogeneity of the hexokinases was recognized there was considerable evidence indicating that the hexokinase activity of rat liver was an adaptive enzyme which was reduced after fasting or in diabetes, but was restored by feeding again or by treatment with insulin (DiPietro and Weinhouse, 1960; Walker and Rao, 1964; Sols, Sillero and Salas, 1965). This effect now appears to be mediated by type II hexokinase, since all rat tissues known to be highly sensitive to insulin, e.g. the diaphragm, epididymal fat pad, muscle and heart, contain large amounts of this isoenzyme. By contrast, type I hexokinase predominates in those, such as the brain and kidney, which are relatively insensitive to the action of insulin (Katzen and Schimke, 1965; Katzen, 1967). The reversible diminution mainly in the type II hexokinase in diabetic or fasted rats is illustrated in Fig. 20, which also demonstrates enzyme restoration on correction of the condition (Hansen et al., 1967b).

Type II hexokinase exists in two electrophoretically distinct forms which can be detected in the fat pad and other insulin-sensitive tissues. The more rapidly migrating of these disappears on fasting or in drug-induced diabetes (Katzen and Schimke, 1965; Katzen, 1967; Katzen, Soderman and Cirillo, 1968). The two forms, however, cannot be distinguished when mercaptoethanol is incorporated into the staining mixture, a result which led Katzen to postulate that the insulin-sensitive form was possibly a disulphide-bridged polymer or a hybridized form of the other.

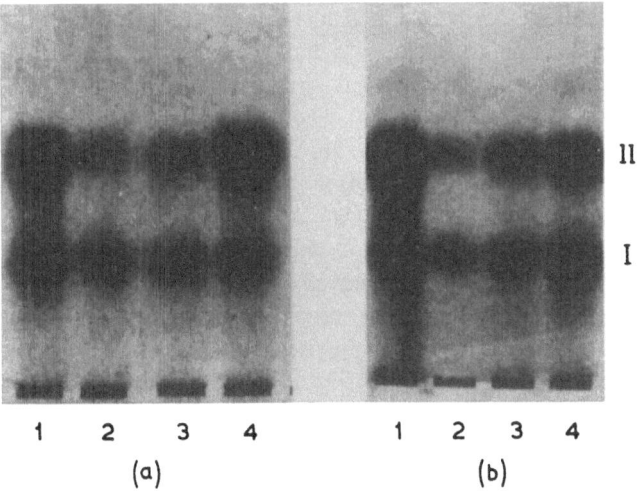

Fig. 20. Effect of nutritional status on adipose tissue hexokinase iso enzymes in: (*a*) fasted and refed rats. (1. normal; 2. fasted 72 hr.; 3. fasted 72 hr. and refed 24 hr.; 4. fasted 72 hr. and refed 48 hr., and (*b*) diabetic and insulin-treated rats (1. normal; 2. diabetic; 3. diabetic + insulin 24 hr.; 4. diabetic + insulin 48 hr.). (Hansen, Pilkis and Krahl, 1967.)

It is further suggested that it is the predominance of type I hexokinase in tissues which are less sensitive to insulin which is responsible for the ability of these tissues to function adequately in the absence of the hormone. Katzen (1967) has put forward a working hypothesis for the mechanism of insulin which relates the multiplicity of the hexokinases with the action of insulin on glucose transport at the cell membrane.

Apart from the erythrocytes, few human tissues have hitherto been extensively studied, but in general the hexokinase patterns appear to resemble those of the rat. In the liver and adipose tissue Brown *et al.* (1967) observed types I to IV, of which band IV was a 'high-K_m

glucokinase'. The latter seemed to be related to the glucose intake, for it was absent from the livers of poorly nourished patients, but was found in all well-nourished individuals. No such difference, however, was found in the distribution of hexokinase types I, II and III. In human- and dog-liver and adipose tissue Brown *et al.* (1967) observed an additional zone of 'low-K_m hexokinase' activity migrating towards the cathode at pH 7·4, but its significance has not yet been established.

Phosphoglucomutase

Comparatively few studies on the heterogeneity of phosphoglucomutase have so far been made, though thanks mainly to the recent contributions of Handler and his co-workers, we now have a substantial knowledge of the general biochemistry of this enzyme (Joshi and Handler, 1964; Hashimoto and Handler, 1966). The first demonstrations of its multiplicity were those of Yankeelov, Horton and Koshland (1964) in rabbit muscle, Tsoi and Douglas (1964) in yeast, and Spencer, Hopkinson and Harris (1964) who reported seven zones of activity after the starch-gel electrophoresis of human haemolysates.

Spencer *et al.* (1964) employed a tetrazolium-staining technique similar to that used for the location of dehydrogenases: the substrate is glucose 1-phosphate, which in the presence of glucose 1,6-diphosphate is converted into glucose 6-phosphate. This product is oxidized by the NADP-dependent glucose 6-phosphate dehydrogenase, a reaction which may be coupled with the tetrazolium–formazan system. Appropriate controls must be performed to ensure that other red-cell enzymes are not responsible for the appearance of the formazan.

In population studies Spencer *et al.* (1964) observed three distinct patterns of phosphoglucomutase isoenzymes: type 1, in which isoenzymes *a, c, e, f* and *g* are present; type 2, in which components *b, d, e, f* and *g* occur; and type 2-1, in which all seven bands may be detected (Fig. 21). Individuals of type 2-1 exhibit smaller amounts of *a* and *c* than those of type 1 and smaller amounts of *b* and *d* than those of type 2. Familial studies indicate that the three types are genetically determined, and calculations based upon estimates of the postulated gene frequencies in the general population show excellent agreement between the expected and observed numbers of the different phenotypes. Similar patterns have also been observed in Caucasian and Negro Americans (Brewer, Bowbeer and Tashian, 1967).

A number of unusual phenotypes have since been reported. Members of one family studied by Hopkinson and Harris (1965) exhibited anomalous *e, f* and *g* components. Usually bands *e* and *f* are much more intensely stained than band g, but in this family band g stained intensely, and a further prominent band migrating faster than g was also found. The authors interpret this as evidence for a second 'structural' locus determining phosphoglucomutase components e, f and g. A similar conclusion was reached by Tsoi and Douglas (1964) on the basis of their studies of the yeast enzyme. Brewer *et al.* (1967) have also reported a variant in whom bands e and f appear to have split to form a pair of e and f bands which migrate faster than the g band.

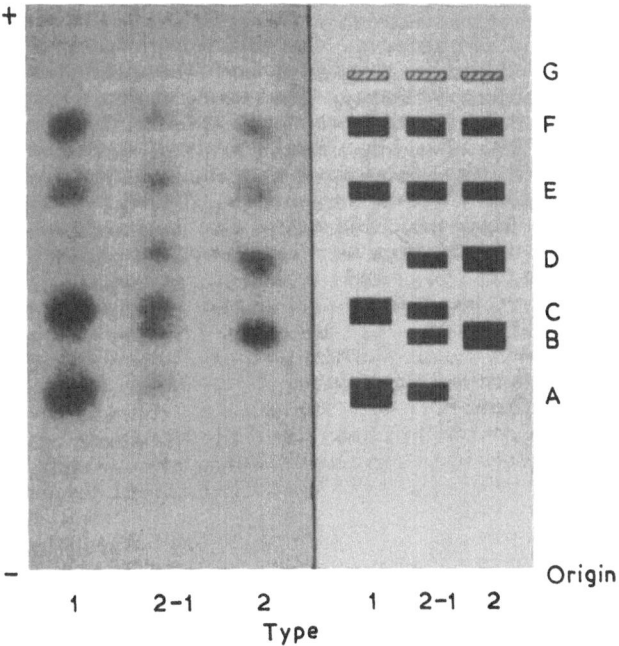

Fig. 21. Phosphoglucomutase isoenzyme patterns obtained by starch-gel electrophoresis of haemolysates from individuals of phenotypes 1, 2-1 and 2. (Spencer, Hopkinson and Harris, 1964.)

Several other uncommon phenotypes were observed by Hopkinson and Harris (1965): these included 3-1 and 3-2, which represent heterozygotes for an uncommon allele with the usual forms. Some thirteen different electrophoretic patterns have now been reported by Harris, Hopkinson and Luffman (1968).

Several studies of the usual gene frequencies have been performed on a variety of racial groups. No very clear-cut differences have been observed, though the gene frequency for type 1 appears to be somewhat higher in Alaskan and Mexican Indians than in other races (Scott, Duncan, Ekstrand and Wright, 1966; Lisker and Giblett, 1967).

The multiple forms of phosphoglucomutase from a variety of sources have also been investigated by column chromatography. Three peaks of activity were obtained from rabbit muscle with a CM-cellulose column (Yankeelov *et al.*, 1964), and up to four peaks were detected in the eluate from CM-Sephadex columns by Joshi, Hooper, Kuwacki, Sakurada, Swanson and Handler (1967). The distribution in some of the tissues studied is summarized in Table 7.

TABLE 7 *The relative distribution of phosphoglucomutases separated by CM-Sephadex chromatography* [Based upon Yankeelov *et al.* (1964) and Joshi *et al.* (1967)]

Source	Phosphoglucomutase eluted in peak (%)			
	I	II	III	IV
Human muscle	80	14	6	—
Rabbit liver	6	54	11	29
Rabbit muscle	20	70	10	—
Rat muscle	64	24	12	—
Rat liver (adult)	92	8	1	—
Rat liver (new-born)	96–97	3–4	—	—
Flounder muscle	99·5	0·5	—	—
Sweet potato*	1–2	60	20	18
E. coli	>99·9	—	—	—

* On DEAE-Sephadex.

Peak III was shown to consist of a dephosphorylated form of peak II, since when peak III was labelled with glucose 1-phosphate-^{32}P and mixed with peaks I and II, then rechromatographed, the radioactivity appeared in peak II (Joshi *et al.*, 1967; Yankeelov *et al.*, 1964). Similarly, the peak IV found in fresh rabbit liver and the sweet potato was shown to be a dephosphorylated form of peak I (Joshi *et al.*, 1967).

It seems therefore that phosphoglucomutase occurs in only two chromatographically separable forms, and further reports from Handler's laboratory on the behaviour of these components on starch-gel electrophoresis are awaited with interest. It is possible that the *a* and *b* components observed by Harris and his co-workers in human haemolysates might be the dephosphorylated forms of the *c* and *d* isoenzymes respectively.

Meanwhile Joshi *et al.* (1967) have investigated the chemical nature of their peaks I and II, and have found peak I to contain more leucine, tyrosine and methionine, and peak II to have more histidine, alanine and valine. 'Fingerprint' maps of tryptic digests, in which 46 of the 65 theoretical spots were distinguished, were consistent with their conclusion from ultracentrifugal studies that the enzyme consists of a single chain with a molecular weight of about 62,000.

They also observed that the phosphoglucomutase activity of rat liver and muscle is reduced by about 50% during starvation and that this is substantially restored on subsequent treatment with insulin or adrenaline. Furthermore, the changes were confined to peak I, peak II being relatively unaffected, but the significance of these observations remains to be established.

Up to the time of writing no figures are available for the kinetic constants of these components.

Glucose 6-phosphate dehydrogenase

The first step in the pentose–phosphate shunt involves the oxidation of glucose 6-phosphate to 6-phosphogluconate or the corresponding lactone, a reaction catalysed by glucose 6-phosphate dehydrogenase, the *zwischenferment* of Warburg and Christian (1931).

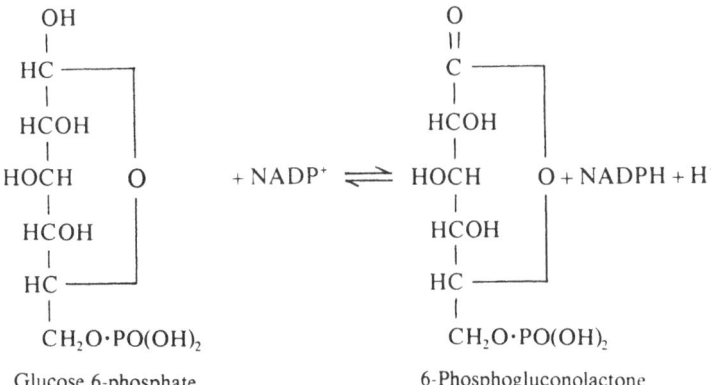

Fig. 22. Reaction catalysed by glucose 6-phosphate dehydrogenase.

This enzyme occurs in yeasts, bacteria and various mammalian tissues, especially the liver and blood cells, but it is to the human erythrocytic enzyme that most attention has recently been paid. This interest arises because of its importance in drug-induced haemolytic anaemias and in congenital non-spherocytic anaemias. These conditions are commonly associated with a deficiency of glucose 6-phosphate dehydrogenase in the erythrocytes of sensitive individuals (Carson, Flanagan, Ickes and Alving, 1956). Several electrophoretically distinct forms of the red-cell enzyme have recently been recognized, both in healthy persons and in patients with the deficiency state (Boyer, Porter and Weilbacher, 1962; Kirkman and Hendrickson, 1962).

The enzyme is particularly important in the metabolism of the erythrocyte which depends upon glucose as its principal source of energy. Normally most of this is metabolized by the Embden–Meyerhof pathway, the various oxidation steps of which are catalysed by enzymes requiring NAD as coenzyme. The main function of the NAD reactions appears to be the production of ATP, the chief high-energy compound of the red cell. The coenzyme of the pentose–phosphate pathway, however, is NADP, whose purpose is believed to be the provision of adequate reducing capacity to protect haemoglobin and other substances from oxidative inactivation. The evolutionary development of the pentose pathway and its functions, especially those of the reaction catalysed by glucose 6-phosphate dehydrogenase, have recently been reviewed by Horecker (1967) and Marks (1967).

Since the mature red cell has no nucleus and possesses little ribonucleic acid, it cannot synthesize enzymes, and consequently its complement of enzymes is complete at the reticulocyte stage. As it ages there is a gradual fall in the activity of various enzymes (Oski, 1965). This is of particular significance in drug-induced anaemias associated with glucose 6-phosphate dehydrogenase deficiency. Older cells are rapidly haemolysed, while young cells remain relatively unaffected. The erythrocyte has no means for correcting the genetic defect. Isoenzyme studies have played a major part in the investigation of such hereditary anomalies.

It is pertinent to point out that red cells and other tissues containing glucose 6-phosphate dehydrogenase also contain 6-phosphogluconate dehydrogenase, the substrate for which is the product of the first enzyme reaction. 6-Phosphogluconate dehydrogenase is also an NADP-dependent enzyme and, unless suitable blank corrections are applied, the observed activity will represent at least some NADP reduced by 6-phosphogluconate dehydrogenase (Glock and McLean, 1953).

Another practical point is the necessity to remove the stromata from haemolysates before studying glucose 6-phosphate dehydrogenase, since stromal NADP-hydrolase removes bound NADP, upon which the stability of the enzyme depends (Carson, Schrier and Kellermeyer, 1959; Kirkman, 1959; Chung and Langdon, 1963; Carson, Ajmar, Hashimoto and Bowman, 1966).

Demonstration of glucose 6-phosphate dehydrogenase heterogeneity

The existence of multiple forms of glucose 6-phosphate dehydrogenase was first recognized by Tsao (1960) in the course of a study of the starch-gel electrophoresis of rat tissues in glycylglycine buffer at pH 7·4. The dehydrogenase patterns were detected by a tetrazolium-staining technique. Two bands of activity which remained close to the origin were found in homogenates of the heart and in erythrocytes. These components also occur in the liver and kidney, which in addition contain a third component migrating towards the anode.

Tsao (1960) extended his studies to the cytologically homogeneous cells produced by tissue culture. The glucose 6-phosphate dehydrogenase bands of mouse fibroblast and ascites lymphoma cells were fast migrating, the former being resolved into two components, while the latter appeared to be homogeneous. Human amnion cells produced a single band of activity with a somewhat lower mobility than the enzyme from the mouse cells, but human cells cultured from a carcinoma of the ovary exhibited two fast bands and a rather diffuse slower component.

Boyer *et al.* (1962) demonstrated two bands of glucose 6-phosphate dehydrogenase activity in normal human red-cell haemolysates after starch-gel electrophoresis. These bands were also present in leucocytes, sometimes accompanied by a number of minor components. Kirkman and Hendrickson (1962) found secondary zones of activity close to the original bands in erythrocyte preparations, but suggested that the minor bands might represent metastable forms of the principal component. Two bands have also been reported by other investigators (Ramot, Bauminger, Brok, Gafni and Shwartz, 1964). In the writer's laboratory normal haemolysates have been shown by acrylamide-gel electrophoresis to contain one major band accompanied by several minor components, and similar experience has been reported by Boivin, Piguet and Galand (1966) and Kirkman and Hanna (1968). Kirkman, McCurdy and Naiman (1964) have suggested that variations in co-enzyme binding might be responsible for at least some of the multiple bands found in normal subjects, since after adding NADP to the sample before electrophoresis they obtained a single band (see p. 67).

Leucocytic glucose 6-phosphate dehydrogenase has been reported to give a different electrophoretic pattern from that of the erythrocytes (Bonsignore, Fornaini, Leoncini, Fantoni and Segni, 1966; Monteleone, Nadler, Justice and Hsia, 1967). However, there is a widespread belief that the white-cell enzyme is identical with that of the erythrocyte, and it is suggested that differences in treatment of the sample prior to electrophoresis might be responsible for the different patterns observed

(Beutler, Mathai and Smith, 1968). This suggestion is supported by the finding that the addition of NADP causes fusion of the bands found in leucocytes (Monteleone *et al.*, 1967).

Population studies showed that on electrophoresis the main zone of glucose 6-phosphate dehydrogenase activity (A) in about 19% of normal American Negro males migrates somewhat faster than that (B) of other American Negro males and Americans of European origin (Boyer *et al.*, 1962). The A component has a mobility very similar to, but slightly slower than, haemoglobin A, while the B enzyme migrates at approximately the same rate as haemoglobin S. In order to differentiate these bands from those of individuals with enzyme deficiency, they are described as A+ and B+ respectively. Similar observations were also made by Beutler, Yeh and Fairbanks (1962) and by Davidson, Nitowsky and Childs (1963). It is now estimated that about twenty different variants of glucose 6-phosphate dehydrogenase have been recognized either by electrophoresis or by study of their enzymatic characteristics (see p. 108).

Chemical structure and properties of human erythrocytic glucose 6-phosphate dehydrogenase

A highly purified crystalline preparation of the normal human red-cell enzyme (B+) has recently been obtained by Yoshida (1966). After preliminary separation from haemoglobin on a DEAE-cellulose column the enzyme was purified by successive chromatography on CM-cellulose, calcium phosphate, DEAE-Sephadex and CM-Sephadex, followed by ammonium sulphate fractionation. The final product had a specific activity of 730 micromolar units per mg, much higher than those obtained by previous investigators (Kirkman, 1962; Balinsky and Bernstein, 1963; Chung and Langdon, 1963). This was apparently due to the precautions taken to avoid losses through destruction of the enzyme protein by proteolytic enzymes such as plasmin. The latter was inactivated by treatment of the eluate from the first column chromatograph with di-isopropyl fluorophosphate and ϵ-aminocaproic acid. Proteolytic denaturation may be a factor in the well-known lability of glucose 6-phosphate dehydrogenase in haemolysates at room temperature (Wilkinson, 1962).

Yoshida (1966) found his crystalline enzyme to have a molecular weight of 240,000 and suggested that it might be a hexamer comprising six identical sub-units. Chung and Langdon (1963) had previously reported a value of 190,000 estimated on the basis of sedimentation studies with a partly purified preparation. Like Kirkman and

Hendrickson (1962) and Tsutsui and Marks (1962), they also observed that removal of NADP causes the enzyme molecule to dissociate into inactive sub-units with a molecular weight about half that of the active enzyme protein. It has recently been established that each enzyme molecule binds two molecules of NADP (Yoshida, 1967a).

The writer is indebted to Dr Yoshida for the information in advance of publication that at alkaline pH, the enzyme molecule dissociates into two sub-units with molecular weights about 120,000. The enzymatically active form (at pH 6–8) also has a molecular weight of 120,000.

A similar fall in the molecular weight of rat-mammary-gland glucose 6-phosphate dehydrogenase on dilution has been reported by Nevuldine and Levy (1965). Gel-filtration gave values of 130,000 and 250,000 for the dilute and more concentrated solutions of the enzyme respectively.

The common Negro variant (A+) has been shown to be very similar to the B+ form in molecular weight, sub-unit size, amino-acid composition, enzymatic properties and immunochemical specificity, but it has a slightly higher pH optimum (Yoshida, 1967a). By the study of fingerprint maps of tryptic digests of the two isoenzymes, Yoshida (1967b, 1968) observed one pair of peptides to differ slightly in their electrophoretic mobilities. On isolation and complete hydrolysis these were found to differ in one respect only. An asparagine residue of the B+ enzyme was replaced by an aspartic acid in the A+ enzyme.

This corresponds to the single amino-acid difference observed between haemoglobin A and the haemoglobin S, and Yoshida suggests that since the m-RNA codons for asparagine are AAU or AAC and those for aspartic acid are GAU or GAC, a single-step transition in the gene could account for the synthesis of the B+ rather than the A+ enzyme. He further suggests that the variant found in Mediterranean populations (see p. 108) may also be due to a single amino-acid substitution.

In addition to D-glucose 6-phosphate, D-galactose 6-phosphate and 2-deoxy-D-glucose 6-phosphate can also serve as substrates for the human red-cell enzyme (Kirkman, 1962; Yoshida, 1966). The Mediterranean mutant form (p. 108) shows unusually high activity with the substrate analogues (Kirkman, Riley and Crowell, 1960; Marks, Banks and Gross, 1962; Kirkman *et al.*, 1964).

The B+ enzyme can also utilize NAD as coenzyme, but its V_{max} is only about 5% of that observed with NADP (Yoshida, 1966). The effect of magnesium ions seems to depend upon the nature of the coenzyme. The reaction of the human erythrocytic enzyme is accelerated when NADP is used as coenzyme (Balinsky and Bernstein, 1963), but Mg^{2+} has an inhibitory effect when NAD is used (Yoshida, 1966). The reaction of the rat-mammary-gland enzyme with NADP is also stimulated by Mg^{2+}, but that with NAD is inhibited (Levy, 1963). More recently a commercially purified enzyme has been shown to be competitively inhibited by Mg^{2+} and Mn^{2+} (Mangiarotti, Garré, Acquarone and Silengo, 1966).

The erythrocytic glucose 6-phosphate dehydrogenase of deficient individuals with congenital nonspherocytic haemolytic anaemia differs from that of normal erythrocytes in being more sensitive to inhibition

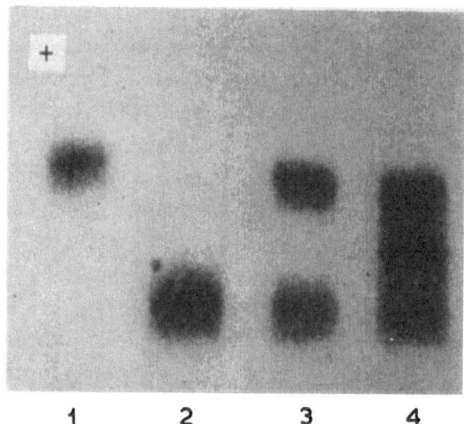

Fig. 23. Effects of *in vitro* hybridization on the starch-gel electrophoresis patterns of human erythrocyte glucose 6-phosphate dehydrogenase. 1. A + enzyme; 2. Seattle variant; 3. mixture of A + enzyme and Seattle variant; 4. mixture of A + enzyme and Seattle variant after hybridization treatment. (Yoshida, Steinmann and Harbert, 1967.)

by neutral salts such as sodium chloride and perchlorate. This observation appears to offer a chemical means for differentiating some of the various forms of the enzyme (Warren and Peterson, 1966).

Since glucose 6-phosphate dehydrogenase has a well-defined sub-unit structure, it is to be expected that *in vitro* hybridization of the enzyme from different sources would be possible (see p. 60). Inter-species hybridization has in fact been demonstrated by Beutler and Collins (1965, 1966), who detected an extra hybrid band after starch-gel electrophoresis of a mixture of the human and rat enzymes, and two extra bands from a mixture of the bovine and rat enzymes.

Early attempts to hybridize the normal human (B+) and the common

Negro variant (A+) were unsuccessful, but recently Yoshida, Steinmann and Harbert (1967) obtained a convincing demonstration. A mixture of the two enzymes was repeatedly precipitated from phosphate buffer (pH 6·8) with sulphuric acid and ammonium sulphate until the enzyme was substantially inactivated. The enzyme mixture was then reactivated by incubation at 30–35°. On subsequent chromatography on a CM-Sephadex column an additional peak of enzyme activity was found between those of the A+ and B+ enzymes (Fig. 23). A similar inactivation–reactivation process led to the formation of an artificial hybrid between the A+ and Seattle variants (p. 109). This was demonstrated by starch-gel electrophoresis.

Genetic variants of human erythrocyte glucose 6-phosphate dehydrogenase

The discovery of the A+ and B+ variants discussed on p. 105 has been followed by the recognition of several other mutant forms. Some twenty or so have now been reported, but it is possible that there may be some duplication. Owing to the relative instability, especially of some of the rarer mutant forms, it is usually not practicable for investigators to exchange specimens, and characterization has had to depend for the most part upon electrophoretic mobility and chemical properties measured in different laboratories in different parts of the world. There is now an extensive literature on the subject, and a detailed review would be beyond the scope of this monograph. Excellent reviews of the haematological aspects have recently been compiled by Boivin *et al.* (1966) and Beutler (1966). Reference must also be made to the City of Hope Symposium on *Hereditary Disorders of Erythrocyte Metabolism* (New York: Grune and Stratton, 1968).

The relatively common occurrence of the A+ phenotype among American Negro males focused attention on its distribution among Negro women. Boyer *et al.* (1962) found three groups of individuals, who possessed either the A or B form or both, but the proportions of women showing both A and B bands was much lower than those expected for a heterozygous group. This suggested an X-linked rather than autosomal inheritance, a conclusion previously reached from studies of glucose 6-phosphate dehydrogenase deficiency (Childs, Zinkham, Browne, Kimbro and Torbert, 1958; Gross, Hurwitz and Marks, 1958).

Evidence in support of the X-linkage was obtained from electrophoretic studies on the wild type of *Drosophila melanogaster*. Young, Porter and Childs (1964) observed three distinct phenotypes, one

TABLE 8 *Variant forms of human erythrocytic glucose 6-phosphate dehydrogenase*

Type	Ethnic group	Activity (Percentage of normal)	Haematological anomaly	Electrophoretic mobility	Affinity for G 6-P	Gal 6-P	2-Deoxy G 6-P	References*
A+	Negro	100	None	A	N	N	N	Boyer et al. (1962)
A−	Negro	15–25	Drug sensitivity	A or slightly faster than A	N	—	—	Marks. Szeinberg and Banks (1961) Boyer et al. (1962) Kirkman and Crowell (1963) Yoshida et al. (1967)
B+	Caucasian	100	None	B	N	N	N	Boyer et al. (1962)
B−	Jewish	2–15	Drug sensitivity	B	Decr.	Incr.	Incr.	Ramot et al. (1964)
	Greek		Favism					
	Italian		CNSHA		Incr.	—	Incr.	Kirkman, Doxiadis et al. (1965) Stamatoyannopoulos et al. (1964) Vecchio et al. (1966)
Baltimore (C)	Negro	100	None	Slower than B	—	—	—	Beutler et al. (1968)
Ibadan	Negro	100	None	Slower than Baltimore	—	—	—	Boyer et al. (1962) Porter et al. (1964)
Sardinia I	Caucasian		None	Slower than Ibadan	—	—	—	Porter et al. (1964)
Sardinia II	Caucasian		None	Slower than Sardinia I	—	—	—	Porter et al. (1964)
Oklahoma	Caucasian	9	CNSHA	B	Decr.	—	N	Kirkman and Riley (1961)

Variant	Race, activity	CNSHA	Electrophoresis				References
Chicago I	Caucasian 9–26	CNSHA	Faster than B	N	—	N	Kirkman et al. (1964); Beutler (1968)
Eyssen	Caucasian very labile	CNSHA		—	—	—	Marks et al. (1961); Boyer et al. (1962)
Duarte	Caucasian 8.5	CNSHA	B	Decr.	—	Incr.	Beutler, Mathai and Smith (1968)
Albuquerque	Caucasian 1 very labile	CNSHA	B	Decr.	—	Decr.	Beutler et al. (1968)
Tubingen	European 100	CNSHA	B	Incr.	—	—	Helge and Borner (1966)
Berlin	German 1	CNSHA	B	? Decr.	—	Decr.	Waller, Löhr and Gayer (1966)
Ohio	Italian 3–16	CNSHA	Similar to A	Decr.	—	N	Pinto, Newton and Richardson (1966)
Seattle	Caucasian 8–21	None	Slower than B	Incr.	Incr.	Incr.	Kirkman, Simon and Pickard (1965)
Tel-Hashomer	Jewish 15–25	None	Slower than B	Incr.	Incr.	—	Ramot et al. (1964)
Austin I	Negro 75	None	Slower than B	N	N	N	Long, Kirkman and Sutton (1965)
Austin II	Negro 90	None	Slower than Austin I	N	N	N	Long et al. (1965)
D	Caucasian 8–16	None	Slower than B	—	—	—	Shows et al. (1964)
Barbieri	Italian 50	None	Faster than A	Decr.	—	—	Marks, Banks and Gross (1962)
Canton	Chinese 4–24	None	Between A and B	Incr.	Incr.	Incr.	Kirkman et al. (1964); Wong, Ling-Yu-Shih, Hsia and Tsao (1965)
Madison	Caucasian	None	Slower than Seattle	—	—	—	McCurdy et al. (1966); Nance and Uchida (1964)

Abbreviations: G 6-P, glucose 6-phosphate; Gal 6-P, galactose 6-phosphate; 2-Deoxy G 6-P, 2-deoxy-D-glucose 6-phosphate; N, normal; CNSHA, congenital non-spherocytic haemolytic anaemia.

* A fuller bibliography is given by Boivin et al. (1966) and Beutler (1966, 1968).

having a single slow-moving band of activity, one showing a single fast band and a third which exhibited both bands. With individual fly homogenates, all three phenotypes were detected in a single strain of females (10 slow, 11 double and 48 fast), while only two phenotypes were found in the males (2 slow and 25 fast). In various combinations of single pair matings it was found that the female progeny of 'fast' males and 'slow' females mated with 'slow' males produced only 'fast' males and 'double' females. All the F_1 males have the maternal phenotype, while all F_1 females whose X-chromosomes are derived from both parents show both bands of enzyme activity.

Further confirmation is provided by studies of the electrophoretic patterns of the erythrocyte glucose 6-phosphate dehydrogenase of the reciprocal hybrids of the horse and the donkey. Female mules and hinnies contain both horse and donkey enzymes, but the male mule with an X-chromosome from its horse mother has the horse enzyme only. Similarly, the red cells of the male hinny contain the donkey enzyme only (Trujillo, Walden, O'Neil and Anstall, 1965).

A summary of the variant forms which have been reported appears in Table 8. It will be seen that the electrophoretic mobility alone is insufficient to characterize the enzyme and that studies of such chemical properties as the total activity in the erythrocytes and the affinities for the natural substrate, glucose 6-phosphate, and the substrate analogues, galactose 6-phosphate and 2-deoxy-D-glucose 6-phosphate, are frequently helpful. Several other properties, however, have also been investigated. The pH curves of some variants, e.g. the Seattle and Canton types, show a bimodal relationship. As compared with the normal (B+) enzyme, these examples show increased affinity (lower K_m values) for NADP, whereas others, e.g. the B- and Oklahoma variants, exhibit decreased affinity for the coenzyme. The Eyssen variant is very labile, and several others are less stable than the normal enzyme.

As mentioned on p. 102, the primary role of glucose 6-phosphate dehydrogenase and other oxidoreductases of the pentose–phosphate pathway in the metabolism of the red cell appears to be the maintenance of sufficient reducing capacity to prevent oxidative denaturation of haemoglobin. NADPH is required by methaemoglobin reductase and also for the reduction of glutathione. It seems likely therefore that the increased lability of glucose 6-phosphate dehydrogenase deficient erythrocytes may be at least partly due to the lack of adequate NADPH. For a more detailed discussion of this aspect the reader is referred to the reviews of Boivin *et al.* (1966), Prankerd (1965, 1967) and Marks (1967).

The clinical implications of glucose 6-phosphate dehydrogenase variants and deficiency have recently been reviewed by Motulsky and Stamatoyannopoulos (1966) and Porter (1968). In susceptible individuals haemolytic episodes may be precipitated by a variety of agencies, including exposure to the fava bean (*Vicia faba*) (Sansone and Segni, 1958), many antimalarial, sulphonamide and analgesic drugs (Carson *et al.*, 1956; Beutler, 1966) and unrelated diseases, such as bacterial infections, diabetic ketoacidosis and viral hepatitis (Burka, Weaver and Marks, 1966; Choremis, Kattamis, Kyriazakous and Gavriilidou, 1966). There appears to be an association between the incidence of hyperbilirubinaemia and glucose 6-phosphate dehydrogenase deficiency in premature Negro babies (Eshaghpour, Oski and Williams, 1967).

Many screening tests, recently summarized by Stamatoyannopoulous, Papayannopoulou, Bakapoulos and Motulsky (1967) and Lubin and Oski (1967), have been devised for the detection of glucose 6-phosphate dehydrogenase deficiency. They include methaemoglobin reduction, dye decolorization, the ascorbate-cyanide test and cyanmethaemoglobin elution.

Glucose 6-phosphate dehydrogenase heterogeneity in other tissues
Glucose 6-phosphate dehydrogenase is distributed throughout most tissues: skeletal muscle probably contains least of this enzyme, and the liver is possibly the richest source. The liver, however, contains NADP glycohydrolase, NADP pyrophosphatase and a third protein, all of which inhibit glucose 6-phosphate dehydrogenase (Carson *et al.*, 1959; Bonsignore, DeFlora, Mangiarotti, Lorenzoni and Alemà, 1968). In a given individual the electrophoretic pattern appears to be identical with that of the erythrocytes.

An isoenzyme of glucose 6-phosphate dehydrogenase having a high affinity for galactose 6-phosphate was detected in the liver of the deer mouse, *Peromyscus maniculatus*, by Shaw and Barto (1965), and has since been found in other mammalian livers (Shaw, 1966; Ohno, Payne, Morrison and Beutler, 1966; Shaw and Koen, 1969). On starch-gel electrophoresis this component migrates cathodally to the sex-linked glucose 6-phosphate dehydrogenase and stains equally well when either hexose 6-phosphate is used as substrate. For this reason it is suggested that it should be known as hexose 6-phosphate dehydrogenase (Ohno *et al.*, 1966; Shaw and Koen, 1968).

In the deer mouse one or three zones of activity may be detected according to whether the animal is homozygous or heterozygous for the autosomal gene controlling the biosynthesis of the enzyme. More complex patterns are seen in the livers of other species, including man. Some human livers show no galactose-enzyme activity, while others have from one to four cathodic bands. Such activity, however, appears to be absent from pig liver (Shaw and Koen, 1968).

Linder and Gartler (1965) have used glucose 6-phosphate dehydrogenase as a cell marker in a study of the origin of uterine leiomyomas in women heterozygous for the A and B variants. Samples of normal myometrium showed both A and B bands on starch-gel electrophoresis, but the enzyme from all the leiomyomas was either of the A or B type – never both. This provides strong support for the hypothesis that in certain cases one of the X-chromosomes may become inactive (Lyon, 1961). Linder and Gartler conclude that the single phenotype tumours arose from a single cell. (This contrasts with these authors' experience with the autosomally inherited phosphoglucomutase, when all tumours from heterozygous individuals exhibited the heterozygous phenotypes.)

A similar conclusion was reached by Beutler (1967), who found a predominance of the A variant in multiple tumours from a heterozygous patient with lymphoma. On the other hand, in a patient with multiple metastases from a carcinoma of the colon, some metastases were mainly of type A, while others were of type B. Despite the fact that only a single primary tumour was found in this patient, it seems that it had originated from more than one cell.

6-Phosphogluconate dehydrogenase
6-Phosphogluconate dehydrogenase catalyses the second reaction in the pentose phosphate pathway, oxidizing 6-phosphogluconate to ribulose 5-phosphate and carbon dioxide:

$$
\begin{array}{c}
\text{COOH} \\
|\\
\text{HCOH} \\
|\\
\text{HOCH} \\
|\\
\text{HCOH} \\
|\\
\text{HCOH} \\
|\\
\text{CH}_2\text{O·PO(OH)}_2
\end{array}
+ \text{NADP}^+ \rightleftharpoons
\begin{array}{c}
\text{CH}_2\text{OH} \\
|\\
\text{C}=\text{O} \\
|\\
\text{HCOH} \\
|\\
\text{HCOH} \\
|\\
\text{CH}_2\text{O·PO(OH)}_2
\end{array}
+ \text{CO}_2 + \text{NADPH} + \text{H}^+
$$

Like glucose 6-phosphate dehydrogenase it is an NADP-dependent enzyme.

Heterogeneity in human erythrocytes

It occurs in several genetically determined forms in most human and animal tissues, but most studies have so far been made on the human erythrocytic enzyme. Fildes and Parr (1963) subjected haemolysates prepared from washed red cells to starch-gel electrophoresis and located the positions of the bands of enzyme activity by a modification of the general procedure for NADP-dependent dehydrogenases. An agar 'overlay' containing 6-phosphogluconate as substrate, together with NADP, MTT and N-methylphenazonium methosulphate in tris buffer at pH 8·0 was applied to the starch gel and incubated at 37° for 1 hr.

In about 95% of a London population these investigators found a single band (A), but a second band (B) with a somewhat lower mobility was detected in about 5% of the sample. It stained with about half the intensity of the A band and, on the basis of familial studies, Fildes and Parr concluded that the variant trait is genetically determined. It appeared to be a heterozygous condition for a gene pair with simple co-dominant autosomal Mendelian inheritance, but in this series no example of the atypical homozygote was detected.

Subsequently, a faint third band (C), migrating more slowly than the B component, was observed in the variant individuals, and it was postulated that the enzyme is composed of two sub-units each synthesized under the control of a separate gene (Fildes and Parr, 1964). Discovery of the corresponding homozygote (Canning) by Parr (1966) provides strong support for this hypothesis. Other anomalies have since been reported, and ten or more different phenotypes of the human erythrocytic enzyme have now been reported (Table 9).

The common variant seems to occur with a somewhat greater frequency among American Negroes, Brazilians and Ugandans than among British and American Caucasians (Dern, Brewer, Tashian and Shows, 1966; Parr, 1966), but the highest incidence so far reported (20–30%) is among the Bantu population of South Africa (Gordon, Keraan and Vooijs, 1967). Using a slightly different electrophoretic technique, however, Bowman, Carson, Frischer and de Garay (1966) found a similar incidence between Americans of both races.

Certain individuals have been shown to have 6-phosphogluconate dehydrogenase with normal electrophoretic mobility but with about half the normal activity (Parr and Fitch, 1964; Brewer and Dern, 1964). In a study of 1,148 individuals Brewer and Dern found the incidence of enzyme deficiency to be about 0·3% in Negroes and about 0·7% in the white population. A similar enzyme deficiency has also been observed in individuals with multiple electrophoretic bands (Shows, Tashian,

TABLE 9 *Human red-cell 6-phosphogluconate dehydrogenase pheno-types* [Based upon Parr (1966) and Parr and Fitch (1967)]

Type	Enzyme activity (percentage of normal)	Incidence in London population	Isoenzyme pattern	Suggested genotype
Usual	100	95	A	PGD^{A}PGD^A
Ilford*	50–60	1	A	PGD^{A}PGD^O
Common variant	80–100	4	A, B, C (trace)	PGD^{A}PGD^C
Canning	70–90	0·05	A (trace), B, C †	PGD^{C}PGD^C
Newham*	40–50	0·02	A (trace), B, C †	PGD^{C}PGD^O
Richmond	100	0·02	A, D, E	PGD^{A}PGD^Y
Hackney	100	0·02	A, B, C	PGD^{A}PGD$^\delta$
Whitechapel*	0–5	—	—	PGD^{W}PGD^W
Dalston	70–75	—	A	PGD^{A}PGD^W

* The Ilford and Newham phenotypes were originally described as 'half activity' and 'half activity Canning' respectively. The Whitechapel form is that previously reported as 'fully deficient' (Parr and Fitch, 1967).

† White-cell extracts from Canning homozygotes give single C bands on electrophoresis. The presence of the A and B bands in the haemolysates is attributed to the superimposition of an additional factor as yet unexplained. (Parr and Fitch, 1967.)

Brewer and Dern, 1964; Parr and Fitch, 1964; Dern *et al.*, 1966; Parr, 1966).

A more complete form of enzyme deficiency was detected by Parr (1966) and described as the 'Whitechapel' (see Table 9) phenotype by Parr and Fitch (1967). Relatives of Whitechapel variants with about 75% of the normal activity in their red cells appear to be heterozygotes for the trait and have been designated 'Dalston' variants (Parr and Fitch, 1967). Since the white cells of the Whitechapel and Dalston variants contained respectively appreciable or full 6-phosphogluconate dehydrogenase activity, the authors conclude that in these individuals the enzyme is unstable. Unlike the white cells, the erythrocytes have no means for replacing an unstable enzyme by synthesis.

All these deficient individuals were apparently symptom-free, but there are reports of two patients with congenital non-spherocytic haemolytic anaemia and an associated moderate deficiency of 6-phosphogluconate dehydrogenase (Lausecker, Heidt, Fischer, Harjleyb and Löhr, 1965; Scialom, Najean and Bernard, 1966). These patients have been reviewed by Boivin *et al.* (1966), who conclude that on the evidence available the possibility cannot be excluded that the hereditary anaemia and the enzyme deficiency may be coincidental.

The variants observed by Parr (1966) are summarized in Table 9. Named after the districts of London in which the individuals concerned resided, the different phenotypes may be distinguished according to their electrophoretic patterns and their enzyme activities. The Canning phenotype is the atypical homozygote corresponding to the heterozygote observed by Fildes and Parr (1963). The finding of a similar electrophoretic pattern accompanied by about half the normal activity led to the recognition of the Newham variant (Parr and Fitch, 1967).

The Richmond variant is of a quite different pattern, for the A isoenzyme was accompanied by two faster components described by Parr (1966) as D and E. The Hackney pattern is similar to that of the common variant, but the B and C bands had slightly different electrophoretic mobilities.

The patterns observed in Americans of both Negro and European origin and in Mexican Lacandón Indians by Bowman *et al.* (1966) differ in certain respects from those reported by the British workers. The major components appear to correspond with the A, B and C bands of Parr and his colleagues, but a number of minor bands were also detected. The usual pattern consisted of an A band accompanied by a minor band in the C position, while the common variant (heterozygote) exhibited major A and B components of equal intensity plus two minor bands, one in the C position and one with a slower mobility. The atypical homozygote pattern consisted of major B and C bands with a minor A band and two other slowly migrating bands. The presence of the minor bands is difficult to interpret on the basis of the genetic mechanism discussed above, but it seems that the atypical homozygote of Carson *et al.* (1966) may well be identical with the Canning variant of Parr (1966).

In this connection it is interesting to note that, although Parr (1966) generally found other tissues to possess the same 6-phosphogluconate dehydrogenase isoenzyme composition as the erythrocyte from the same individual, in certain cases referred to by Parr and Fitch (1967), minor bands occur in the red-cell pattern which do not appear in extracts of white cells (see footnote to Table 9).

It is possible that the minor bands might have been caused by dissociation of the enzyme molecule as the result of dilution or removal of NADP, as observed by Kirkman and Hendrickson (1962) and Yoshida (1966) with glucose 6-phosphate dehydrogenase. Such changes might take place during or prior to electrophoresis (see pp. 104–105).

6-Phosphogluconate dehydrogenase isoenzymes in other species

Variants of 6-phosphogluconate dehydrogenase have been reported in other species. In laboratory rats Parr (1966) found a single 'slow' band which migrated much faster than the human A band, but in wild rats he also found two additional phenotypes. One of these exhibited a single 'fast' band, while the other appeared to be the heterozygote having both 'slow' and 'fast' bands and a third more intense band of intermediate mobility.

Similar polymorphism has been observed in the domestic cat by Thuline, Morrow, Norby and Motulsky (1967), and in the pigeon by Childs, Kazazian and Young (cited by Parr, 1966).

Structure of 6-phosphogluconate dehydrogenase isoenzymes

The structure of 6-phosphogluconate dehydrogenase isoenzymes has not yet been extensively studied, but the genetic data summarized above are consistent with the presence of at least two sub-units. Indeed, at the present time relatively little is known about the chemical nature of the enzyme, but summaries of its kinetic properties have been collected by Glock (1961) and Noltmann and Kuby (1963).

There is evidence, however, that the common variant and Canning enzymes are less stable than the usual form (Parr and Parr, 1965; Carson *et al.*, 1966). These variants are also more susceptible to inhibition by red-cell stromata (Carson *et al.*, 1966).

Phosphoglucose isomerase

The occurrence of phosphoglucose isomerase in multiple forms has recently been reported. Nakagawa and Noltmann (1967) resolved the crystalline yeast enzyme into three chromatographically distinct components which also behaved differently on cellulose acetate electrophoresis. All three isoenzymes were obtained from a live, genetically homogeneous yeast culture, as well as from the enzyme purified from brewer's and baker's yeast, thus eliminating the possibility that they resulted from a mixture of strains.

Phosphoglucose isomerase heterogeneity has also been detected by the starch-gel electrophoresis of haemolysates and muscle extracts of wild and laboratory mice (Carter and Parr, 1967). A tetrazolium staining technique was used with fructose 6-phosphate as substrate. The glucose 6-phosphate formed by the isomerase reacted with glucose 6-phosphate dehydrogenase coupled to the tetrazolium system.

Some animals exhibited a single electrophoretically slow band of activity, whereas others showed a single much faster band. When a pair of such animals was crossed, all the progeny, male and female alike, exhibited a pattern consisting of both fast and slow bands of the parents accompanied by a third zone of intermediate mobility (Fig. 24). These results, the authors suggest, indicate that the enzyme probably has a dimeric structure and that it is controlled by an autosomal inheritance mechanism.

Most human haemolysates contain a single major isomerase band which, on starch-gel electrophoresis at pH 8·0, migrates towards the cathode. Several genetic variants, however, have recently been reported.

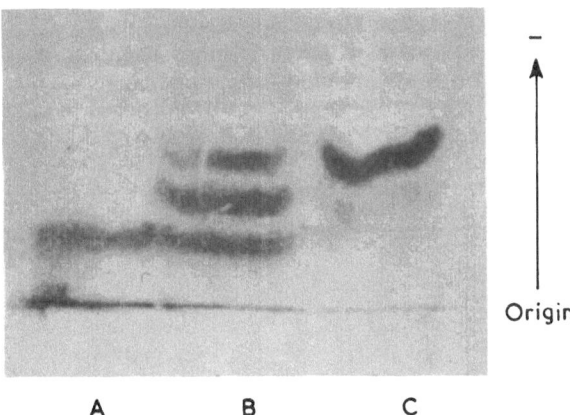

Origin

A   B   C

Fig. 24. Starch-gel electrophoresis patterns of mouse hae molysates stained for phosphoglucose isomerase. Pattern (*b*) is that from one of the offspring of a mating pair which gave patterns (*a*) and (*c*). (Carter and Parr, 1967.)

Most of these exhibit three major isoenzymes, one of which is identical with the usual component (Detter, Ways, Giblett, Baughan, Hopkinson, Povey and Harris, 1968; Fitch, Parr and Welch, 1968). Fitch *et al.* (1968) conclude that the human enzyme, like that of the mouse, is probably a dimer.

Aldolase
Study of the molecular heterogeneity of aldolase (ketose 1-phosphate aldehyde-lyase, E.C.4.1.2.7, and fructose 1,6-diphosphate D-glyceraldehyde 3-phosphate lyase, E.C.4.1.2.13) is remarkable in that it has led to reconsideration of the sub-unit structure of this enzyme. C-terminal analysis (Drechsler, Boyer and Kowalsky, 1959; Spolter, Adelman and Weinhouse, 1965), N-terminal analysis (Edelstein and Schachman, 1966; Sine and Hass, 1967), determination of substrate

binding sites (Ginsburg and Mehler, 1966; Castellino and Barker, 1966; Kobashi, Lai and Horecker, 1966), amino-acid composition and tryptic digest studies (Shimizu and Ozawa, 1967) and molecular-weight determinations of the enzyme and its sub-units (Stellwagen and Schachman, 1962; Deal, Rutter and Van Holde, 1963; Schachman and Edelstein, 1966) all indicated that the enzyme molecule is composed of three sub-units. More recent electrophoretic studies, however, have produced results which are more easily explained by a four sub-unit structure. Nevertheless, Chan, Morse and Horecker (1967) have recently reported evidence suggesting that rabbit-muscle aldolase contains sub-units of two different types combined in triads of the AAB pattern.

Aldolase catalyses the cleavage of fructose 1,6-diphosphate into D-glyceraldehyde phosphate and dihydroxyacetone phosphate:

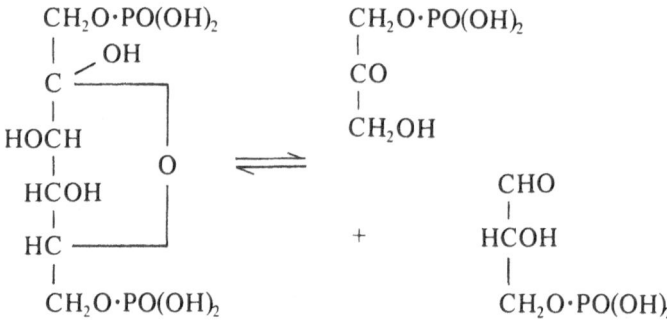

The interconversion of the products of the forward reaction is catalysed by triose phosphate isomerase, in the presence of which either substance may be used as substrate for the reverse aldolase reaction.

Early work on aldolase has been reviewed by Rutter (1962).

Classification of the aldolases

The numerous aldolases have been grouped into two main classes by Rutter (1964) according to whether a lysine amino group of the enzyme combines with a molecule of the substrate, dihydroxyacetone phosphate, to form a Schiff's base (Grazi, Cheng and Horecker, 1962). Those of class I, the lysine aldolases, occur in animals, plants and a limited number of micro-organisms, while those of class II are found in bacteria and fungi. Most of the class I animal aldolases can be further classified as A, B or C aldolases which have been isolated from muscle, liver and brain respectively (Rutter, Woodfin and Blostein, 1963; Blostein and Rutter, 1964).

The liver enzyme exhibits strong action, relative to that with fructose diphosphate, on fructose 1-phosphate, against which the muscle enzyme shows only feeble activity (Hers and Kusaka, 1953; Leuthardt, Testa and Wolf, 1953; Peanasky and Lardy, 1958). The brain enzyme has an activity ratio intermediate between those of liver and muscle.

Electrophoretic studies
On electrophoresis, the mammalian- and avian-muscle, heart and spleen aldolases show single zones of activity. Liver and kidney aldolases, however, show multiple bands: the main bands of the A and B enzymes are accompanied by three zones of intermediate mobility, while the brain enzyme (C) exhibits a quite distinct, more anodic, pattern of five bands (Penhoet, Rajkumar and Rutter, 1966; Foxwell, Cran and Baron, 1966; Christen, Rensing, Schmid and Leuthardt, 1966; Anstall, Lapp and Trujillo, 1966; Herskovits, Masters, Wassarman and Kaplan, 1967; Foxwell, Buck and Baron, 1968; Nicholas and Bachelard, 1969).

Penhoet, Kochman, Valentine and Rutter (1967) demonstrated that artificial five-membered sets of isoenzymes could be prepared from various binary combinations of the A, B and C enzymes. Moreover, they were able to show that on dissociation and re-association the hybrid enzymes gave a series of patterns containing all five components (Fig. 25). They also found the A–C hybrids to have similar molecular weights to the parent enzymes. By means of dissociation and recombination experiments with (^{3}H)-leucyl-aldolase A and aldolase C, they showed the specific radioactivities of the set of five isoenzymes obtained to be in the proportions 1, 0·75, 0·5, 0·25 and 0, indicating that the individual sub-units comprise a quarter of the enzyme molecule. This conclusion was supported by electron microscopy, by means of which the enzyme molecule appears either as a square or a tetrahedron (Rutter, Rajkumar, Penhoet, Kochman and Valentine, 1968).

Chan *et al.* (1967), however, produced further evidence for the 3 sub-unit theory and point out that random hybridization of trimeric enzyme molecules would be expected to give rise to a considerably greater number of hybrids. They suggest that possibly some of the isoenzymes in the five-membered pattern may not be homogeneous, i.e. that different hybrids may have the same electrophoretic mobilities, or that there may be some preferential mode of recombination.

The five-membered pattern has also been obtained by Foxwell, Buck and Baron (1968), who partially purified rabbit-brain aldolase by ammonium sulphate precipitation and separated the isoenzymes by electrophoresis on Pevikon (p. 241). Except for isoenzyme 3, the five components displayed a continuous increase in the ratio of the activity with fructose 1,6-diphosphate to that with fructose 1-phosphate, which the authors consider to be consistent with the tetrameric hypothesis (see also Baron, Buck and Foxwell, 1969).

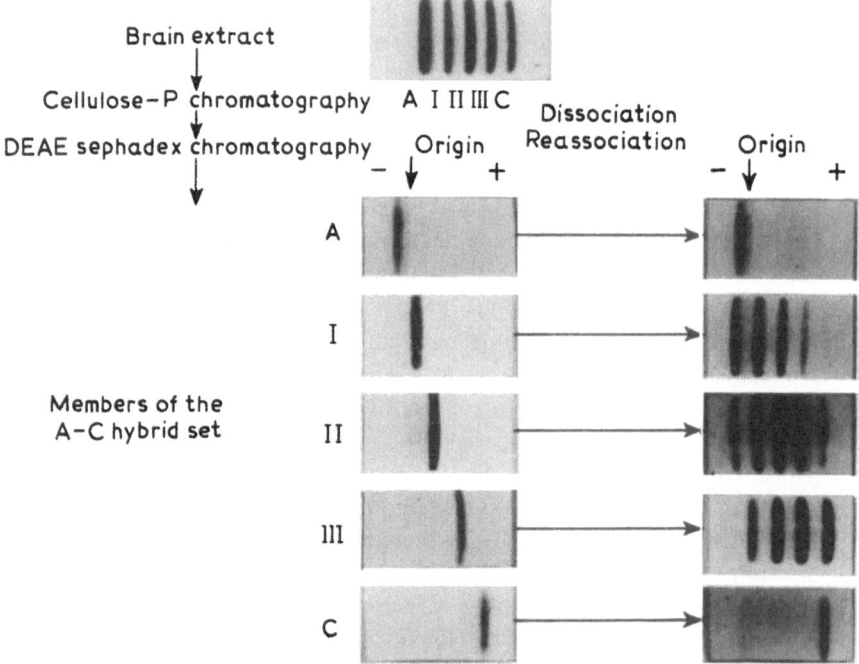

Fig. 25. Cellulose acetate electrophoretic patterns of the A–C aldolase hybrid iso-enzymes after separation and exposure to dissociation and reassociation. Each of the hybrid isoenzymes gives up to five components after such treatment, but the pure A and C components appear to contain sub-units of a single type only. (Penhoet, Kochman, Valentine and Rutter, 1967.)

Similar conclusions were reached by Nicholas and Bachelard (1969) who separated aldolase isoenzymes from guinea-pig cerebral cortex by chromatography on DEAE-cellulose. Each of their five components was demonstrated by starch-gel electrophoresis to be free from contamination with other forms. Further confirmation has recently been reported by Gracy, Lacko and Horecker (1969), who observed rabbit-liver aldolase and its sub-units to have molecular weights of about 158,000 and 39,000 respectively.

Baron *et al.* (1969) suggested that aldolase A may exist in two different forms, Aα and Aβ. This has now been confirmed by subjecting the highly purified enzyme to isoelectric focusing when Susor, Kochman and Rutter (1969) obtained a set of five isoenzymes. These appear to have the sub-unit compositions: $A\alpha_4$; $A\alpha_3 A\beta$; $A\alpha_2 A\beta_2$; $A\alpha A\beta_3$ and $A\beta_4$ (Buck and Baron, personal communication).

Cellulose acetate was used as electrophoretic medium by Penhoet *et al.* (1966, 1967), who stained the aldolase protein with Amido black, but similar results were reported by Anstall *et al.* (1966), who located

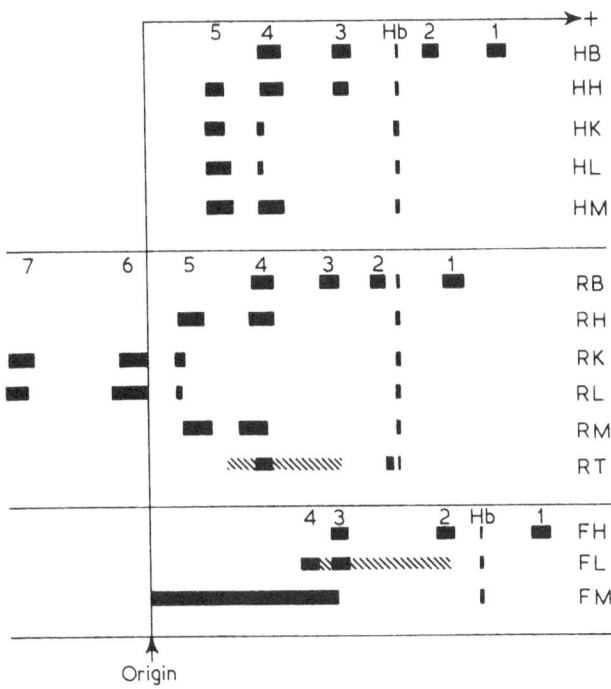

Fig. 26. Diagrams representing the aldolase isoenzyme patterns obtained by starch-gel electrophoresis of human brain, heart, kidney, liver and muscle; rat brain, heart, kidney, liver, muscle and testis; and frog heart, liver and muscle. (Anstall, Lapp and Trujillo, 1966.)

the bands of aldolase separated by starch-gel electrophoresis by incubation with fructose diphosphate and hydrazine to trap the dihydroxyacetone phosphate formed. The gels were then stained with 2,4-dinitrophenylhydrazine, and after treatment with sodium hydroxide the zones of enzyme activity appeared as red spots on a yellow background.

Anstall *et al.* (1966) compared the patterns obtained from various human, rat and frog tissues, and in all three species found the brain enzyme to consist of three or four components migrating rapidly towards the anode. The human liver and kidney enzymes showed two slowly migrating anodic bands, while the corresponding tissues from the rat migrated towards the cathode (Fig. 26).

By contrast, Herskovits *et al.* (1967) found avian liver and kidney aldolases to migrate faster than those of the heart and brain. This has been further investigated by Masters (1967), who, on the basis of the results obtained when purified chicken aldolases were pre-incubated with liver or intestinal supernatant fractions, attributes the unusual mobilities to modifications *in situ*. The mixed aldolases appear as a single band with a mobility intermediate between those of the individual components. The change in mobility is accompanied by reduction of activity with fructose diphosphate and in the fructose diphosphate/fructose 1-phosphate activity ratio. Masters suggests that such modification might occur *in vivo* and might explain some of the difficulties experienced in elucidating the chemical nature of this enzyme.

Fructose 1-phosphate aldolase
Interest in the multiple molecular forms of aldolase has been further stimulated by the recognition of the enzyme defect in patients with the rare metabolic disorder, hereditary fructose intolerance. The biochemical abnormalities are due to the absence of liver fructose 1-phosphate aldolase (Hers and Joassin, 1961; Perheentupa, Pitkänen, Nikkilä, Somersalo and Hakosala, 1962; Froesch, Wolf, Baitsch, Prader and Labhart, 1963). A similar deficiency has recently been demonstrated in the kidney (Morris, Ueki, Loh, Eanes and McLin, 1967).

Although in hereditary fructose intolerance the fructose 1,6-diphosphate aldolase activity is 50% or more of the normal value while fructose 1-phosphate aldolase activity is undetectable, there is no evidence that the two substrates are acted upon by different enzyme proteins. Indeed it now seems that the enzyme abnormality is the production of an anomalous aldolase B, while the aldolase A is normal (Schapira, Nordmann and Dreyfus, 1968).

Glyceraldehyde 3-phosphate dehydrogenase
Few reports have so far appeared concerning the isoenzymes of glyceraldehyde 3-phosphate dehydrogenase. Lebherz and Rutter (1967) compared the electrophoretic patterns on cellulose acetate of tissue extracts from a variety of animals, plants, fish, birds, insects and microorganisms. The trout, perch, turtle, spinach and yeast enzymes display multiple forms, but those from the rat, rabbit, chicken, frog, honey-bee,

Euglena gracilis and *Escherichia coli* appear to be electrophoretically homogeneous. Most of the species displaying multiple forms exhibit five-membered patterns, which the authors interpret as suggesting that, like lactate dehydrogenase, this enzyme is a tetramer comprising monomers of two different types (pp. 58–63).

Papadopoulos and Velick (1967), however, found the rabbit-muscle and -liver enzymes to differ in their electrophoretic mobilities and certain kinetic properties. Although both enzymes were powerfully inhibited by the products of the forward reaction, they differed markedly in their sensitivity to product inhibition by NAD, in the reverse reaction: values for the $K_{i(NAD)}$ for the liver and muscle enzymes of 2,000 μM and 125 μM respectively were observed at 37 and pH 7·4. Papadopoulos and Velick suggest therefore that the liver enzyme is well adapted for gluconeogenesis.

α-Glycerophosphate dehydrogenase

A recent report indicates that the α-glycerophosphate dehydrogenase of chicken pectoralis muscle differs from that of chicken heart in their optimum substrate concentrations, pH stability curves and certain other properties (Rouslin, 1967). So far no electrophoretic studies of this enzyme appear to have been described, and it remains uncertain whether this enzyme occurs in multiple forms. Rouslin, however, reports that he found some correlation between the types of α-glycerophosphate dehydrogenase and lactate dehydrogenase in chicken breast muscle, brain, liver and heart.

Phosphopyruvate hydratase (enolase)

The first report of the occurrence of multiple forms of enolase was that of Malmström (1957), who separated them from yeast but later considered them to be artifacts. Cory and Wold (1965), however, succeeded in separating three components from the crystalline enzyme of rainbow-trout muscle. More recently Pfleiderer and his colleagues have re-investigated the enolases of yeast and animal tissues and have established their heterogeneity beyond doubt.

Three major anionic bands were detected in *Saccharomyces cerevisiae* 24-2c, by electrophoresis on cellulose acetate, starch gel and acrylamide gel at pH 8·6, but the slower components, II and III, appeared as double bands on starch gel (Pfleiderer, Neufahrt-Kreiling, Kaplan and Fortnagel, 1966; Dave, Kaplan and Pfleiderer, 1966). Rat-skeletal muscle, on the other hand, exhibits three cationic enolase components on starch-gel electrophoresis at pH 8·6, while under the same conditions beef muscle shows only two isoenzymes.

Subsequently Pfleiderer, Neufahrt-Kreiling, Kaplan and Fortnagel (1968) separated the three main isoenzymes from yeast by column chromatography on DEAE-Sephadex A50, but these were found to have almost identical catalytic and immunochemical properties. Pfleiderer *et al.* (1968) therefore conclude that yeast enolase isoenzymes have little metabolic significance.

Pyruvate kinase

Interest in pyruvate kinase has been greatly stimulated by recent recognition of its role as a regulatory enzyme in glycolysis and gluconeogenesis (reviewed by Weber, Lea, Hird Convery and Stamm, 1967) and by its congenital deficiency in the erythrocytes of patients with certain congenital haemolytic anaemias (Valentine, Tanaka and Miwa, 1961; Oski and Diamond, 1963; Bowman and Procopio, 1963).

It catalyses the reversible conversion of phosphoenolpyruvate to pyruvate:

$$
\begin{array}{ccc}
\mathrm{CH_2} & & \mathrm{CH_3} \\
| & & | \\
\mathrm{CO\cdot PO(OH)_2 + ADP} & \rightleftharpoons & \mathrm{CO + ATP} \\
| & & | \\
\mathrm{COOH} & & \mathrm{COOH}
\end{array}
$$

Under physiological conditions, however, the reaction is essentially unidirectional in favour of pyruvate formation. During gluconeogenesis some blockade of pyruvate kinase, possibly mediated by fatty acids, appears to allow parallel synthetic pathways to become effective in terms of glucose production (Weber *et al.*, 1967). The reaction is activated by K^+ ions, but this effect is antagonized by Na^+ ions.

The first observation that pyruvate kinase occurs in multiple forms was that of Von Fellenberg, Richterich and Aebi (1963), who found the rat-kidney and -liver enzymes to have different electrophoretic mobilities from those of other rat tissues. They suggested that these forms might be concerned with gluconeogenesis rather than with glycolysis. Subsequently Tanaka, Harano, Morimura and Mori (1965) found two different forms of the enzyme in rat liver, one of which is identical with the muscle enzyme (Boyer, 1962). The activity of the liver-specific isoenzyme was shown to be markedly reduced by fasting or in experimental diabetes, but it is restored by insulin treatment. The muscle enzyme, on the other hand, is not diet-dependent (Tanaka *et al.*, 1965).

Pyruvate kinase heterogeneity has recently been described in human tissues by Bigley, Stenzel, Jones, Campos and Koler (1968), who on starch-gel electrophoresis found three zones of activity. The erythrocytic enzyme appeared as a single fast-migrating band (I), while that from the brain, heart, leucocytes and skeletal muscle remained as a single band (III) near the origin. The liver contained bands I and III, with most activity in the latter, but the kidney enzyme was largely concentrated in a band of intermediate mobility (II) accompanied by minor quantities of band I (Fig. 27). Bands I and III can also be separated by DEAE-cellulose chromatography, and all three isoenzymes can be differentiated by fractional precipitation with ammonium sulphate.

The three isoenzymes can be distinguished antigenically. The rat-liver isoenzyme obtained by Tanaka *et al.* (1965) is not affected by anti-rat-muscle pyruvate kinase antiserum, nor are the human type III isoenzymes of Bigley *et al.* (1968) significantly inhibited by anti-type I serum. Isoenzyme II, however, is partly inhibited by the antiserum to the type I isoenzyme. Both groups report almost complete inhibition by the homologous antisera.

Two slow-moving bands of pyruvate kinase activity (PK I and PK II) have recently been separated from normal erythrocytes by high-voltage electrophoresis on Cellogel, but various members of a family with pyruvate kinase deficiency showed a complete absence of PK II in their erythrocytes. Very similar patterns were obtained with the red-cell glutathione reductase: one of the two normal bands was absent in the cells of patients with a deficiency of this enzyme (Rüdiger, Blume, Löhr and Schalhorn, 1968).

Since the inhibition of pyruvate kinase by fatty acids is believed to be one of the regulatory mechanisms controlling glycolysis and gluconeogenesis, it is significant that octanoate exerts a greater inhibitory effect on isoenzyme III than on isoenzyme I (Bigley *et al.*, 1968).

The finding of two interconvertible forms of pyruvate kinase after cellulose acetate electrophoresis of the rat epididymal fat-pad is also of interest in this connection (Pogson, 1968). One of these is fast migrating and is activated by fructose diphosphate, while the other migrates at a slower rate and is insensitive to fructose diphosphate. The fast isoenzyme is converted into the slow form when incubated with low concentrations of fructose diphosphate, while EDTA reverses the process.

Further evidence of the heterogeneity of pyruvate kinase is provided by determination of the K_m values for phosphoenolpyruvate of the erythrocytic enzyme (Campos, Koler and Bigley, 1965; Wiemann and Tönz, 1966; Boivin *et al.*, 1966). Using pyruvate kinase partly purified

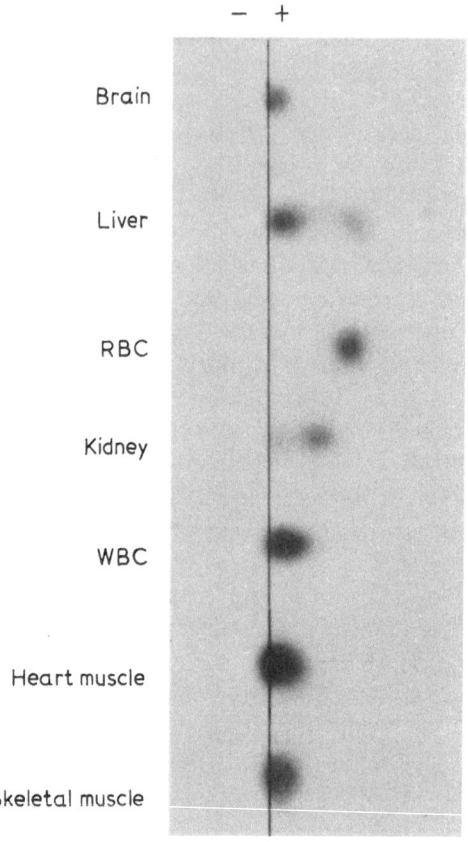

Fig. 27. Pyruvate kinase isoenzymes of human tissues separated by starch-gel electrophoresis at pH 8·6. (Bigley, Stenzel, Jones, Campos and Koler, 1968.)

by DEAE-cellulose chromatography, Boivin and Galand (1967) obtained a mean value of 1·5 mM for the enzyme of normal red cells, while the corresponding figure for that of a pyruvate-kinase-deficient patient was 0·31 mM. Substrate affinities, however, seem to vary in different individuals, as Sachs, Wicker, Gilcher, Conrad and Cohen (1968) found the abnormal enzyme to have a higher K_m for phospho-enolpyruvate (0·104–0·127 mM) than the normal enzymes (0·02–0·037 mM).

Bigley and Koler (1968), on the other hand, have described a patient with a congenital deficiency of red-cell pyruvate kinase, whose liver was also deficient in the 'erythrocytic' isoenzyme. This isoenzyme appeared to be qualitatively normal in both liver and red cells as judged by its chromatographic, electrophoretic, kinetic and antigenic properties, and its heat stability.

REFERENCES

ANSTALL, H. B., LAPP, C. and TRUJILLO, J. M. (1966), *Science*, **154**, 657.

BALINSKY, D. and BERNSTEIN, R. E. (1963), *Biochim. biophys. Acta* **67**, 313.

BARON, D. N., BUCK, G. M. and FOXWELL, C. J. (1969), *Adv. Enzyme Reg.* **7**, 325.

BEUTLER, E. (1966), *The Metabolic Basis of Inherited Disease*, ed. STANBURY, J. B., WYNGAARDEN, J. B. and FREDRICKSON, D. S., 2nd ed., New York: McGraw Hill, p. 1060.

BEUTLER, E. (1967), *Am. J. clin. Path.* **47**, 303.

BEUTLER, E. and COLLINS, Z. (1965), *Science*, **150**, 1306.

BEUTLER, E. and COLLINS, Z. (1966), *Experientia* **22**, 827.

BEUTLER, E., MATHAI, C. K. and SMITH, J. E. (1968), *Blood* **31**, 131.

BEUTLER, E., YEH, M. and FAIRBANKS, V. F. (1962), *Proc. Natl. Acad. Sci., Wash.* **48**, 9.

BIGLEY, R. H. and KOLER, R. D. (1968), *Ann. Hum. Genet., Lond.* **31**, 383.

BIGLEY, R. H., STENZEL, P., JONES, R. T., CAMPOS, J. O. and KOLER, R. D. (1968), *Enzymol. biol. clin.* **9**, 10.

BLOSTEIN, R. and RUTTER, W. J. (1963), *J. biol. Chem.* **238**, 3280.

BOIVIN, P. and GALAND, C. (1967), *Rev. franç. Étud. clin. biol.* **12**, 372.

BOIVIN, P., PIGUET, H. and GALAND, C. (1966), *Nouv. Rev. franç. Hématol.* **6**, 769.

BONSIGNORE, A., DE FLORA, A., MANGIATROTTI, M. A., LORENZONI, I. and ALEMA, S. (1968), *Biochem. J.* **106**, 147.

BONSIGNORE, A., FORNAINI, G., LEONCINI, G., FANTONI, A. and SEGNI, P. (1966), *J. clin. Invest.* **45**, 1865.

BOWMAN, H. S. and PROCOPIO, F. (1963), *Ann. int. Med.* **58**, 567.

BOWMAN, J. E., CARSON, P. E., FRISCHER, H. and DE GARAY, A. L. (1966), *Nature, Lond.* **210**, 811.

BOYER, P. D. (1962), *The Enzymes*, ed. BOYER, P. D., LARDY, H. and MYRBÄCK, K., 2nd ed., Vol. 6, New York: Academic Press, p. 95.

BOYER, S. H., PORTER, I. H. and WEILBACHER, R. G. (1962), *Proc. Natl. Acad. Sci., Wash.* **48**, 1868.

BREWER, G. J., BOWBEER, D. R. and TASHIAN, R. E. (1967), *Acta genet., Basel* **17**, 97.

BREWER, G. J. and DERN, R. J. (1964), *Am. J. hum. Genet.* **16**, 472.

BREWER, G. J. and KNUTSEN, C. A. (1968), *Science*, **159**, 650.

BREWER, G. J. and POWELL, R. D. (1963), *Nature, Lond.* **199**, 704.

BREWER, G. J., POWELL, R. D., SWANSON, S. H. and ALVING, A. S. (1964), *J. Lab. clin. Med.* **64**, 601.

BROWN, J., MILLER, D. M., HOLLOWAY, M. T. and LEVE, G. D. (1967), *Science*, **155**, 205.

BURKA, E. R., WEAVER, Z. and MARKS, P. A. (1966), *Ann. int. Med.* **64**, 817.

CAMPOS, J. O., KOLER, R. D. and BIGLEY, R. H. (1965), *Nature, Lond.* **208**, 194.

CARSON, P. E., AJMAR, F., HASHIMOTO, F. and BOWMAN, J. E. (1966), *Nature, Lond.* **210**, 813.

CARSON, P. E., FLANAGAN, C. L., ICKES, C. E. and ALVING, A. S. (1956), *Science,* **124**, 484.

CARSON, P. E., SCHRIER, S. L. and KELLERMEYER, R. W. (1959), *Nature, Lond.* **184**, 1292.

CARTER, N. D. and PARR, C. W. (1967), *Nature, Lond.* **216**, 511.

CASTELLINO, F. J. and BARKER, R. (1966), *Biochem. biophys. Res. Commun.* **23**, 182.

CHAN, W., MORSE, D. E. and HORECKER, B. L. (1967), *Proc. Natl. Acad. Sci., Wash.* **57**, 1013.

CHILDS, B., ZINKHAM, W. H., BROWNE, E. A., KIMBRO, E. L. and TORBERT, J. V. (1958), *Johns Hopkins Hosp. Bull.* **102**, 21.

CHOREMIS, C., KATTAMIS, C. A., KYRIAZAKOU, M. and GAVRIILIDOU, E. (1966), *Lancet* **i**, 269.

CHRISTEN, P., RENSING, U., SCHMID, A. and LEUTHARDT, F. (1966), *Helv. chim. Acta* **49**, 1872.

CHUNG, A. E. and LANGDON, R. G. (1963), *J. biol. Chem.* **238**, 2317.

CORY, R. P. and WOLD, E. (1965), *Federation Proc.* **24**, 594.

CRANE, R. K. and SOLS, A. (1953), *J. biol. Chem.* **203**, 273.

DARROW, R. A. and COLOWICK, S. P. (1961), *Methods in Enzymology,* ed. COLOWICK, S. P. and KAPLAN, N. O., Vol. 5, New York: Academic Press, p. 226.

DAVE, P., KAPLAN, R. W. and PFLEIDERER, G. (1966), *Biochem. Z.* **345**, 440.

DAVIDSON, R. G., NITOWSKY, H. M. and CHILDS, B. (1963), *Proc. Natl. Acad. Sci., Wash.* **50**, 481.

DEAL, W. C., RUTTER, W. J. and VAN HOLDE, K. E. (1963), *Biochemistry* **2**, 246.

DERECHIN, M., RAMEL, A. H., LAZARUS, N. R. and BARNARD, E. A. (1966), *Biochemistry* **5**, 4017.

DERN, R. J., BREWER, G. J., TASHIAN, R. E. and SHOWS, T. B. (1966), *J. Lab. clin. Med.* **67**, 255.

DETTER, J. C., WAYS, P. O., GIBLETT, E. R., BAUGHAN, M. A., HOPKINSON, D. A., POVEY, S. and HARRIS, H. (1968), *Ann. hum. Genet., Lond.* **31**, 329.

DRECHSLER, E. R., BOYER, P. D. and KOWALSKY, A. G. (1959), *J. biol. Chem.* **234**, 2627.

DIPIETRO, D. L. and WEINHOUSE, S. (1960), *J. biol. Chem.* **235**, 2542.

EATON, G. M., BREWER, G. J. and TASHIAN, R. E. (1966), *Nature, Lond.* **212**, 944.

EDELSTEIN, S. J. and SCHACHMAN, H. K. (1966), *Federation Proc.* **25**, 412.

ESHAGHPOUR, E., OSKI, F. A. and WILLIAMS, M. (1967), *J. Pediat.* **70**, 595.

FILDES, R. A. and PARR, C. W. (1963), *Nature, Lond.* **200**, 890.

FITCH, L. I., PARR, C. W. and WELCH, S. G. (1968), *Biochem. J.* **110**, 560.

FOXWELL, C. J., BUCK, G. M. and BARON, D. N. (1968), *Biochem. J.* **110**, 721.

FOXWELL, C. J., CRAN, E. J. and BARON, D. N. (1966), *Biochem. J.* **100**, 44P.

FROESCH, E. R., WOLF, H. P., BAITSCH, H., PRADER, A. and LABHART, A. (1963), *Am. J. Med.* **34**, 151.

GAZITH, J., SCHULZE, I. T., GOODING, R. H., WOMACK, F. C. and COLOWICK, S. P. (1968), *Ann. N.Y. Acad. Sci.* **151**, 307.

GINSBERG, A. and MEHLER, A. H. (1966), *Federation Proc.* **25**, 407.

GLOCK, G. E. (1961), *Biochemists' Handbook,* ed. LONG, C., London: Spon, p. 345.

GLOCK, G. E. and MCLEAN, P. (1953), *Biochem. J.* **55**, 400.

GONZALES, C., URETA, T., SANCHEZ, R. and NIEMEYER, H. (1964), *Biochem. biophys. Res. Commun.* **16**, 347.

GORDON, H., KERAAN, M. M. and VOOIJS, M. (1967), *Nature, Lond.* **214**, 466.

GRACY, R. W., LACKO, A. G. and HORECKER, B. L. (1969), *J. biol. Chem.* **244**, 3913.

GRAZI, E., CHENG, T. and HORECKER, B. L. (1962), *Biochem. biophys. Res. Commun.* **7**, 250.

GROSS, R., HURWITZ, R. and MARKS, P. A. (1958), *J. clin. Invest.* **37**, 1176.

GROSSBARD, L. and SCHIMKE, R. T. (1966), *Federation Proc.* **25**, 220.

HANSEN, R., PILKIS, S. J. and KRAHL, M. E. (1967a), *Federation Proc.* **26**, 257.

HANSEN, R., PILKIS, S. J. and KRAHL, M. E. (1967b), *Endocrinology* **81**, 1397.

HARRIS, H., HOPKINSON, D. A. and LUFFMAN, J. (1968), *Ann. N.Y. Acad. Sci.* **151**, 232.

HASHIMOTO, T. and HANDLER, P. (1966), *J. biol. Chem.* **241**, 3940.

HELGE, H. and BORNER, K. (1966), *Dtsch. med. Wschr.* **91**, 1584.

HERS, H. G. and JOASSIN, G. (1961), *Enzymol. biol. clin.* **1**, 4.

HERS, H. G. and KUSAKA, T. (1953), *Biochim. biophys. Acta* **11**, 427.

HERSKOVITS, J., MASTERS. C. J., WASSARMAN, P. M. and KAPLAN, N. O. (1967), *Biochem. biophys. Res. Commun.* **26**, 24.

HOLMES, E. W., JR, MALONE, J. I., WINEGRAD, A. I. and OSKI, F. S. (1967), *Science,* **156**, 646.

HOLMES, E. W., JR, MALONE, J. I., WINEGRAD, A. I. and OSKI, F. S. (1968), *Science,* **159**, 651.

HOPKINSON, D. A. and HARRIS, H. (1965), *Nature, Lond.* **206**, 410.

HORECKER, B. L. (1967), *Am. J. clin. Path.* **47**, 269.

JOSHI, J. G. and HANDLER, P. (1964), *J. biol. Chem.* **239**, 2741.

JOSHI, J. G., HOOPER, J., KUWACKI, T., SAKURADA, T., SWANSON, J. R. and HANDLER, P. (1967), *Proc. Natl. Acad. Sci., Wash.* **57**, 1482.

KAJI, A., TRAYSER, K. A. and COLOWICK, S. P. (1961), *Ann. N.Y. Acad. Sci.* **94**, 798.

KAPLAN, J.-C. and BEUTLER, E. (1968), *Science*, **159**, 215.

KATZEN, H. M. (1967), *Advances in Enzyme Regulation*, ed. WEBER, G., Vol. 5, Oxford: Pergamon Press, p. 335.

KATZEN, H. M. and SCHIMKE, R. T. (1965), *Proc. Natl. Acad. Sci., Wash.* **54**, 1218.

KATZEN, H. M., SODERMAN, D. D. and CIRILLO, V. J. (1968), *Ann. N.Y. Acad. Sci.* **151**, 351.

KIRKMAN, H. N. (1959), *Nature, Lond.* **184**, 1291.

KIRKMAN, H. N. (1962), *J. biol. Chem.* **237**, 2364.

KIRKMAN, H. N. and CROWELL, B. B. (1963), *Nature, Lond.* **197**, 286.

KIRKMAN, H. N., DOXIADIS, S. A., VALAES, T., TASSOPOULOS, N. and BRINSON, A. G. (1965), *J. Lab. clin. Med.* **65**, 212.

KIRKMAN, H. N. and HANNA, J. E. (1968), *Ann. N.Y. Acad. Sci.* **151**, 133.

KIRKMAN, H. N. and HENDRICKSON, E. M. (1962), *J. biol. Chem.* **237**, 2371.

KIRKMAN, H. N., MCCURDY, P. R. and NAIMAN, J. L. (1964), *Cold Spring Harb. Symp. quant. Biol.* **29**, 391.

KIRKMAN, H. N. and RILEY, H. D. (1961), *Am. J. Dis. Child.* **102**, 313, 584.

KIRKMAN, H. N., RILEY, H. D. and CROWELL, B. B. (1960), *Proc. Natl. Acad. Sci., Wash.* **46**, 938.

KIRKMAN, H. N., SIMON, E. R. and PICKARD, B. M. (1965), *J. Lab. clin. Med.* **68**, 834.

KOBASHI, K., LAI, C. Y. and HORECKER, B. L. (1966), *Arch. Biochem. Biophys.* **117**, 437.

KUNITZ, M. and MCDONALD, M. R. (1946), *J. gen. Physiol.* **29**, 393.

LAUSECKER, C., HEIDT, P., FISCHER, D., HARJLEYB, H. and LÖHR, G. W. (1965), *Arch. Franç. Pédiat.* **22**, 789.

LAZARUS, N. R., RAMEL, A. H., RUSTUM, Y. M. and BARNARD, E. A. (1966), *Biochemistry* **5**, 4003.

LEBHERZ, H. G. and RUTTER, W. J. (1967), *Science*, **157**, 1198.

LEUTHARDT, F., TESTA, E. and WOLF, H. P. (1954), *Helv. chim. Acta* **36**, 227.

LEVY, H. R. (1963), *J. biol. Chem.* **238**, 775.

LINDER, D. and GARTLER, S. M. (1965), *Science*, **150**, 67.

LISKER, R. and GIBLETT, E. R. (1967), *Am. J. hum. Genet.* **19**, 174.

LONG, W. K., KIRKMAN, H. N. and SUTTON, H. E. (1965), *J. Lab. clin. Med.* **65**, 81.

LUBIN, B. H. and OSKI, F. A. (1967), *J. Pediat.* **70**, 788.

LYON, M. P. (1961), *Nature, Lond.* **190**, 372.

MCCURDY, P. R., KIRKMAN, H. N., NAIMAN, J. L., JIM, R. T. S. and PICKARD, B. M. (1966), *J. Lab. clin. Med.* **67**, 374.

MALMSTRÖM, B. G. (1957), *Arch. Biochem.* **70**, 58.

MANGIAROTTI, G., GARRÉ, C., ACQUARONE, M. E. and SILENGO, L. (1966), *Ital. J. Biochem.* **15**, 67.

MARKS, P. A. (1967), *Am. J. clin. Path.* **47**, 287.

MARKS, P. A., BANKS, J. and GROSS, R. T. (1962), *Nature, Lond.* **194**, 454.

MARKS, P. A., SZEINBERG, A. and BANKS, J. (1961), *J. biol. Chem.* **236**, 10.

MASTERS, C. J. (1967), *Biochem. biophys. Res. Commun.* **28**, 978.

MONTELEONE, P. L., NADLER, H. H., JUSTICE, P. and HSIA, D. Y.-Y. (1967), *Clin. chim. Acta* **18**, 275.

MORRIS, R. C., JR, UEKI, I., LOH, D., EANES, R. Z. and MCLIN, P. (1967), *Nature, Lond.* **214**, 920.

MOTULSKY, A. G. and STAMATOYANNOPOULOS, G. (1966), *Ann. int. Med.* **65**, 1329.

MURRAY, R. F., JR. and BALL, J. A. (1967), *Science*, **156**, 81.

NAKAGAWA, Y. and NOLTMANN, E. A. (1967), *J. biol. Chem.* **242**, 4782.

NANCE, W. E. and UCHIDA, I. (1964), *Am. J. hum. Genet.* **16**, 380.

NEVULDINE, B. H. and LEVY, H. R. (1965), *Biochem. biophys. Res. Commun.* **21**, 28.

NICHOLAS, P. C. and BACHELARD, H. S. (1969), *Biochem. J.* **112**, 587.

NOLTMANN, E. A. and KUBY, S. A. (1963), *The Enzymes*, ed. BOYER, P. E., LARDY, H. and MYRBACK, K., Vol. 7, New York: Academic Press, p. 238.

OHNO, S., PAYNE, H. W., MORRISON, M. and BEUTLER, E. (1966), *Science*, **153**, 1015.

OSKI, F. A. (1965), *Pediat. Clin. N. Amer.* **12**, 687.

OSKI, F. A. and DIAMOND, L. K. (1963), *New Engl. J. Med.* **269**, 763.

PAPADOPOULOS, C. S. and VELICK, S. F. (1967), *Federation Proc.* **26**, 557.

PARR, C. W. (1966), *Nature, Lond.* **210**, 487.

PARR, C. W. and FITCH, L. I. (1954), *Biochem. J.* **93**, 28c.

PARR, C. W. and FITCH, L. I. (1967), *Ann. hum. Genet.* **30**, 339.

PARR, C. W. and PARR, I. B. (1965), *Biochem. J.* **95**, 16p.

PEANASKY, R. T. and LARDY, H. A. (1958), *J. biol. Chem.* **233**, 365.

PENHOET, E., KOCHMAN, M., VALENTINE, R. and RUTTER, W. J. (1967), *Biochemistry* **6**, 2940.

PENHOET, E., RAJKUMAR, T. V. and RUTTER, W. J. (1966), *Proc. Natl. Acad. Sci., Wash.* **56**, 1275.

PERHEENTUPA, J., PITKÄNEN, E., NIKKILÄ, E. A., SOMERSALO, O. and HAKOSALA, J. (1962), *Ann. Paedat. Fenn.* **8**, 221.

PILKIS, S. J. and HANSEN, R. J. (1968), *Biochim. biophys. Acta* **159**, 189.

PILKIS, S. J., HANSEN, R. J. and KRAHL, M. E. (1968), *Federation Proc.* **27**, 589.

PILKIS, S. J. and KRAHL, M. E. (1966), *Federation Proc.* **25**, 523.

PINTO, P. V. C., NEWTON, W. A. J. and RICHARDSON, K. E. (1966), *J. clin. Invest.* **45**, 823.

PFLEIDERER, G., NEUFAHRT-KREILING, A., KAPLAN, R. W. and FORTNAGEL, P. (1966), *Biochem. Z.* **346**, 269.

PFLEIDERER, G., NEUFAHRT-KREILING, A., KAPLAN, R. W. and FORTNAGEL, P. (1968), *Ann. N.Y. Acad. Sci.* **151**, 78.

POGSON, C. I. (1968), *Biochem. biophys. Res. Commun.* **30**, 297.

PORTER, I. H. (1968), *Ann. intern. Med.* **68**, 251.

PORTER, I. H., BOYER, S. H., WATSON-WILLIAMS, E. J., ADAM, A., SZEINBERG, A. and SINISCALCO, M. (1964), *Lancet* **i**, 895.

PRANKERD, T. A. J. (1965), *Br. med. J.* **ii**, 1017.

PRANKERD, T. A. J. (1967), *Am. J. clin. Path.* **47**, 282.

RAMOT, B., BAUMINGER, S., BROK, F., GAFNI, D. and SHWARTZ, J. (1964), *J. Lab. clin. Med.* **64**, 895.

ROUSLIN, W. (1967), *Federation Proc.* **26**, 557.

RÜDIGER, H. W., BLUME, K. G., LÖHR, G. W. and SCHALHORN, A. (1968), *Klin. Wschr.* **46**, 397.

RUTTER, W. J. (1962), *The Enzymes*, ed. BOYER, P. D., LARDY, H. A. and MYRBÄCK, K., 2nd ed., Vol. 5, New York: Academic Press, p. 341.

RUTTER, W. J. (1964), *Federation Proc.* **23**, 1248.

RUTTER, W. J., RAJKUMAR, T., PENHOET, E., KOCHMAN, M. and VALENTINE, R. (1968), *Ann. N.Y. Acad. Sci.* **151**, 102.

RUTTER, W. J., WOODFIN, B. M. and BLOSTEIN, R. E. (1963), *Acta Chem. Scand.* **17**, 226.

SACHS, J. R., WICKER, D. J., GILCHER, R. O., CONRAD, M. E. and COHEN, R. J. (1968), *J. Lab. clin. Med.* **72**, 359.

SANSONE, G. and SEGNI, G. (1958), *Boll. Soc. Ital. Biol. Sper.* **15**, 327.

SCHACHMAN, H. K. and EDELSTEIN, S. J. (1966), *Biochemistry* **5**, 2681.

SCHAPIRA, F., NORDMANN, Y. and DREYFUS, J.-C. (1968), *Rev. franç. Étud. clin. biol.* **13**, 267.

SCHIMKE, R. T. and GROSSBARD, L. (1968), *Ann. N.Y. Acad. Sci.* **151**, 332.

SCHULZE, I. T., GAZITH, J. and GOODING, R. H. (1967), *Methods in Enzymology*, ed. WOOD, W. A., Vol. 9, New York: Academic Press, p. 376.

SCIALOM, C., NAJEAN, Y. and BERNARD, J. (1966), *Nouv. Rev. franç. Hématol.* **6**, 452.

SCOTT, E. M., DUNCAN, I. W., EKSTRAND, V. and WRIGHT, R. C. (1966), *Am. J. hum. Genet.* **18**, 408.

SHARMA, C., MANJESHWAR, R. and WEINHOUSE, S. (1963), *J. biol. Chem.* **238**, 3840.

SHAW, C. R. (1966), *Science*, **153**, 1013.

SHAW, C. R. and BARTO, E. (1965), *Science*, **148**, 1099.

SHAW, C. R. and KOEN, A. L. (1968), *Ann. N.Y. Acad. Sci.* **151**, 149.

SHIMIZU, H. and OZAWA, H. (1967), *Biochim. biophys. Acta* **133**, 195.

SHOWS, T. B., TASHIAN, R. E. and BREWER, G. J. (1964), *Science*, **145**, 1056.

132 · Isoenzymes

SHOWS, T. B., TASHIAN, R. E., BREWER, G. J. and DERN, R. J. (1964). *Science*, **145**, 1056.

SINE, H. E. and HASS, L. F. (1967), *J. Am. chem. Soc.* **89**, 1749.

SOLS, A., SILLERO, A. and SALAS, J. (1965), *J. Cell Comp. Physiol.* **66** (Suppl. 1), 23.

SPENCER, N., HOPKINSON, D. A. and HARRIS, H. (1964), *Nature, Lond.* **204**, 742.

SPOLTER, P. D., ADELMAN, R. C. and WEINHOUSE, S. (1965), *J. biol. Chem.* **240**, 1327.

STAMATOYANNOPOULOS, G., PANAYOTOPOULOS, A. and PAPAYAN-NOPOULOU, T. (1964), *Lancet* ii, 932.

STAMATOYANNOPOULOS, G., PAPAYANNOPOULOS, T., BAKOPOULOS, C. and MOTULSKY, A. G. (1967), *Blood* **29**, 87.

STELLWAGEN, E. and SCHACHMAN, H. K. (1962), *Biochemistry* **1**, 1056.

SUSOR, W. A., KOCHMAN, M. and RUTTER, W. J. (1969), *Science*, **165**, 1260.

TANAKA, T., HARANO, Y., MORIMURA, H. and MORI, H. (1965), *Biochem. biophys. Res. Commun.* **21**, 55.

THULINE, H. C., MORROW, A. C., NORBY, D. E. and MOTULSKY, A. G. (1967), *Science*, **157**, 431.

TRAYSER, K. A. and COLOWICK, S. P. (1961), *Arch. Biochem. Biophys.* **94**, 177.

TRUJILLO, J. M., WALDEN, B., O'NEIL, P. and ANSTALL, H. B. (1965). *Science*, **148**, 1603.

TSAO, M. U. (1960), *Arch. Biochem. Biophys.* **90**, 234.

TSOI, A. and DOUGLAS, H. C. (1964), *Biochim. biophys. Acta* **92**, 513.

TSUTSUI, E. A. and MARKS, P. A. (1962), *Biochem. biophys. Res. Commun.* **8**, 338.

VALENTINE, W. N., OSKI, F. A., PAGLIA, D. E., BAUGHAN, M. A., SCHNEIDER, A. S. and NAIMAN, J. L. (1967), *New Engl. J. Med.* **276**, 1.

VALENTINE, W. N., TANAKA, K. R. and MIWA, S. (1959), *Trans. Assoc. Am. Physicians* **74**, 100.

VECCHIO, F., SCHETTINI, F., DI FRANCESCO, L., MELONI, T. and RUSSINO, G. (1966), *Acta Haematol.* **35**, 46.

VINUELA, E., SALAS, M. and SOLS, A. (1963), *J. biol. Chem.* **238**, 1175.

VON FELLENBERG, R., RICHTERICH, R. and AEBI, H. (1963), *Enzymol. biol. clin.* **3**, 240.

WALKER, D. G. (1963), *Biochim. biophys. Acta* **77**, 209.

WALKER, D. G. (1966), In *Essays in Biochemistry*, ed. CAMPBELL, P. N. and GREVILLE, G. D., Vol. 2, London: Academic Press, p. 33.

WALKER, D. G. and RAO, S. (1964), *Biochem. J.* **90**, 360.

WALLER, H. D., LÖHR, G. W. and GAYER, J. (1966), *Klin. Wschr.* **44**, 122.

WARBURG, O. and CHRISTIAN, W. (1931), *Biochem. Z.* **242**, 206.

WARREN, J. C. and PETERSON, D. M. (1966), *Science, N.Y.* **152**, 1245.

WEBER, G., LEA, M. A., HIRD CONVERY, H. J. and STAMM, N. B. (1967), *Advances in Enzyme Regulation*, ed. WEBER, G., Vol. 5, Oxford: Pergamon Press, p. 257.

WIEMANN, U. and TÖNZ, O. (1966), *Nature, Lond.* **209**, 612.

WILKINSON, J. H. (1962), *An Introduction to Diagnostic Enzymology*, London: Arnold, p. 220.

WONG, P. W. K., LING-YU-SHIH, D., HSIA, D. Y.-Y. and TSAO, Y. C. (1965). *Nature, Lond.* **208**, 1323.

YANKEELOV, J. A., HORTON, H. R. and KOSHLAND, D. E. (1964), *Biochemistry* **3**, 349.

YOSHIDA, A. (1966), *J. biol. Chem.* **241**, 4966.

YOSHIDA, A. (1967a), *Biochem. Genet.* **1**, 81.

YOSHIDA, A. (1967b), *Proc. Natl. Acad. Sci., Wash.* **57**, 835.

YOSHIDA, A. (1968), *Ann. N.Y. Acad. Sci.* **151**, 145.

YOSHIDA, A., STEINMANN, L. and HARBERT, P. (1967), *Nature, Lond.* **216**, 275.

YOUNG, W. J., PORTER, J. E. and CHILDS, B. (1964), *Science, N.Y.* **143**, 140.

CHAPTER 6

Lactate Dehydrogenase Isoenzymes

The heterogeneity of lactate dehydrogenase was first recognized when Neilands (1952) demonstrated activity in each of the two electrophoretically distinct proteins, previously separated by Meister (1950) from the crystalline ox-heart enzyme. Subsequently Vesell and Bearn (1957) found three distinct components when human sera were subjected to starch-block electrophoresis, and Sayre and Hill (1957) obtained similar results with paper electrophoresis and gradient-elution chromatography. About the same time Wieland and Pfleiderer (1957) found that most organs contain up to five protein fractions each exhibiting lactate dehydrogenase activity. Soon afterwards it was appreciated that the isoenzyme composition of the serum was of considerable diagnostic value (Vesell and Bearn, 1958a, 1958b; Hess, 1958; Wieme, 1959), and this gave a tremendous impetus to further study of variations in the isoenzyme content of the blood serum.

Meanwhile it became apparent that lactate dehydrogenase isoenzymes differed from each other in a variety of respects other than in their electrophoretic and chromatographic properties. Differences in chemical composition, substrate specificities, thermal stabilities, abilities to utilize coenzyme analogues and susceptibilities to inhibitors have been reported. In addition, a high degree of immunochemical specificity has been observed: antisera have been prepared which inhibit the action of particular isoenzymes without affecting that of others.

More recently certain facts concerning the biological significance of lactate dehydrogenase isoenzymes have become apparent, and these will be considered in this chapter along with the factors listed above. Methods for the isolation and detection of isoenzymes have been reviewed in Chapters 2 and 3, and their chemical structures have been discussed in Chapter 4. As already mentioned, the five isoenzymes will be referred to as LD_1, LD_2, LD_3, LD_4 and LD_5, LD_1 carrying the highest negative charge and migrating farthest towards the anode during electrophoresis.

Distribution of lactate dehydrogenase isoenzymes
Though it has been clearly established by a variety of different techniques that five distinct isoenzymes usually occur in human and animal

tissues, their relative distribution varies considerably, not only from tissue to tissue but also from species to species. The results of some quantitative studies are summarized in Table 10, from which it will be noted that in certain cases there are considerable discrepancies between the values obtained for the same tissue by different investigators. These may be partly due to variations in technique and in the treatment of the specimen before electrophoresis, and also possibly to disproportionate loss of the more labile slow-moving isoenzymes, LD_4 and LD_5. Other likely causes of variation, especially in the case of the human specimens, include post-mortem autolysis and contamination of certain tissues with blood. Despite these anomalies, there is broad agreement that in heart, erythrocytes and kidney, the fast-moving isoenzymes LD_1 and LD_2 predominate, whereas in liver and skeletal muscle the principal isoenzymes are LD_4 and LD_5. LD_3 appears to be the most abundant fraction in many other human tissues, including the spleen, pancreas, thyroid, adrenals and lymph nodes.

Several investigators have reported the absence of LD_5 from red blood cells, an observation which is not at first sight consistent with random reassociation of the enzyme sub-units. However, it has now been established that erythrocytes lose their complement of this isoenzyme during ageing (Vesell and Bearn, 1962a; Rosa and Schapira, 1964; Starkweather *et al.*, 1965). Since LD_5 is less stable than the other lactate dehydrogenase isoenzymes, it disappears more rapidly, and since there is no mechanism for protein synthesis in mature erythrocytes, it cannot be replaced (Schapira and Rosa, 1967).

Palmer and Kjellberg (1967) separated connective tissue from rat-liver homogenates and determined the isoenzyme composition of the liver-cell preparation. They concluded that liver cells probably contain LD_5 only and that the small amounts of other isoenzymes originate in other cells. However, Nemchinskaya, Ganelina and Braun (1968) have recently obtained evidence suggesting a difference between the relative amounts of isoenzymes present in rat-liver nuclei and in the cytoplasmic fraction. In both cases LD_5 was the principal component, but the nuclear fraction was found to contain appreciably more LD_3 and LD_4.

The relative electrophoretic mobilities of the heart lactate dehydrogenases from several different species are shown in Fig. 28, from which it will be observed that in the chicken the principal heart isoenzyme has a markedly different mobility from those of the other species. Care must therefore be exercised in extrapolating experimental results to animals of a different species. Though the isoenzyme pattern may vary considerably from species to species, it remains remarkably constant,

Species	Tissue	Percentage distribution of LD isoenzymes					
		LD_1	LD_2	LD_3	LD_4	LD_5	References
Human	Heart	35	36	12	16	11	[1, 2]
		67	29	4	< 1	< 1	[3]
		71	27	2	—	—	[4]
		49	45	6	—	—	[5]
		53	29	16	2	< 1	[6]
	Kidney	12	14	24	25	23	[1, 2]
		42	48	9	1	—	[4]
		30	50	15	5	—	[5]
		52	28	16	4	< 1	[7]
	Erythrocytes	39	56	—	5	—	[1, 2]
		39	46	15	—	—	[3]
		46	39	11	4	—	[4]
		42	36	15	5	2	[7]
		36	44	14	4	2	[14]
	Brain	21	26	26	20	8	[2, 8]
		25	25	34	15	1	[9]
	Adrenal	3	20	75	—	2	[5]
	Lung	10	20	30	25	15	[5]
	Lymph node	10	25	60	—	5	[5]
	Pancreas	30	15	50	—	5	[5]
	Placenta	12	18	15	30	25	[1, 2]
	Thymus	10	11	30	28	21	[1, 2]
	Thyroid	12	25	55	—	8	[5]
	Spleen	6	11	35	28	20	[3]
		10	25	40	20	5	[5]
	Leucocytes	11	58	27	4	—	[4]
		8	12	50	18	12	[7]
		6	19	28	19	22	[15]
	Skeletal muscle	4	7	21	27	41	[1, 2]
	(*see* p. 144)	—	—	2	1	97	[3]
		—	—	17	10	73	[4]
		16	18	18	12	36	[5]
		< 1	15	38	36	11	[6]
		1	4	8	9	78	[10]
	Liver	2	4	11	27	56	[1,2]
		—	6	12	6	76	[3]
		8	10	33	13	36	[4]
		2	5	12	—	81	[5]
		< 1	10	19	17	54	[6]
		2	2	3	12	80	[10]
	Uterus	5	25	44	22	4	[13]
	Uterus (gravid)	2	5	28	45	20	[13]

TABLE 10—*continued*

		Percentage distribution of LD isoenzymes					
Species	Tissue	LD_1	LD_2	LD_3	LD_4	LD_5	References
Rat	Heart	50	30	12	6	2	[10, 11]
	Brain	40	20	15	20	5	[10, 11]
		39	20	18	18	5	[12]
	Kidney	20	15	10	20	35	[10, 11]
	Skeletal muscle	1	1	3	5	90	[10, 11]
	Liver	1	1	3	5–10	85–90	[10, 11]
Rabbit	Heart	100	—	—	—	—	[3]
		94	2	1	3	< 1	[10]
	Erythrocytes	100	—	—	—	—	[3]
	Kidney	77	11	7	3	2	[3]
		63	10	8	9	9	[10]
	Spleen	43	43	14	< 1	< 1	[3]
	Brain	43	20	27	6	4	[3]
	Lung	23	35	21	4	17	[3]
	Liver	2	4	24	59	11	[3]
		1	3	17	39	40	[10]
	Skeletal muscle	—	—	—	—	100	[3]
	(*see* p. 144)	< 1	< 1	< 1	3	95	[10]
Monkey	Heart	35	20	13	15	16	[2]
	Erythrocytes	29	49	15	7	—	[2]
	Lung	18	24	18	17	23	[2]
	Spleen	13	15	15	23	34	[2]
	Liver	7	8	13	28	43	[2]

The above figures have been calculated from results published by the following authors:

[1] Wieme (1959)
[2] Wieme and Van Maercke (1961)
[3] Plagemann, Gregory and Wróblewski (1960a)
[4] Vesell and Bearn (1961)
[5] Wróblewski and Gregory (1961)
[6] Plummer, Elliott, Cooke and Wilkinson (1963)
[7] Withycombe (1965)
[8] Löwenthal, Van Sande and Karcher (1961)
[9] Gerhardt, Clausen, Christensen and Riishede (1963)
[10] Wieland, Pfleiderer, Haupt and Wörner (1959)
[11] Wieland and Pfleiderer (1961)
[12] Bonavita, Ponte and Amore (1962)
[13] Geyer (1968)
[14] Starkweather, Cousineau, Schoch and Zarafonetis (1965)
[15] Malasková and Holeyšovská (1969)

in the absence of disease, between different individuals of the same species.

There have been several reports that a greater number than five lactate dehydrogenase isoenzymes may be detected, especially when starch-gel electrophoresis is employed, though Wieland and Pfleiderer (1961) refer to their occasional observations of extra bands of activity in rat tissues separated by paper or cellulose acetate electrophoresis. Allen (1961) found a total of seven anodic fractions and two cathodic fractions in various mouse tissues, and Blanchaer (1962) observed that

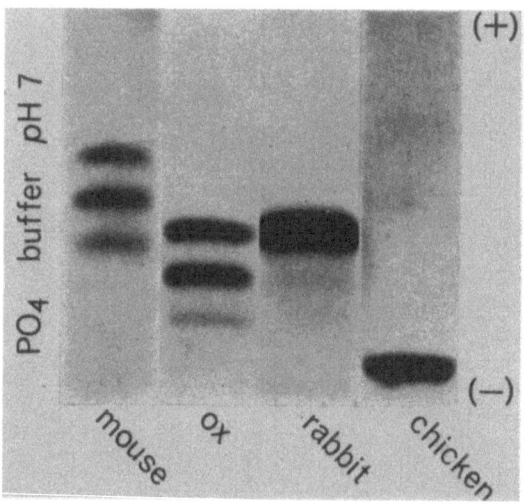

Fig. 28. Starch-gel electrophoresis of the lactate dehydrogenase isoenzymes of mouse, ox, rabbit and chicken
heart (Markert and Appella, 1961).

in various human tissue homogenates and pathological human sera as many as eight fractions could be detected on starch-gel electrophoresis. Other investigators have also observed additional zones of activity in pathological human sera and tissue extracts (Beautyman, 1962; Latner, 1964; Soetens, Karcher, Van Sande and Löwenthal, 1965; Fujimoto, Nazarian and Wilkinson, 1968) (see p. 183).

The observation, first made by Fritz and Jacobson (1963), that the main bands of lactate dehydrogenase activity consist of a number of sub-bands, however, is of a quite different nature. These investigators found that the five major bands of enzymic activity in extracts of mouse tissues could be split into a total of fifteen bands during electrophoresis on polyacrylamide gels when 0.005M-β-mercaptoethanol was previously incorporated into the gel. Since a similar effect was observed with starch gel, the possibility of a polyacrylamide artifact could be eliminated. Others have since confirmed these findings, and various theories have been proposed to explain such multiplication of the main bands. These have been considered in Chapter 4.

Multiple sub-band formation also occurs in the frog (Moyer, Speaker and Wright, 1968) and in the horse (Rauch, 1968). In the former considerable inter-species variation in the absolute electrophoretic mobilities of the major bands is found, e.g. one of the LD_1 sub-bands of *Rana septentrionalis* has a similar mobility to that of the LD_5 of *R. catesbiana*, but the relative rates of migration of corresponding isoenzymes are constant. Moyer *et al.* (1968) found 18–25 isoenzymes in two species, though most frogs exhibit 6–8 bands on electrophoresis.

Nine zones of activity have been detected in various tissues of the trout, an observation which Goldberg (1965) suggests may be attributed to the presence of a third type of sub-unit.

Mention has also been made (p. 69) of the sixth isoenzyme (LD_x) found in post-pubertal human testis (Blanco and Zinkham, 1963; Goldberg, 1963). This has an electrophoretic mobility midway between LD_3 and LD_4. Since LD_x does not occur in pre-pubertal testis, while it is the principal form of spermatozoal lactate dehydrogenase, it seems that this isoenzyme is associated in some way with spermatogenesis.

The testes of other species also contain characteristic isoenzymes (Zinkham, Blanco and Kupchyk, 1963). After electrophoresis the LD_x fractions of the rabbit and the dog appear between LD_4 and LD_5, while in the mouse LD_x has a lower (anodic) mobility than LD_5. Two separate LD_x fractions occur in homogenates of guinea-pig and rat testes: these appear between LD_3 and LD_4 and between LD_4 and LD_5. In bull testis three LD_x components are found, one between LD_3 and LD_4 and two between LD_4 and LD_5, but there is no electrophoretic evidence of an LD_x in the testes of the duck, chicken, hog or cat.

An LD_x band has also been demonstrated in rat kidney, by Ressler, Cook, Olivero and Joseph (1965), but it is not certain whether this is identical with the testicular isoenzyme.

In the mouse LD_x activity becomes apparent when the animals are about 22 days old and reaches a maximum about three weeks later. It is not detectable until primary and secondary spermatocytes can be seen (Goldberg and Hawtrey, 1967; Hawtrey and Goldberg, 1968). Evidence relating the synthesis of LD_x to spermatogenesis has also been obtained in the guinea-pig. LD_x is not found in the testes of this species when spermatogenesis is immunologically suppressed (Zinkham, 1968).

Zinkham and his co-workers have described a multiplicity of testicular isoenzymes in the pigeon. Three distinct phenotypes were observed: type I, which consists of a major and a minor band, both considerably more cathodal than LD_5; type II, consisting of five bands, the fastest of which migrates between LD_4 and LD_5; and type III, which exhibits a single band between LD_4 and LD_5 (Fig. 29) (Blanco,

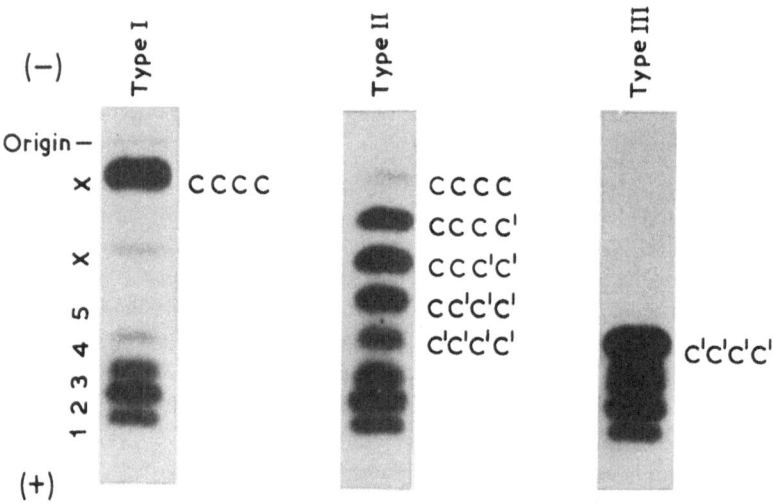

Fig. 29. The three types of lactate dehydrogenase isoenzyme pattern found in the testes of pigeons, showing the multiple LD_x isoenzymes. The minor LD_x of Type I is probably a hybrid of the H and C sub-units. (Zinkham, 1968.)

Zinkham and Kupchyk, 1964; Zinkham, Blanco and Kupchyk, 1964). The presence of the three types is attributed to the synthesis of two C type sub-units. The type I variant, it is suggested, is a polymer of four C sub-units, while the type III isoenzyme comprises four C′ sub-units. Combination of both types of sub-unit leads to the formation of the five LD_x isoenzymes of the type II phenotype.

This mechanism is supported by population studies and by breeding experiments (Zinkham, Kupchyk, Blanco and Isensee, 1965, 1966; Zinkham, 1968).

Relatively few studies have so far been made of the isoenzymes of the mammalian ova. Only one band of activity is detectable in homogenates of fertilized mouse ova. This is identical in its electrophoretic and catalytic properties with mouse LD_1 isolated from other tissues (Auerbach and Brinster, 1967; Rapola and Koskimies, 1967). According to Rapola and Koskimies (1967), this pattern persists throughout the early stages of development from unfertilized egg to the blastocyst stage, but after implantation there is a sudden change in the isoenzyme pattern. As discussed on p. 164, LD_5 predominates in the implanted embryo. Before implantation therefore the egg enzyme consists exclusively of the H sub-unit, while that of the embryo consists mainly of the M sub-unit. This suggests an abrupt change in the activity of the corresponding genes about the time of implantation.

These results also indicate that the spermatozoal LD_x activity is lost immediately upon fertilization. It seems therefore that LD_x activity is concerned solely with spermatozoal metabolism.

Chemical and kinetic properties

Substrate specificities and affinities
Classical studies of mammalian lactate dehydrogenases showed them to be specific for L(+)-lactate* (Meyerhof and Lohmann, 1926; Green and Brosteaux, 1936; Kubowitz and Ott, 1943), but Ottolenghi and Denstedt (1958a) reported that the rabbit-erythrocyte enzyme reacts equally well with both D and L forms. Plagemann *et al.* (1960b) were unable to confirm this, but they demonstrated that the D form is not inhibitory. Green and Brosteaux (1936) showed that α-hydroxybutyrate was an effective substrate for lactate dehydrogenase, and several other α-hydroxy acids have since been shown to be of value in the study of isoenzymes. Though the use of optically active substrates other than lactate does not appear to have been investigated, the observations of Plagemann *et al.* (1960b) suggest that results obtained with DL forms might be interpreted in terms of the L form. Since the existence of isoenzymes was recognized, Allen (1961) has used DL-α-hydroxybutyrate, DL-α-hydroxycaproate and DL-α-hydroxyvalerate in the study of mouse LD isoenzymes.

Similar substrate specificity is found in the reverse reaction. Meister (1950) compared a series of α-keto and α-γ-diketo acids as substrates for crystalline ox-heart lactate dehydrogenase and found that only

---

* The cytochrome *c*-dependent lactate dehydrogenases of yeast also exhibit stereo-specificity in their substrate requirements, but the enzyme requiring D-lactate appears to be quite distinct from that oxidizing the L-form (see review by Nygaard, 1963). Tarmy and Kaplan (1967a, 1967b) have recently isolated and characterized a stereospecific D-lactate dehydrogenase from *Escherichia coli* which requires NAD as coenzyme. It is activated by pyruvate, apparently mediated through a change in conformation.

TABLE 11 *Michaelis constants (K_m) of crystalline lactate dehydrogenases*

Species	Tissue	Principal isoenzyme	Substrate	pH	Coenzyme concentration	K_m	References
Human	Heart	LD_1	Pyruvate	7·4	$0·43–1·7 \times 10^{-4}$M	$1·18 \times 10^{-4}$M	[1]
Human	Heart	LD_1	Lactate	8·7	$4·76 \times 10^{-4}$M	$0·44 \times 10^{-2}$M	[1]
Human	Heart	LD_1	Lactate	8·7	$2·38 \times 10^{-4}$M	$0·80 \times 10^{-2}$M	[1]
Human	Heart	LD_1	Lactate	8·7	$0·95 \times 10^{-4}$M	$1·65 \times 10^{-2}$M	[1]
Human	Liver	LD_5	Lactate	8·8	$0·44 \times 10^{-4}$M	$2·82 \times 10^{-2}$M	[2]
Human	Liver	LD_5	Lactate	8·8	$0·88 \times 10^{-4}$M	$2·56 \times 10^{-2}$M	[2]
Human	Liver	LD_5	Lactate	8·8	$1·75 \times 10^{-4}$M	$2·23 \times 10^{-2}$M	[2]
Human	Liver	LD_5	Lactate	8·8	$3·50 \times 10^{-4}$M	$1·96 \times 10^{-3}$M	[2]
Ox	Heart	LD_1	Pyruvate	6·98	$0·48 \times 10^{-4}$M	$1·4 \times 10^{-4}$M	[3]
Ox	Heart	LD_1	Lactate	9·44	$2·6 \times 10^{-4}$M	11×10^{-2}M	[3]
Ox	Muscle	LD_5	Pyruvate	7·5	$1·0 \times 10^{-4}$M	$5·2 \times 10^{-4}$M	[4]
Ox	Muscle	LD_5	Lactate	8·9	$1·0 \times 10^{-4}$M	$2·5 \times 10^{-2}$M	[4]
Chicken	Heart	LD_1	Pyruvate	7·5	$1·0 \times 10^{-4}$M	$0·8 \times 10^{-4}$M	[4]
Chicken	Heart	LD_1	Lactate	8·9	$1·0 \times 10^{-4}$M	$0·7 \times 10^{-2}$M	[4]
Chicken	Muscle	LD_5	Pyruvate	7·5	$1·0 \times 10^{-4}$M	$3·2 \times 10^{-4}$M	[4]
Chicken	Muscle	LD_5	Lactate	8·9	$1·0 \times 10^{-4}$M	$4·0 \times 10^{-2}$M	[4]

References: [1] Nisselbaum and Bodansky (1961a)
[2] Gibson, Davisson, Bachhawat, Ray and Vestling (1953)
[3] Winer and Schwert (1958)
[4] Pesce, Fondy, Stolzenbach, Costello and Kaplan (1967)

pyruvate and α-oxobutyrate were readily reduced in the presence of NADH, though certain of the diketo acids reacted at a much slower rate.

Numerous figures for the Michaelis constants (K_m) for pyruvate and lactate have been published, but owing to variations in the purity of the enzyme preparations and in the conditions used, it is difficult to compare directly the different values reported. Table 11 lists values for the purified heart enzyme, which consists almost exclusively of LD_1, and for the liver or muscle enzymes composed very largely of LD_5.

The K_m values for the fast isoenzymes are lower than the corresponding figures for the slow isoenzymes, i.e. LD_1 and LD_2 have greater affinities for their substrates than have LD_4 and LD_5.

The results summarized in Table 11 have been confirmed by studies on electrophoretically separated isoenzymes from crude tissue extracts. It should be borne in mind, however, that the figures reported were determined on relatively impure preparations and that they would be modified by the possible presence of activators or inhibitors or of non-lactate dehydrogenase proteins with which the substrate might become associated. The relative differences, nevertheless, are of considerable interest, and some of the published results appear in Table 12.

TABLE 12 *Apparent Michaelis constants (K_m) for electrophoretically separated lactate dehydrogenase isoenzymes*

Species	Substrate	pH	Coenzyme concentration	K_m		Ref.
				LD_1	LD_5	
Rabbit	Pyruvate	7·4	$1·5 \times 10^{-4}$M	$6·7 \times 10^{-5}$M	$3·5 \times 10^{-4}$M	[1]
Human	Pyruvate	7·4	$1·2 \times 10^{-4}$M	$7·7 \times 10^{-5}$M	$8·3 \times 10^{-4}$M	[2]
Human	2-Oxo-butyrate	7·4	$1·2 \times 10^{-4}$M	$6·3 \times 10^{-4}$M	$2·7 \times 10^{-3}$M	[2]

References: [1] Plagemann *et al.* (1960*b*)
[2] Withycombe (1965)

Lactate dehydrogenase is strongly inhibited by excess of pyruvate (Kubowitz and Ott, 1943; Ottolenghi and Denstedt, 1958*b*; Winer and Schwert, 1958; Rosalki and Wilkinson, 1960), and a technique based upon this effect has been devised by Plagemann *et al.* (1960*b*) to determine the relative proportions of LD_1 and LD_5 in a mixture of the two isoenzymes. At pH 7·0 a pyruvate concentration of 1·2mM was optimal for human LD_5, but permitted only 70% of the optimal activity of LD_1, while LD_1 showed greatest activity with 0·15mM-pyruvate, at which concentration LD_5 exhibited only 70% of its optimal activity. Similar results were obtained with rabbit isoenzymes. If the reaction rates with these substrate concentrations are described as V_1 and V_2 respectively, the isoenzymes, LD_2, LD_3 and LD_4, are found to have $V_1 : V_2$ ratios intermediate between those of LD_1 and LD_5 (Table 13). When the $V_1 : V_2$ ratios are plotted logarithmically against the relative composition of a series of mixtures of LD_1 and LD_5 a direct relationship is found and

TABLE 13 *The ratio of lactate dehydrogenase activities at concentrations of 1·2mM-pyruvate (V_1) and 0·15mM-pyruvate (V_2) for various rabbit tissues*

(Plagemann, Gregory and Wróblewski, 1960*b*)

Tissue	Predominant isoenzyme	$V_1 : V_2$
Heart	LD_1, LD_2	0·38 + 0·04
Erythrocytes	LD_1, LD_2	0·44 + 0·02
Kidney	LD_1, LD_2	0·41 + 0·04
Liver	LD_5	1·5 + 0·2
Skeletal muscle	LD_5	2

there is excellent agreement between the observed and calculated values. The practical value of the $V_1 : V_2$ ratio is somewhat diminished by the fact that the ratio tends to increase on storage, especially in the case of LD_1. This process is accelerated by the addition of trypsin.

A similar method based upon the difference in the relative affinities for lactate has been devised by Stambaugh and Post (1966*a*) for the determination of H and M sub-units in biological fluids. Another related technique is that of Krieg, Rosenblum and Henry (1967), who compared the relative rates for the conversion of pyruvate to lactate with those for the reverse reaction.

The difference in the degree of inhibition of LD isoenzymes by excess pyruvate appears to be of considerable metabolic significance. Since the reduction of pyruvate to lactate by LD_1 is strongly inhibited by quite low concentrations of pyruvate, it has been suggested that rapid accumulation of lactate could not occur in a tissue rich in this isoenzyme, such as the heart, and complete oxidation of glucose via the citric acid cycle is therefore to be expected. LD_5, on the other hand, functions more efficiently when exposed to concentrations of pyruvate inhibitory to LD_1, and is inhibited only by much higher concentrations. It appears therefore that tissues rich in LD_5, e.g. skeletal muscle, would allow the rapid conversion of pyruvate into lactate, and hence the establishment of an oxygen debt under anaerobic conditions (Cahn, Kaplan, Levine and Zwilling, 1960).

Certain skeletal muscles, however, such as the soleus, exhibit lactate dehydrogenase isoenzyme patterns similar to that of heart muscle. The anomaly may be explained by the fact that such muscles consist principally of red fibres, whose main physiological concern is with posture and support, and thus they remain in more or less continuous contraction (Dawson, Goodfriend and Kaplan, 1964).

Direct evidence in support of the concept that LD_1 is concerned with aerobic oxidation while LD_5 is associated with anaerobic glycolysis is provided by the observation of Goodfriend, Sokol and Kaplan (1966) that exposure of cells (originally derived from monkey heart) in tissue culture to low oxygen tensions provokes increased synthesis of the LD_5 isoenzyme. This increase of M sub-unit synthesis is blocked by exposure to certain antibiotics and to cold, but its occurrence suggests a possible control mechanism for sub-unit synthesis. Exposure of human lymphocytes cultured with phytohaemagglutinin to low oxygen tensions also leads to a relative increase in the synthesis of the M sub-unit (Hellung-Larsen and Andersen, 1968), while exposure to high oxygen tensions reduces the extent of the shift towards synthesis of the M sub-unit in calf-kidney cortex cells grown in tissue culture (Güttler and Clausen, 1969).

Substrate inhibition studies with pyruvate have been applied to the investigation of the functions of the two forms of lactate dehydrogenase in the breast muscle of birds (Wilson, Cahn and Kaplan, 1963). The flight habits of birds correlate with the sensitivity of their breast muscle enzyme to inhibition by pyruvate. The muscle enzyme of birds such as the pheasant, grouse and domestic fowl, which fly only occasionally, resembles that of human skeletal muscle in not being inhibited by pyruvate, and when the birds are forced to fly their breast muscles soon become fatigued, due to the accumulation of an oxygen debt. Other birds, such as the humming-bird and storm petrel, which normally spend much of their time in flight, have breast muscle lactate dehydrogenase which, like that of human heart, is strongly inhibited by pyruvate. Thus, accumulation of lactate does not occur, and the muscle does not become fatigued.

The lactate dehydrogenases of malarial parasites consist predominantly of cathodic isoenzymes which are metabolically more efficient than the anodic isoenzymes of the host erythrocytes. It is suggested that this may account for the ability of the parasite to grow at the expense of the host red cells (Sherman, 1962).

Impressive though the evidence adduced by Kaplan and his co-workers undoubtedly is, the theory has been criticized because the temperatures and concentrations employed have been outside physiological ranges. Vesell (1965a) found purified human LD_1 and LD_5 to exhibit similar substrate inhibition at $37°$, though he confirmed that LD_1 was much more susceptible to inhibition by increasing pyruvate at $25°$. Moreover, at $25°$ the lactate dehydrogenase of human tissue homogenates did not show the difference in substrate inhibition expected on the basis of their isoenzyme composition. Vesell therefore questions whether the differences observed at $25°$ are of any physiological significance at $37°$, and in support refers to similar observations made at $40°$ by Plagemann, Gregory and Wróblewski (1961).

On the other hand, Latner, Siddiqui and Skillen (1966) found the enzyme from human heart, kidney and erythrocyte homogenates (mainly LD_1) to be much more sensitive to pyruvate inhibition at both $25°$ and $37°$ than that in liver and skeletal muscle homogenates (mainly LD_5). This conflict of evidence has not yet been resolved, but it is conceivable the pH tolerances ($0.2–0.3$) may have some relevance, since Fritz (1967) has shown that LD_5 is extremely sensitive to pH changes. Kaplan, Everse and Admiraal (1968), however, observed well-marked differences between the sensitivities to pyruvate inhibition of the chicken H_4 and M_4 isoenzymes at 37°, not only at pH 7·5 but also at pH 6·0.

Kaplan's theory has also been criticized by Stambaugh and Post (1966b) on the grounds that substrate inhibition is not detectable with the concentrations of pyruvate likely to be reached under physiological conditions. These investigators, however, found LD_1 to be more sensitive than LD_5 to product inhibition by physiological concentrations of L-lactate. It is probable therefore that Kaplan's views on the biological significance of lactate dehydrogenase isoenzymes may be substantially correct, but their differences in catalytic activity may be due to inhibition by the product rather than by the substrate. Another possibility is that inhibition may be mediated by a relatively stable inactive ternary complex comprising enzyme + NAD + pyruvate (Kaplan *et al.*, 1968).

It has recently been shown by Wuntch, Chen and Vesell (1970) that at high enzyme concentrations (more than $3·5 \times 10^{-7}$M), LD_1 is not sensitive to inhibition by excess pyruvate. Since the presence of intracellular enzyme concentrations of this order has been confirmed (Criddle *et al.*, 1968), the physiological significance of the inhibition of LD_1 by high pyruvate concentrations must still be regarded as uncertain.

It is well established that the rate-determining steps in reactions catalysed by lactate dehydrogenase are the equilibria between the bound and free substrates (*vide, inter alia*, Criddle, McMurray and Gutfreund, 1968).

TABLE 14 *Ratios of the activities of human tissue lactate dehydrogenase isoenzymes with 3·3mM-2-oxobutyrate and 0·7mM-pyruvate at 25°*

(Plummer, Elliott, Cooke and Wilkinson, 1963)

	Activity with 2-oxobutyrate: activity with pyruvate				
	Isoenzyme				
Tissue	1	2	3	4	5
Heart	1·0	1·0	0·67	0·29	—
Skeletal muscle	0·84	0·80	0·45	0·28	0·10
Liver	—	0·53	0·53	0·29	0·16

A marked difference in the relative activities with pyruvate and α-oxobutyrate between the fast and slow isoenzymes has been demonstrated (Rosalki and Wilkinson, 1960). LD_1 and LD_2 exhibit comparable activities with 0·7mM-pyruvate and 3·3mM-2-oxobutyrate (these substrate concentrations are optimal for normal human serum), whereas LD_4 and LD_5 exhibit a marked preference for pyruvate (Table 14). Tissues rich in either fast or slow isoenzymes have characteristic ratios of activity with the two substrates (Wilkinson, Cooke, Elliott and Plummer, 1961; Rosalki, 1962; Plummer *et al.*, 1963), and some examples are shown in Table 15 and Fig. 30.

TABLE 15 *Ratios of dehydrogenase activities at* 25° *with* $3 \cdot 3mM$-2-*oxobutyrate and* $0 \cdot 7mM$-*pyruvate of tissue extracts*

(Based upon Plummer, Elliott, Cooke and Wilkinson, 1963)

Species	Tissue	Activity with 2-oxobutyrate: activity with pyruvate
Rabbit	Heart	1·03
	Kidney	0·89
	Liver	0·34
	Skeletal muscle	0·19
Human	Heart	1·08
	Kidney	0·83
	Erythrocytes	0·93
	Liver	0·34
	Skeletal muscle	0·37
Mouse	Heart	0·35
	Liver	0·15
	Skeletal muscle	0·14

Similar relationships are found when the relative activities of human heart and liver preparations are studied by means of the reverse reactions, i.e. the oxidation of lactate or DL-2-hydroxybutyrate with NAD as coenzyme (Plummer and Wilkinson, 1963).

The activity with 2-oxobutyrate has found certain clinical applications which are discussed later (p. 188), and in this connection it has been found convenient to describe such activity as '2-hydroxybutyrate dehydrogenase' (Elliott and Wilkinson, 1961). The possibility that 2-oxobutyrate and pyruvate are reduced by separate enzymes arises because Rosalki and Wilkinson (1960) observed differences in temperature coefficients, pH curves and relative activities in human sera, but since no electrophoretically separated fraction has been found to display activity exclusively against one or other of these substrates, it seems that '2-hydroxybutyrate dehydrogenase' and lactate dehydrogenase are identical. This conclusion is supported by the observation that partial purification of human-heart lactate dehydrogenase yielded several fractions, all of which had similar ratios of activity with the two substrates, and that commercial samples of the crystalline ox-heart enzyme show equal activity with either substrate. Moreover, crystalline rabbit-muscle lactate dehydrogenase shows the same ratio of activities with the two substrates as a crude extract of rabbit skeletal muscle (Plummer *et al.*, 1963).

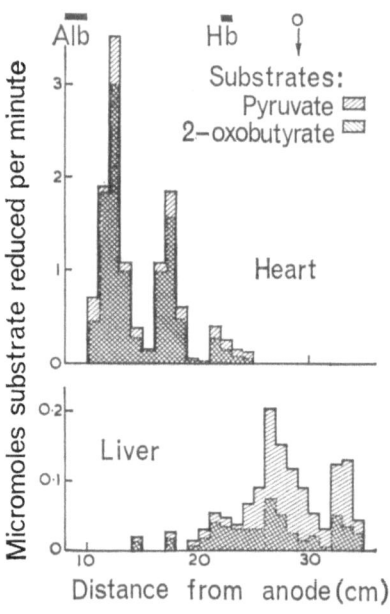

Fig. 30. Starch-block electrophoresis of human heart and human liver lactate dehydrogenase isoenzymes showing the relatively high activity of LD_1 and LD_2 with 2-oxobutyrate.

Interesting differences in the abilities of lactate dehydrogenase isoenzymes to utilize substrate analogues may also be demonstrated among those of various mouse tissues (Allen, 1961). As in human and rabbit tissues, the fast (anodic) components are active with DL-α-hydroxybutyrate, but not with DL-α-hydroxycaproate or DL-α-hydroxyvalerate. In contrast with the results obtained with human and rabbit isoenzymes, the most cathodic fraction found in the mouse (in the testis) shows activity with all three α-hydroxyacids similar to that with L-lactate. This is probably due to the special properties of the LD_x component found in the testis, but proof of this must await further study. Meanwhile, however, the LD_x isoenzyme of human spermatozoa has been found to exhibit relatively high activity with 2-oxobutyrate, as judged by activity ratios and K_m values (Tables 16 and 17) (Withycombe and Wilkinson, 1964; Wilkinson and Withycombe, 1965; Stambaugh and Buckley, 1967).

TABLE 16 *Relative activities of human lactate dehydrogenase isoenzymes with* $3\cdot3$*mM-2-oxobutyrate and* $0\cdot7$*mM-pyruvate*
(Wilkinson and Withycombe, 1965)

Isoenzyme	Source	Ratio of dehydrogenase activities (2-oxobutyrate/pyruvate)
LD_1	Heart	0·90
LD_2	Heart	0·77
LD_3	Skeletal muscle	0·68
LD_x	Spermatozoa	1·17
LD_4	Skeletal muscle	0·57
LD_5	Liver	0·32

TABLE 17 *Optimum pyruvate and 2-oxobutyrate concentrations and Michaelis constants* (K_m) *for human lactate dehydrogenase isoenzymes*
(Wilkinson and Withycombe, 1965)

Isoenzyme	Optimum concentration (mM)		K_m (mM)	
	Pyruvate	2-Oxobutyrate	Pyruvate	2-Oxobutyrate
LD_1	0·40	3·3	0·08	0·84
LD_2	0·48	5·7	0·11	1·24
LD_3	0·62	12·5	0·17	1·90
LD_x	0·62	1·65	0·05	0·18
LD_4	0·77	50	0·18	3·47
LD_5	1·00	90	0·83	10·0

Effect of pH variation

LD_5 differs from LD_1 in its sensitivity to variation in pH. With pyruvate as substrate, Fritz (1967) found rabbit-skeletal-muscle LD_5 to have a sharp peak of activity at pH 6·8 (37°), while rabbit-heart LD_1 was scarcely affected by pH changes in the range 6·2–8·0 (Fig. 31). These results have been confirmed in the writer's laboratory by Mr Y. Fujimoto. Fritz (1967) also showed that LD_5 behaved like LD_1 after it had been pre-incubated with the various buffer solutions before adding the substrate. At higher pH values LD_1 was less sensitive to substrate inhibition by pyruvate than at lower pH values, an observation which Fritz regards as confirming the conclusion of Vesell (1966) that the differences between the responses of LD_1 and LD_5 to substrate inhibition diminish when the reactions are performed under physiological conditions (p. 145). It seems that insufficient attention has hitherto been paid to the effects of small pH changes on isoenzyme activities when using pyruvate as substrate.

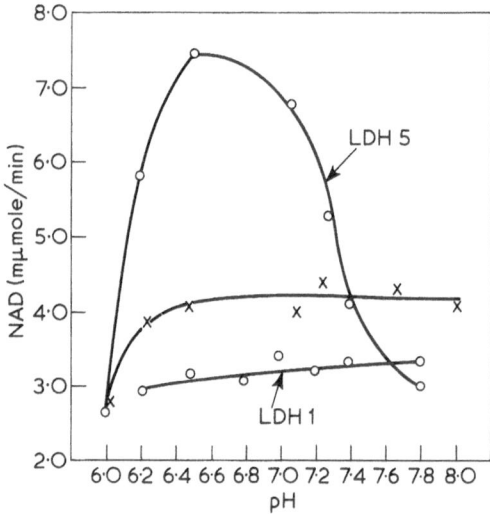

Fig. 31. Effect of pH variation on rabbit-heart LD_1 and rabbit-muscle LD_5 with 0·224mM-pyruvate as substrate in 0·03M-phosphate buffer at 37°. The results shown in curve (X) were determined at pH 7·4 after LD_5 had first been incubated for 3 min. at the pH indicated. (Fritz, 1967.)

Boyd (1967) has also found similar pH sensitivity in sheep isoenzymes. He reported that sheep LD_5 is relatively stable in phosphate buffer at pH 7·4, but is rapidly inactivated when incubated with sheep serum, a result which is ascribed to the increase in the serum pH from 7·4 to 7·8.

When lactate is employed as substrate, however, the corresponding human isoenzymes show little difference in their pH optima (LD_1, 8·4; LD_5, 8·2) (Fujimoto, unpublished observations). This is to be expected if the suggestion of Winer and Schwert (1958) is correct. From the results of a kinetic study, these investigators postulated that during the reduction of pyruvate the imidazole ring of a histidine residue at the active centre of the enzyme molecule becomes protonated, whereas it must be unprotonated during the reverse reaction. This theory has recently been confirmed and extended by Schwert, Miller and Peanasky (1967).

The effect of pH on isoenzyme activity has recently been shown to be related to the nature and concentration of the buffer solution employed. At pH 6·9, LD_1 is markedly inhibited by concentrations of tris buffer greater than 0·25M, but both isoenzymes are relatively insensitive to increasing concentrations of phosphate. Similar results are obtained at pH 7·4 with phosphate buffer, but both isoenzymes are sensitive to high concentrations of tris buffer at this pH (Vesell, Fritz and White, 1968).

Fritz (1965) reported that several of the tricarboxylic acid cycle intermediates apparently activated LD_5 at low concentrations, but inhibited at high concentrations, while LD_1 is inhibited at all the concentrations studied. Subsequently he suggested that these effects were due to small variations in the pH and that only oxaloacetate possessed any activating effect (Fritz, 1967). However, Orleans-Harding and Mahler (1968) have since found that citrate activates the M_4 isoenzyme of rat uterus at concentrations ranging from 3×10^{-6}M to $1·5 \times 10^{-4}$M, while having no effect on the activity of the H_4 isoenzyme. This effect was enhanced by prolonged administration of oestradiol-17β to the animals, and supports the view, originally postulated by Fritz (1965), that LD_5 may act as a regulatory enzyme.

Effect of inhibitors
Although lactate dehydrogenase is inhibited by compounds such as *p*-chloromercuribenzoate, which react with sulphydryl groups, the first reports of differential inhibition were those of Pfleiderer and Jeckel (1957) and Wieland, Pfleiderer and Ortanderl (1959), who found that sulphite preferentially inhibits the fast-migrating isoenzymes of the rat (cf. Bonavita and Guaneri, 1962).

Subsequently, the observation by Novoa, Winer, Glaid and Schwert (1959) that, with pyruvate as substrate, crystalline lactate dehydrogenase is competitively inhibited by oxamate but non-competitively by oxalate, was confirmed and extended to human-heart and -liver preparations by Plummer and Wilkinson (1961, 1963). The effects of increasing concentrations of oxamate on the activity of human-heart and -liver extracts with pyruvate and 2-oxobutyrate as substrates are shown in Fig. 32. The inhibitory effect is expressed as the ratio v/v_i, where v is the activity in the absence of the inhibitor and v_i the activity in its presence. It will be noted that the reduction of 2-oxobutyrate is inhibited to a much greater extent than that of pyruvate. The more weakly bound substrate appears to be more easily displaced by oxamate from the enzyme surface. By contrast, oxalate inhibits the reduction of both substrates equally with any given enzyme preparation (Fig. 33). Thus, while oxamate differentiates between substrates, the non-competitive inhibitor, oxalate, exerts different effects according to the source of the enzyme. Preparations of human heart containing LD_1 and LD_2 are affected to a greater extent than those of human liver containing LD_4 and LD_5. A relationship between the electrophoretic mobility of the LD isoenzymes and their reciprocal inhibitor constants for oxalate has now been established (Emerson, Wilkinson and Withycombe, 1964).

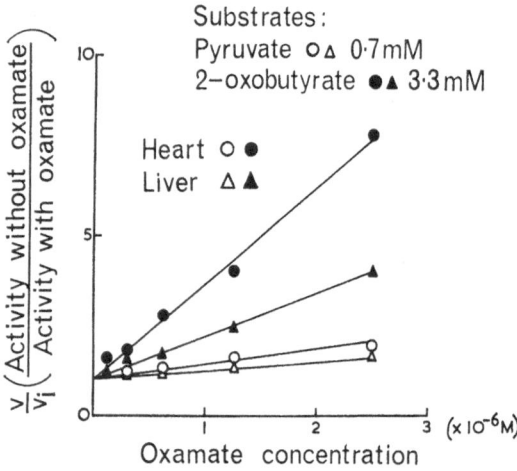

Fig. 32. The effect of oxamate on lactate dehydrogenase activity of human-heart and human-liver extracts. (Plummer and Wilkinson, 1963.)

The dissociation of lactate dehydrogenase into sub-units by high concentrations of urea has already been discussed (p. 60), but at lower concentrations urea behaves as a competitive inhibitor of the enzyme. It can be used to differentiate between the fast and slow isoenzymes, but unlike oxalate, its effect is greatest on LD_4 and LD_5 (Richterich, Burger and Weber, 1962; Schindler and Richterich, 1962; Richterich and Burger, 1963a; Plummer, Wilkinson and Withycombe, 1963; Konttinen and Lindy, 1967a).

Two quite distinct effects are observed when, with pyruvate as substrate, v/v_i ratios are plotted against urea concentrations (Fig. 34). With increasing urea concentration there is at first a slow increase in the degree of inhibition until a critical concentration, which varies according to the isoenzyme composition, is reached. At this point a sharp inflexion of the curve occurs, and thereafter inhibition is virtually com-

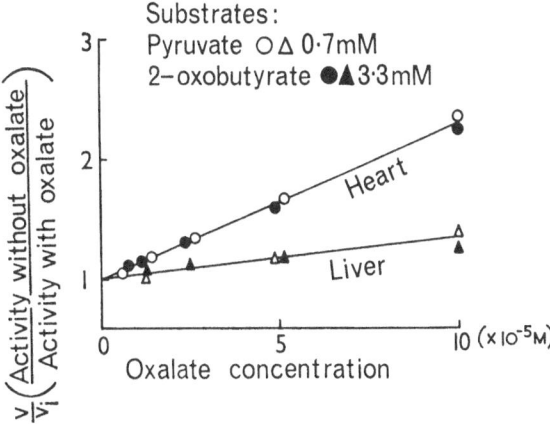

Fig. 33. The effect of oxalate on the lactate dehydrogenase activity of human heart and human-liver extracts. (Plummer and Wilkinson, 1963.)

plete. Inflexion occurs at a concentration of about 4M with LD_1 and LD_2, but with LD_4 and LD_5 it takes place at a much lower concentration (about 1M). Thus when 2M-urea is incorporated into the reaction mixture it is possible to distinguish lactate dehydrogenase preparations containing LD_1 and LD_2 from those containing LD_4 and LD_5, since the latter are completely inhibited, while the former retain about 80% of their original activity (Withycombe, Plummer and Wilkinson, 1965).

Similar results are obtained when 2-oxobutyrate is used as substrate, and also when the reverse reactions, i.e. the oxidation of lactate or 2-hydroxybutyrate, are investigated.

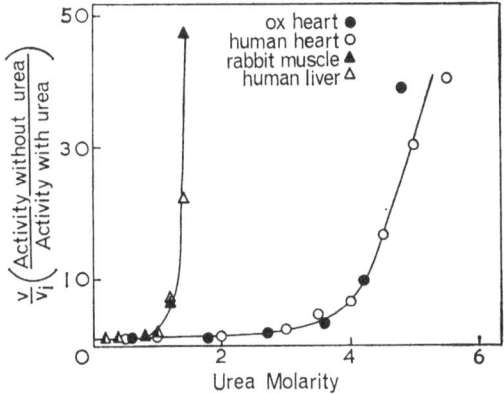

Fig. 34. The inhibition of the lactate dehydrogenase activity of various tissue extracts by urea. Substrate: 0·7mM-pyruvate. (Withycombe, Plummer and Wilkinson, 1965.)

The two effects exerted by urea appear to be quite different, for at low concentrations inhibition is competitive, indicating that urea can displace substrate from the active centre, whereas at higher levels such competition seems to be obscured by changes in the secondary or tertiary structure of the enzyme protein due to the rupture of hydrogen bonds (Withycombe *et al.*, 1965). In this connection it is interesting to note that the promotion of freeze–thaw hybridization of lactate dehydrogenase by sub-inhibitory concentrations of urea (less than $0 \cdot 1M$) observed by Massaro (1967) is attributed to denaturation brought by concentration during freezing, followed by renaturation as the result of dilution during thawing.

Several diagnostic techniques dependent upon the inhibition of the serum LD_5 have been introduced. These are discussed on p. 193.

The LD_x isoenzyme of human testes and spermatozoa behaves like LD_1 in being strongly inhibited by $0 \cdot 2mM$-oxalate and in being relatively resistant to inhibition by urea (Withycombe and Wilkinson, 1964; Wilkinson and Withycombe, 1965).

An interesting application of urea inhibition which offers an approach to isoenzyme histochemistry has been suggested by Brody (1964*a*), who incorporated $2 \cdot 61M$-urea into the reagents used to stain electropherograms of human gastrocnemius muscle extracts. LD_3, LD_4 and LD_5 were selectively inhibited. These findings have been confirmed by Emery (1967), who reports that there is good agreement between the results obtained with urea inhibition and those obtained by starch-gel electrophoresis. Brody (1964*b*, 1968) has described the use of selective inhibitors in a histochemical study of lactate dehydrogenase isoenzyme distribution in the light and dark fibres of normal and dystrophic human muscle.

Rat-liver nuclear lactate dehydrogenase is reported to be more resistant to inhibition by urea than the enzyme from the corresponding cytoplasmic fraction (Nemchinskaya *et al.*, 1968) (see p. 152).

Differentiation of lactate dehydrogenase isoenzymes by means of urea inhibition is rendered more clear-cut by varying the substrate concentration. The inhibitory effect of increased pyruvate concentration on LD_1 activity may be abolished or even reversed by carrying out the reaction in the presence of urea. For example, with 5mM-pyruvate as substrate, $2 \cdot 0M$-urea activates human LD_1 by 25%, but almost completely suppresses LD_5 (Lindy and Konttinen, 1966, 1967*a*, 1967*b*; Konttinen and Lindy, 1967*a*, 1967*b*, 1967*c*).

A similar procedure, based upon the oxidation of lactate, has been devised by Babson (1967), who found LD_1 and LD_5 to be best differentiated by comparing their activities in $1\cdot0$M-lactate with those in $0\cdot01$M-lactate + $1\cdot0$M-urea. Human-heart extracts give ratios about $0\cdot4$, whereas the corresponding figures for liver extracts are about 8. A number of chemical analogues of urea also exert differential inhibition. Hydantoic acid is effective as a concentration of $0\cdot3$M, but methylurea proved to be the most potent compound of this series, for at a final concentration of $0\cdot2$M it completely inhibited LD_5 (Withycombe *et al.*, 1965). Neither of these substances, however, appears to have any practical advantage over urea as a differential inhibitor.

The inhibitory effect of urea on lactate dehydrogenase is greatly increased by preliminary treatment with urease. Owing to the increase in alkalinity during the reverse reaction, this effect cannot readily be investigated with pyruvate as substrate, but when lactate is used at pH $9\cdot3$ concentrations of urea, normally sub-inhibitory ($<0\cdot1$M), selectively suppress LD_5 activity. This appears to be due to the production of ammonium bicarbonate, which is a potent inhibitor of the cathodic isoenzyme under the conditions described above (Fujimoto and Wilkinson, 1970).

Under certain conditions the carbonate ion also behaves as a selective inhibitor of LD_5. A comparison of the effects of varying concentrations of tris-HCl and bicarbonate–carbonate buffers on the activity of rabbit-muscle LD_5 is shown in Fig. 35. At pH $9\cdot3$ and with $0\cdot04$M-L-lactate as substrate, LD_5 activity is very sensitive to changes in the bicarbonate–carbonate buffer concentration between $0\cdot01$ and $0\cdot1$M, but is unaffected by similar changes when tris buffer is used. The inhibition is not due to a non-specific osmotic effect, since similar results are obtained when the buffers are prepared in $0\cdot1$M-sodium chloride. It is tempting to speculate on the biological significance of the dramatic response of LD_5 to variations in the bicarbonate concentration within the normal physiological range, and it appears that this naturally occurring buffer system may act as a regulator of lactate dehydrogenase activity. The effect, however, appears to be due to the carbonate ion, since it is greatest at pH $9\cdot3$, diminishing as the pH is reduced, until at pH $7\cdot4$ it can no longer be detected. Moreover, though bicarbonate–carbonate buffer also inhibits human LD_5, its effect is maximal at about $0\cdot3$M and is not observed at physiological concentrations (Fujimoto and Wilkinson, 1970).

Among other reagents reported to behave as differential inhibitors are phenazine methosulphate (PMS) and 2-mercaptoethanol. In view of the widespread use of PMS as a hydrogen carrier in the tetrazolium technique for isoenzyme staining (pp. 48–50), the observation by Benitez and Fischer (1966) that PMS disproportionately inhibits LD_1 in histological preparations suggests that most isoenzyme measurements employing this technique are likely to be under-estimated with respect to LD_1. The authors, however, did not observe any inhibitory effect *in vitro*. 2-Mercaptoethanol inhibits lactate dehydrogenase isoenzymes, but the activity of the more stable testicular LD_x fraction is unaffected (Ressler, Olivero and Joseph, 1965).

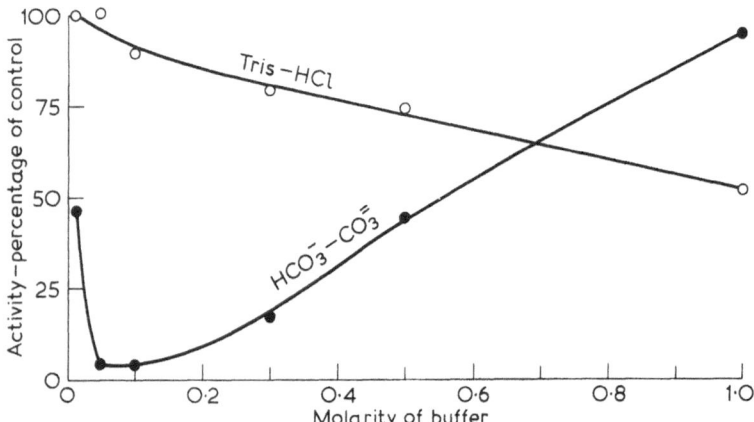

Fig. 35. Effect of carbonate ion on the activity of rabbit-muscle lactate dehydrogenase with lactate as substrate at pH 9·3. The open circles show the results obtained with tris-HCl buffers and the closed circles the results with bicarbonate–carbonate buffer. The inhibitory effect was observed when the enzyme was incubated with the buffer solution for 10 min. before adding the coenzyme, but not when NAD was incorporated into the buffer solution (Fujimoto and Wilkinson, 1970).

Certain solvents have been reported to precipitate or inhibit differentially the slow isoenzymes. The incorporation of chloroform into the reaction mixture has been recommended as a means for improving the diagnostic specificity of the determination of the serum lactate dehydrogenase (Warburton, Smith and Laing, 1963; Warburton and Smith, 1963). This solvent, however, is reported to inhibit the red-cell enzyme to a greater extent than those of sera from patients with heart and liver disease, an observation which appears to conflict with modern views on isoenzyme structure. This anomaly is related to the effects of chloroform on protein precipitation and to the concentrations of protein in the haemolysates, on the one hand, and the sera, on the other (F. G. Warburton, personal communication). Precipitation of the serum proteins with acetone allows a proportion of the LD_1 and LD_2 isoenzymes to remain in solution in an active form, and this technique has also been suggested as a diagnostic test in myocardial infarction (Latner and Turner, 1963).

Reactions with coenzyme analogues
Kaplan and his colleagues have prepared several chemical analogues of nicotinamide-adenine dinucleotide (NAD) which have found application

TABLE 18 *Analogues* of nicotinamide-adenine dinucleotide*
(Kaplan and Ciotti, 1961*a*)

Substituents in the pyridine ring

$X = \overset{S}{\underset{\parallel}{-C}} - NH_2$	3-Thionicotinamide-AD (TNAD)
$X = \overset{O}{\underset{\parallel}{-C}} - H$	Pyridine-3-aldehyde-AD (Py-3-Al AD)
$X = \overset{O}{\underset{\parallel}{-C}} - CH_3$	3-Acetylpyridine-AD (APAD)
$X = \overset{O}{\underset{\parallel}{-C}} - CH_2 \cdot CH_3$	3-Propionylpyridine-AD (PrPAD)

Substituents in the purine ring

Nicotinamide-hypoxanthine dinucleotide (NHXD)

Nicotinamide-6-(2-hydroxyethyl) purine dinucleotide (N6HEPD)

---

* The abbreviations used by the original authors have been altered to bring them into line with those used in this book.

in the study of the isoenzymes of NAD- and NADP-dependent dehydrogenases. Two series of analogues may be used: (*a*) those with substituents in the pyridine ring, and (*b*) those in which adenine is replaced by other purine groups. Some examples are listed in Table 18. When the ratios of activities with APAD and TNAD are determined, striking differences between the results with the heart muscle and skeletal muscle lactate dehydrogenases of various mammals occur (Kaplan, Ciotti, Hamolsky and Bieber, 1960). The values reported range from 0·20 to 0·48 in heart muscle and from 0·72 to 2·00 in skeletal muscle. In the same species the figures for different tissue lactate dehydrogenases correspond closely with their isoenzyme composition. With the human-heart and -kidney enzymes, both of which consist predominantly

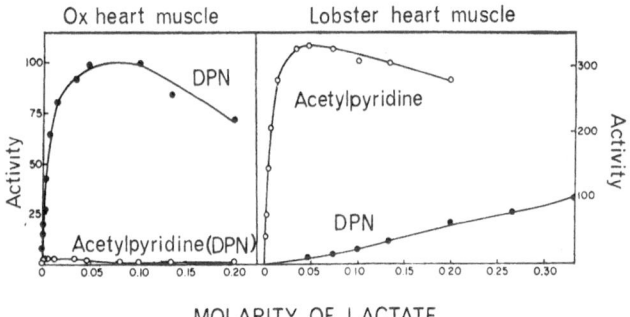

Fig. 36. The effect of increasing substrate concentrations on the rates of oxidation of lactate by ox-heart muscle and lobster-heart muscle lactate dehydrogenases using DPN (NAD) and acetylpyridine-DPN (APAD) as coenzyme. (Kaplan *et al.*, 1960.)

of LD_1 and LD_2, Kaplan and Ciotti (1961*a*) obtained mean values of 0·25 and 0·24 respectively, while the corresponding figures for the liver enzyme, mostly LD_5, and for the skeletal muscle enzyme, consisting largely of LD_4 and LD_5, were 0·97 and 0·66 respectively.

Greater affinity for the analogue than for the naturally occurring coenzyme is sometimes observed, and in some cases there is a marked difference between the effects of the coenzyme and its analogue on the substrate affinity. Figure 36 shows the remarkable differences between the Michaelis curves for lactate with ox-heart and lobster-heart lactate dehydrogenases observed when NAD and APAD are used as coenzyme. The ox-heart enzyme shows very little activity with the coenzyme analogue, while the lobster-heart preparation shows little activity with NAD. Transposition of coenzyme and analogue, however, leads to high activity with both preparations (Kaplan *et al.*, 1960). Lactate dehydrogenase from the skeletal muscle of several other crustaceans and from certain annelida also shows a greater affinity for APAD than for NAD, and the muscle enzyme from most chelicerata can utilize NHXDH more effectively than NADH.

Dennis and Kaplan (1960) have shown that, although the two stereospecific lactate dehydrogenases of *Lactobacillus plantarum* cannot easily be separated either by electrophoresis or by column chromatography, they can be recognized by their activities with NAD analogues. For example, while the $D(-)$-enzyme can utilize NHXD, the $L(+)$-form is completely inactive with this analogue.

Cahn *et al.* (1962) have extended these studies by determining the ratios for the rates of reduction of TNAD and APAD with lactate by extracts of heart or skeletal muscle from a variety of species. While there is a remarkable constancy for the ratio in heart muscle, which ranges from 5·0 to 5·3 in birds, 5·4 to 5·8 in mammals, 6·2 to 6·8 in amphibia and 8·0 in fish, corresponding figures for skeletal muscle are more variable, ranging from 0·22 in fish, 0·48 to 0·63 in birds and 1·60 to 1·62 in amphibia to 0·73 to 2·2 in mammals. The extreme values for mammalian skeletal muscle occur in the rabbit (0·73) and in the man (2·2). This is of considerable interest in view of the marked difference between the ratios of the activities with 2-oxobutyrate and with pyruvate of the skeletal muscle enzyme of these species (Table 15).

TABLE 19 *Ratios of activities of lactate dehydrogenase isoenzymes with NHXDH and NADH as coenzymes* *

(Cahn, Kaplan, Levine and Zwilling, 1962)

Isoenzyme	Ox skeletal muscle	Ox heart
LD_1	2·78	2·50
LD_2	1·83	1·74
LD_3	1·23	1·29
LD_4	0·70	0·78
LD_5	0·53	—

* Substrate concentrations: with NHXDH, $3\cdot3 \times 10^{-4}$M-pyruvate; with NADH, $1\cdot0 \times 10^{-2}$M-pyruvate.

The gradation in the properties of the lactate dehydrogenase isoenzymes is reflected in their abilities to utilize coenzyme analogues. Table 19 shows that a similar progression in the ratios of activities with NHXDH and NADH is found for the five isoenzymes obtained from

TABLE 20 *Ratios of activities of chicken lactate dehydrogenase isoenzymes with NHXDH and NADH as coenzymes*

(Kaplan, 1963)

	Ratios (NHXDH/NADH)	
Isoenzyme	Found	Predicted
LD_1	3·22	—
LD_2	2·50	2·54
LD_3	1·98	1·86
LD_4	0·99	1·18
LD_5	0·50	—

ox-skeletal muscle and for the four from ox heart (Cahn *et al.*, 1962). The close correlation between the observed and predicted values for the activity ratios with NHXDH and NADH for chicken isoenzymes is shown in Table 20 (Kaplan, 1963).

Another application of analogue studies is the investigation of the metabolic functions of muscle enzymes. The light and dark skeletal muscle of certain fish contain lactate dehydrogenases which differ considerably in their relative affinities: some of the results obtained by

TABLE 21 *Relative activities with coenzyme analogues of lactate dehydrogenases of heart and light and dark skeletal muscle of fish*

(Kaplan and Ciotti, 1961a)

		Ratios of LD activities with coenzyme analogues	
Fish	Muscle	$\dfrac{APAD}{NAD}$	$\dfrac{APAD}{Py\text{-}3\text{-}alAD}$
Mackerel	Heart	0·1	0·6
	Dark	0·1	0·5
	Light	0·7	2·7
Trout	Heart	0·2	0·8
	Dark	0·3	1·2
	Light	1·7	11·8
Sea robin	Heart	0·2	1·5
	Dark	1·0	3·6
	Light	1·0	2·4
Salamander	Heart	0·2	0·8
	Dark	0·4	2·3
	Light	0·4	2·2

Kaplan and Ciotti (1961*a*) are listed in Table 21. These show that in the mackerel and the trout the dark muscle enzyme resembles that of the heart muscle in its ability to use coenzyme analogues, while the enzyme of the light muscle is quite different. In the sea robin and salamander, on the other hand, the enzymes of the light and dark muscle appear to be identical with one another.

Effect of temperature on lactate dehydrogenase isoenzymes
It is frequently found that the recovery of LD_4 and LD_5 after electrophoretic separation is less complete than that of LD_1 and LD_2, an observation which suggests that the slow isoenzymes are less stable than the fast isoenzymes. This is reflected in the relative lability of LD_4

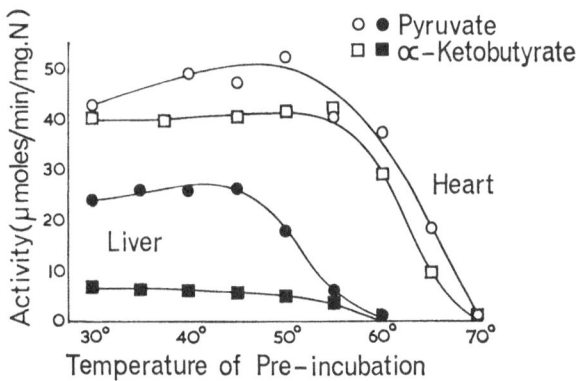

Fig. 37. Effect of pre-incubation on the dehydrogenase activity of heart and liver extracts with pyruvate and α-oxobutyrate as substrate. (Plummer and Wilkinson, 1963.)

and LD_5 when exposed to elevated temperatures. Figure 37 shows the effect of pre-incubation at various temperatures for 30 min. on the rates of reduction of pyruvate and 2-oxobutyrate by extracts of human heart and human liver before determining enzyme activity at 25°. Preincubation at temperatures below 50° has little effect on either preparation with either substrate. At higher temperatures the lactate dehydrogenase of the liver extract, consisting principally of LD_4 and LD_5, is appreciably less stable than that of the heart extract, chiefly LD_1 and LD_2. At 60° the activity of the former is completely destroyed, whereas the latter retains about 75% of its activity. It is interesting to note, however, that each preparation loses its ability to reduce pyruvate and

2-oxobutyrate at the same temperature (Plummer and Wilkinson, 1963).

Hill (1958) observed that the LD_1, LD_2 and LD_3 of human serum show progressively decreasing stabilities when exposed to elevated temperatures, and suggested that the presence of serum proteins, especially albumin, might exert a stabilizing effect. Plagemann *et al.* (1961), however, failed to find any evidence for the protective effects of protein impurities during an investigation of the thermal stabilities of individual rabbit LD isoenzymes of high specific activity. They found that LD_5 was completely inactivated by heating at $53°$ for 6 min., while LD_1 retained its full activity even after 40 min. at this temperature.

Determination of the temperature coefficients of the various isoenzymes is difficult owing to the fact that the optimal pyruvate concentration for both LD_1 and LD_5 increases with increasing temperature. The apparent temperature coefficients decrease as the temperature rises, and consequently nearly straight lines rather than exponential curves are obtained when reaction velocities are plotted against temperature (Plagemann *et al.*, 1961; Plummer and Wilkinson, 1963).

Such problems do not arise during the oxidation of lactate with NAD, for excess substrate can be used without risk of inhibition. From the results of such experiments, Plagemann *et al.* (1961) found values for the temperature coefficients of $2·21 \pm 0·03$ for LD_1 and $1·60 \pm 0·05$ for LD_5 over the range, $15–25°$. They also calculated the energy of activation by means of the Arrhenius equation, and obtained values of $13,188 \pm 74$ calories for LD_1 and $8,285 \pm 400$ calories for LD_5, from which they deduced that LD_5 has the greater catalytic efficiency.

Several investigators have employed the difference in sensitivities at elevated temperatures as a means of assessing the relative amounts of LD_1 and LD_5 in a raised serum lactate dehydrogenase for diagnostic purposes (p. 192).

Although it is generally agreed that LD_5 is much more sensitive to heat than LD_1, unanimity has yet to be reached on a really satisfactory procedure for determining their relative contents by heat-inactivation techniques. Kreutzer and Kreutzer (1965) showed that the incubation temperature has a marked effect upon the proportions of the various human-serum isoenzymes detected after agar-gel electrophoresis. As the temperature is raised from 30 to $60°$ there is a gradual increase in the proportions of LD_1 and LD_2 detected while LD_5, LD_4 and LD_3 are progressively one by one inactivated (Table 22). A temperature about $37°$ appears to be optimal for staining electropherograms. Vesell, Fritz and White (1968), however, have shown that at 39° thermal inactivation of rat-heart LD_1 and rat-skeletal-muscle LD_5 is dependent not only on the time of exposure but also inversely on the enzyme concentration.

TABLE 22 *Effect of increasing incubation temperature on the relative proportions of lactate dehydrogenase in human sera*
(Based on Kreutzer and Kreutzer, 1965)

	Percentage of total enzyme activity				
	Incubation temperature (deg.)				
Isoenzyme	30	40	50	60	70
LD_1	27	23	27	37	100
LD_2	29	32	37	47	0
LD_3	19	21	23	6	0
LD_4	11	12	10	0	0
LD_5	14	12	3	0	0

The thermal stability is also influenced by the coenzyme concentration: the protecting effect of NADH is optimal at a concentration of about 10^{-4}M (Vesell and Yielding, 1966). Certain other metabolically important substances such as fructose 1,6-diphosphate, oxaloacetate and malate protect against the thermal inactivation of frog LD_1 and LD_5 (Vesell and Yielding, 1968).

The lactate dehydrogenase isoenzymes of human sera differ also in their stabilities at low temperatures. LD_1 is relatively unaffected by exposure at 0–4° or at $-10°$, but LD_4 and LD_5 are rapidly inactivated under these conditions, though they retain much of their activities at room temperature (Kreutzer and Fennis, 1964).

Lactate dehydrogenase isoenzymes in developing tissues
The patterns of distribution of LD isoenzymes discussed earlier in this chapter refer to those occurring in adult tissues, but those found in embryonic or new-born tissues are not always identical with their adult counterparts. The first observation of such differences was that of Markert and Møller (1959), who detected a larger number of components in embryonic pig heart than in the adult heart. These authors reject the suggestion that the differences between the adult and embryonic patterns might be due to changes in cell population, since the observed population changes during development do not appear to be sufficiently great to account for such a possibility, and prefer to consider the

isoenzyme pattern of a tissue as a parameter of the state of differentiation of its cells. Recent observations strongly support this view.

Among the early observations is that of Flexner, Flexner, Roberts and De la Haba (1960), who reported differences between the brain lactate dehydrogenases of adult and new-born mice during chromatography on DEAE-cellulose, and between the electrophoretic patterns of the liver enzymes of adult and new-born guinea-pigs. The liver enzymes of adult and new-born rats have been shown to differ in their capacities to reduce coenzyme analogues (Kaplan and Ciotti, 1961a).

The changes in the relative activities of the mouse-kidney isoenzymes during development have been demonstrated by Markert (1962) by starch-gel electrophoresis. The patterns observed one day before and one day after birth are similar in that most activity is associated with the slow isoenzymes, LD_3, LD_4 and LD_5, but over a period of 3–4 weeks an initial rapid then a more gradual change to the adult pattern (principally LD_1, LD_2 and LD_3) takes place. Similar changes also occur in the isoenzyme patterns of other tissues, but the rates of change vary considerably. Whereas in the kidney the pattern found in 3-day-old animals is essentially that of the adult, full development of the characteristic enzyme pattern of the stomach requires about 3 weeks, while the isoenzymes of the tongue are still of the embryonic type at this age. In the skeletal muscle of this species LD_5 predominates in the embryonic form, but measurable amounts of the other isoenzymes occur, while in the adult form LD_1, LD_2 and LD_3 are not usually detectable (Markert, 1963b).

In the mouse the most abundant embryonic isoenzyme in all tissues is LD_5, but in certain other species, including the chicken and man, other isoenzymes predominate in the embryo. With the aid of a quantitative complement-fixation technique Cahn *et al.* (1962) observed that the lactate dehydrogenase of breast muscle from a 6-day-old chick embryo was identical with that of an adult chick heart, whereas that of an 8-day-old chicken was identical with the normal adult skeletal muscle enzyme. They demonstrated that the most rapid change occurred about the time of hatching. These conclusions were confirmed by precipitation studies with anti-chick heart and anti-chick muscle lactate dehydrogenases, and by coenzyme analogue studies. These authors interpret their findings as indicating that the embryonic tissues of the chick depend upon aerobic metabolism of pyruvate (see p. 143), and in this connection point out that lactate is toxic to the early chick embryo (Grabowski, 1961).

Similar conclusions were reached by Philip and Vesell (1962), who compared the isoenzyme patterns obtained by starch-gel electrophoresis of chick-embryo tissues at various stages of development. Soon after fertilization they found LD_1 to be the most abundant isoenzyme in liver, muscle and heart, though LD_2 and LD_3 are also present in smaller amounts, while in the liver LD_1 is accompanied by two faster-migrating bands of low activity. During development there is a gradual transfer of activity from the anodic to the cathodic isoenzymes, and appreciable amounts of LD_4 and LD_5 are detectable in adult liver and muscle. On the other hand, little change in the isoenzyme pattern occurs during development of the heart.

Somewhat different findings, however, have been reported by Nebel and Conklin (1964), who found LD_3 and LD_4 to be the first isoenzymes to appear in most chick embryonic tissues. The discrepancy may in part be due to differences in numbering the isoenzymes, since both groups describe more than five bands of activity: Nebel and Conklin report duplication of the LD_5 band, while Philip and Vesell refer to additional bands in the LD_1 region in the liver. Lindsay (1963) also disagrees with Philip and Vesell and with Cahn *et al.* (1962) in reporting a shift from cathodic to anodic isoenzymes during development. Two additional bands anodic to LD_1 have been reported in adult chicken tissues by Croisille (1964), who also found two even faster (E) bands in the embryonic liver and kidney. The E bands, however, disappeared before hatching occurred.

Because of its wholly ectodermal origin and the absence of connective tissue and a blood supply, the lens of the eye is of particular ontogenetic interest. Genis-Galvez and Maisel (1967) have recently reported on the changes in the lactate dehydrogenase electropherograms during differentiation of the lens in the chick. The cationic isoenzymes, LD_3, LD_4 and LD_5, predominate in the lens of the 12-day-old embryo and this pattern persists until hatching, though there is a slight increase in LD_1, particularly in the epithelium. After hatching there is a marked change, especially in the cortical fibre cells, but to a lesser extent in the nuclear fibre cells, leading to a predominance of LD_1. Three additional components, all anodal to LD_1 and similar to those described by Croisille (1964) in various tissues of the adult chick, appear in the fibre cells, where they persist throughout adult life. In 1-day-old chicks the epithelial cell enzyme consists very largely of LD_5 and LD_3, a pattern, though less well marked, which is also found in the adult. Genis-Galvez and Maisel confirmed their electrophoretic findings by comparing enzyme activities with high and low pyruvate concentrations: the fibre-cell enzyme is inhibited by the high substrate concentration, while that of the epithelial cells is enhanced. They suggest that the sudden change in isoenzyme pattern on hatching is probably related to the lens becoming functional.

It is interesting to note in passing that additional fast-migrating lactate dehydrogenase isoenzymes are found in the eyes of many fish (Markert and Faulhauber, 1965; Goldberg, 1966). Fieldhouse and Masters (1968) have recently described the occurrence of an extra slow-moving band in the guinea-pig embryo and of several extra bands of intermediate mobility in the liver of the duck.

Chick tissues lose their characteristic isoenzyme patterns after tissue culture *in vitro* or on chorioallantoic membrane (Philip and Vesell, 1962; Nebel and Conklin, 1964). Muscle, skin, heart and liver all show a decrease in the anodic fractions, a finding which is interpreted as indicating a return to the random assortment found in immature tissues (Nebel and Conklin, 1964).

In the rat heart, however, changes during development resemble those occurring in the mouse, for LD_5 appears first (Kaplan and Ciotti, 1961*b*; Fine, Kaplan and White, 1962), but soon after birth LD_1 replaces LD_5 as the most abundant form. Similar changes have been observed during the development of rat brain, which requires 50–60 days to achieve the adult pattern (Bonavita *et al.*, 1962). Thus, in the rat and mouse, anaerobic glycolysis leading to the accumulation of lactate appears to be the principal metabolic pathway *in utero*, and it has been suggested that excess lactate may be removed via the placenta (Cahn *et al.*, 1962).

In higher animals the developmental changes in the lactate dehydrogenase isoenzyme pattern are somewhat different from those found in rats and mice. During the embryonic development of porcine skeletal muscle genetic activity at first favours synthesis of the H sub-unit, but as gestation proceeds there is a slight increase in M sub-unit synthesis, leading to the formation of LD_5. This process accelerates considerably at term, and the normal adult enzyme consists very largely of LD_5. By contrast, the anodic isoenzymes predominate in embryonic pig kidney and heart, which in this respect closely resemble the corresponding adult tissues. The lung, which in both foetus and adult shows an intermediate isoenzyme pattern with LD_3 the most abundant form, is interesting, since at term there is a temporary shift towards the cathodic isoenzymes (Fieldhouse and Masters, 1966, 1968).

LD_3, LD_4 and LD_5 predominate at an early stage in the cat embryo. During gestation there is a slight shift towards the anodic isoenzymes, and the characteristic adult patterns appear shortly after birth. The change is most marked in the case of the heart, where the gradual transition towards anodic isoenzymes during foetal development is greatly accelerated in the new-born kitten. In the adult cat skeletal muscle has by far the highest specific activity of all tissues. This is largely due to the marked increase in LD_5 which occurs during the first three months of life (Hinks and Masters, 1966).

Most investigators are agreed that on prolonged tissue culture de-differentiation is accompanied by loss of LD_1 and a corresponding increase in LD_5 (Vesell, Philip and Bearn, 1962; Nitowsky and Soder-man, 1964). Blanco, Rifé and Larson (1967) studied the effects of repeated tissue culture on the isoenzyme patterns of mammary tissue cells from a lactating goat. The original tissue showed a marked predominance of LD_1, but there was a gradual change such that after six days all five isoenzymes were present in approximately equal amounts. Subsequently there was a marked increase in LD_5, accompanied by dedifferentiation as evidenced by loss of the ability of the cells to synthesize milk components.

On repeated passage in the hamster pouch, human diploid embryonic lung cells show no change in their isoenzyme patterns until about the 17th passage, after which there is progressive loss of LD_1. Heteroploid cells exhibit a similar pattern to that found in the early passage of the diploid cells (Childs and Legator, 1965).

The human embryo contains all five lactate dehydrogenase isoen-zymes, LD_3 being the most abundant, but during development the proportions of LD_1 and LD_5 increase so that normal tissue patterns appear soon after birth (Pfleiderer and Wachsmuth, 1961; Latner and Skillen, 1964). The relative proportions of the various isoenzymes in embryonic skeletal muscle, heart muscle and oesophageal muscle are close to the ideal binomial distribution expected as the result of random association of A and B monomers, and Wachsmuth (1964) has recently shown that the isoenzyme distribution remains constant during the mitotic phase of the development of human heart muscle, but during the post-mitotic (amitotic) phase there is a gradual increase in the proportions of LD_1 and LD_2.

Differentiation of the isoenzyme pattern is especially well marked in skeletal muscle (Fine *et al.*, 1962; Fine, Kaplan and Kuftinec, 1963), but in progressive muscular dystrophy of the pseudohypertrophic (Duchenne) type such a change remains substantially incomplete and LD_1, LD_2 and LD_3 remain the principal isoenzymes (Fig. 38) (Rich-terich, Gautier, Zuppinger, Egli and Rossi, 1961; Wieme and Herpol, 1962; Dreyfus, Demos, Schapira and Schapira, 1962; Wieme and Lauryssens, 1962; Emery, 1964, 1968; Pearson, Kar, Peter and

Munsat, 1965). Similar though less marked patterns are found in the skeletal muscle of carriers of the disease trait, and may also occur in some unaffected siblings (Schapira and Demos, 1962). Johnston, Wilkinson, Withycombe and Raymond (1966) compared the HBD/LD ratios* (see p. 188) of muscle biopsy specimens of members of a

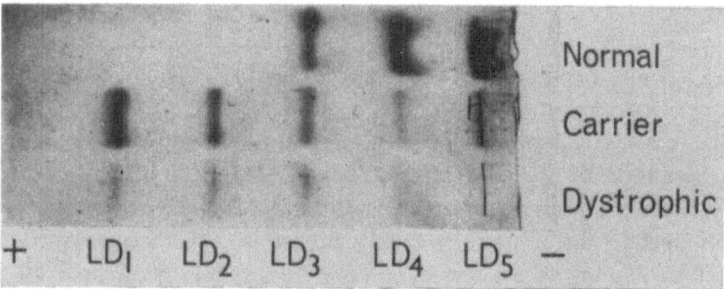

Fig. 38. Acrylamide-gel electrophoresis of the lactate dehydrogenase isoenzymes of normal and dystrophic human muscle. (Johnston, Wilkinson, Withycombe and Raymond, 1966.)

dystrophic family, and found that abnormal values were present not only in the patients but also in their mother and one of their siblings (Table 23). The HBD/LD ratios correlated well with the electrophoretic patterns which indicate a preponderance of LD_1, LD_2 and LD_3. It seems therefore that the genetically transmitted abnormality involves failure to synthesize sufficiently large amounts of the M sub-unit (see p. 60),

TABLE 23 *Dehydrogenase activity in human dystrophic muscle*
(Johnston, Wilkinson, Withycombe and Raymond, 1966)

| Patient | Enzyme activities (μmolar units/mg protein) | | |
	LD	HBD	HBD/LD ratio
(1) Dystrophic	0·48	0·38	0·79
(2) Dystrophic	0·74	0·56	0·76
(3) Unaffected sibling	1·00	0·55	0·55
(4) Mother of (1), (2) and (3)	2·01	1·39	0·69
Normal muscle	5·45	2·03	0·37
	8·00	3·60	0·45

* Ratios of $\dfrac{\text{Activity with } 3\cdot3\text{mM-2-oxobutyrate}}{\text{Activity with } 0\cdot7\text{mM-pyruvate}}$.

but since similar loss of isoenzyme specificity has been observed in skeletal muscle after experimental nerve section (Dawson *et al.*, 1964; Brody, 1965), it is possible that the change may be a non-specific effect.

In view of the close parallel between the changes in the human and avian isoenzyme patterns during development, Kaplan and Cahn (1962) have further investigated this possibility in dystrophic chickens. With the exception of *M. Latissimus dorsi, pars anterior*, all muscles of dystrophic chickens contain a greater proportion of the heart-type lactate dehydrogenase (LD_1) than the corresponding muscles of normal chickens. This difference is observed in 1-week-old as well as in 8-week-old chickens, but both normal and dystrophic groups show a fall in the proportion of LD_1 in their muscles with increasing age. The altered pattern in the dystrophic chickens, as compared with the normals, is accompanied by a diminution in the total enzyme activity of the muscle, but Kaplan and Cahn found no evidence to support the view that this is due to infiltration of the dystrophic muscles by connective tissue. They prefer to conclude that in dystrophic chickens there is failure of the gene controlling the synthesis of M sub-units.

The total lactate dehydrogenase activity of human dystrophic muscle is also much less than that of normal human muscle (Table 23), and as the serum activities of enzymes found in muscle (lactate dehydrogenase, aspartate aminotransferase, creatine kinase, aldolase) are much increased in progressive muscular dystrophy, it seems that the fall in the lactate dehydrogenase activity of dystrophic muscle may be partly due to an increase in membrane permeability (Wieme and Lauryssens, 1962; Heyck, Laudahn and Lüders, 1963; Dawson *et al.*, 1964; Lauryssens, Lauryssens and Zondag, 1964; Brody, 1965; Shepard, Gordon and Wollenweber, 1965; Johnston *et al.*, 1966).

It is not yet known whether the failure of LD_5 to replace LD_1 is a precipitating factor in the disease, but there is no doubt about the failure of dystrophic muscle to sustain anaerobic metabolism. Rosalki and Sinclair (1965) have shown that in dystrophic patients lactate production after exercise is subnormal.

Changes of a rather different type occur during the development of the snail. Goldberg and Cather (1963) separated five zones of lactate dehydrogenase activity by acrylamide-disc electrophoresis from the eggs of the snail, and observed that prior to blastulation two much faster bands are present.

Genetic variants of lactate dehydrogenase isoenzymes

Since synthesis of the H and M sub-units is controlled by separate genes, mutations involving either of these will lead to isoenzyme polymorphism. Several examples of such variants have been described in man and other species.

Almost simultaneously, Shaw and Barto (1963) described a genetic variant of lactate dehydrogenase isoenzymes in the deer-mouse (*Peromyscus maniculatus*), and Boyer, Fainer and Watson-Williams (1963) reported the occurrence of a similar anomaly affecting the synthesis of the H sub-unit in the erythrocytes of a 25-year-old Nigerian male. On starch-gel electrophoresis five LD_1 bands, four LD_2, three LD_3 and two LD_4 bands were observed. LD_5 was not detected, but its absence is not uncommon in normal haemolysates (Vesell and Bearn, 1962) (Table 10). Boyer *et al.* (1963) interpret their findings as reflecting a mutant allele at the genetic locus producing the H sub-unit. Using Appella and Markert (1961) nomenclature, they describe the normal LD_1 sub-unit as B, and the variant sub-unit as β. Accordingly, the variant isoenzymes may have the structures shown in Table 24.

TABLE 24 *Suggested structure of lactate dehydrogenase variants*
(Boyer, Fainer and Watson-Williams, 1963)

Isoenzyme	Normal	Variant
LD_1	B_4 (H_4)	B_4
		$B_3\beta_1$
		$B_2\beta_2$
		$B_1\beta_3$
		β_4
LD_2	B_3A_1 (H_3M_1)	B_3A_1
		$B_2\beta_1A_1$
		$B_1\beta_2A_1$
		β_3A
LD_3	B_2A_2 (H_2M_2)	B_2A_2
		$B_1\beta_1A_2$
		β_2A_2
LD_4	B_1A_3 (H_1M_3)	B_1A_3
		β_1A_3
LD_5	A_4 (M_4)	A_4

The M sub-unit may also be involved in genetic polymorphism. Nance, Claflin and Smithies (1963) reported a Brazilian family, four members of which exhibited such a mutant phenotype. The red cell haemolysates were characterized by a single LD_1 isoenzyme, accompanied by double LD_2 and LD_3 components, a distribution which the authors attribute to a non-random association of sub-units.

Both types of polymorphism were observed by Kraus and Neely (1964) in the course of a study of 940 hospital employees, eight of whom exhibited variant patterns. These were of four different types, described as Memphis 1, 2, 3 and 4 after the city in which the survey was made. Memphis 3 appears to be identical with the pattern previously observed by Boyer *et al.* (1963). The Memphis 1 variant consists of a single LD_2 and three LD_3 bands. Memphis 2 is similar except that the additional LD_2 bands move towards the anode at a faster rate than those of Memphis 1. Memphis 4 is also characterized by two LD_2 components, one of which has the normal mobility, while the other migrates more slowly. The Memphis 1, 2 and 4 variants appear to be due to mutation affecting synthesis of the M sub-unit, while a variant of the H sub-unit would account for Memphis 3.

Kraus and Neely suggested an autosomal codominant inheritance for the unusual isoenzyme patterns as the result of family studies in which it was found that all affected relatives of affected individuals possessed identical abnormalities. They concluded that all the individuals discovered to have the variant isoenzymes are heterozygous.

Two individuals with a variant erythrocytic lactate dehydrogenase isoenzyme pattern identical with the Memphis 4 phenotype have been detected among 1,015 English people studied by Davidson, Fildes, Glen-Bott, Harris, Robson and Cleghorn (1965), who found a further twenty-nine examples of the variant phenotype during subsequent family studies. These suggest that the individuals affected are heterozygous for a comparatively rare autosomal gene.

An individual, believed to be homozygous for a variant form of the H sub-unit, has been described by Vesell (1965*b*), who in the course of screening 600 white and 600 Negro patients found one of the former and three of the latter to have variant isoenzyme patterns. The three Negro patients exhibited the Memphis 1 phenotype of Kraus and Neely (1964), but the white patient, an elderly woman with lymphoblastic sarcoma, appeared to be a homozygote for an unusual form of the H sub-unit. In this patient the variant LD_1 and LD_2 forms were homogeneous and had electrophoretic mobilities greater than those of the general population. The mobility of the main LD_3 band was normal, but a secondary component migrated somewhat faster. The isoenzyme pattern was unrelated to her disease state, and other blood elements showed the same electrophoretic anomaly which, Vesell suggests, is due to homozygous involvement at the *b*-locus. Unfortunately, as the patient had no close relatives, it was impossible for him to confirm this hypothesis. Nor can any satisfactory explanation be given for the normal LD_3 mobility or for the occurrence of the secondary component in this region of the electropherogram.

The incidence of unusual genes has been discussed by Davidson *et al.* (1965), who deduced a gene frequency of about 0·001 for the unusual phenotype in a British population. This is appreciably lower than that found in Negroes by Kraus and Neely (1964) and by Vesell (1965), but comparable with the frequency found by these authors in white populations. Davidson *et al.* (1965) found a variant similar to the Memphis 1 or Memphis 2 variants in two out of 345 Turkish Cypriots and in one of twenty-three Nigerians.

Lactate dehydrogenase mutations have also been reported in a wide variety of animal species. In addition to the variant form of the H sub-unit detected in the deer-mouse by Shaw and Barto (1963), polymorphism of the M sub-unit has been observed in laboratory mice (Costello and Kaplan, 1963) and the baboon (Syner and Goodman, 1966). The most striking examples, however, seem to occur in fish. Goldberg (1966) showed that tissues of the speckled trout (*Salvelinus fontinalis*) and the lake trout (*S. namaycush*) contain dissimilar five-membered isoenzyme patterns, but the hybrid splake, produced by fertilizing lake trout eggs with speckled trout sperm, exhibits nine isoenzymes. The LD_1 of the two homozygotes has the same electrophoretic mobility, but the LD_5 bands differ considerably in this respect. The speckled trout LD_2 migrated at the same speed as lake trout LD_3, and the speckled trout LD_3 was coincident with lake trout LD_5. Goldberg suggests that the gene locus (*b*) controlling synthesis of the LD_1 (H) sub-unit is common to both species, but formation of the LD_5 (M + M´) sub-unit is governed by two different gene loci (*a* and *a´*). The speckled trout would then have the genotype *bba´a´*, the lake trout *bbaa* and the hybrid splake *bbaa´*, giving the isoenzyme patterns shown in Fig. 39.

The fact that only nine of the theoretically possible fifteen isoenzymes can be seen is attributed to several different components having the same electrophoretic mobilities. In confirmation of this conclusion, Goldberg obtained a similar pattern of nine isoenzymes by *in vitro* freeze–thaw hybridization of a mixture of tissue extracts from the two homozygous species, but Hochachka (1966) has proposed an alternative explanation. He suggests that some control mechanism may prevent the reassociation of certain sub-unit combinations.

Similar complex heterozygous isoenzyme patterns have also been observed in the whiting (Markert and Faulhauber, 1965), the cod

(*Gadus morrhua*) and the coalfish (*Gadus virens*) (Lush and Cowey, 1968). The coalfish, like the trout, exhibits genetic variation involving the M sub-unit, while in the whiting and cod the H sub-unit is polymorphic.

A mutation involving the H sub-unit has recently been detected in the horse by Rauch (1968), who separated the unusual LD_1 and prepared a series of five artificial hybrids with the normal equine LD_1.

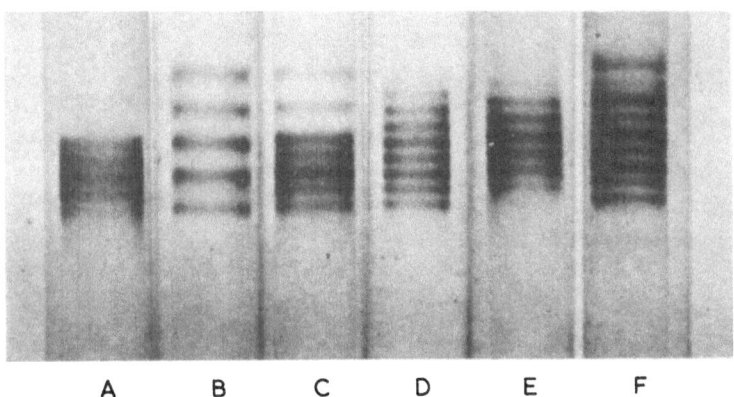

A B C D E F

Fig. 39. Lactate dehydrogenase isoenzymes of the trout separated by acrylamide-gel electrophoresis. The anode is at the bottom. A, lake trout; B, speckled trout; C, mixture of A and B; D, hybrid splake; E, mixture of A and B after freezing; F, mixture of B and E. (Goldberg, 1966.)

An extensive phylogenetic study by Baur and Pattie (1968) of the lactate dehydrogenase isoenzyme patterns of a wide variety of rodents has revealed that of thirty-four species examined, eighteen belonging to the sub-orders Myomorpha and Geomyoidea showed a complete absence of all but LD_5 in their enucleate erythrocytes, though all five isoenzymes occurred in other tissues. The electrophoretic findings in the case of *Rattus norvegicus* were confirmed by the observation that the red-cell enzyme was highly sensitive to heat and to inhibition by urea in the presence of a high concentration of pyruvate (see p. 154). The complete suppression of the H sub-unit in mature erythrocytes in these species appears to have important taxonomical consequences.

Another notable application of lactate dehydrogenase isoenzymes is the observation by Neaves and Gerald (1968) that two diploid parthenogenetic species of *Cnemidophorus* lizards were found to produce two different H sub-units, one of which also occurred in two triploid parthenogenetic species and two sexual species. The other was also found in a third sexual species. The authors suggest that interspecific hybridization between sexual species carrying different *b*-alleles may be responsible for the heterozygosity at the *b*-locus in the diploid parthenogenetic species and may be a factor in the development of parthenogenesis.

Isoenzyme studies have also provided results of major genetic interest in birds. Reference has already been made on p. 140 to the occurrence in the pigeon of three phenotypes characterized by variations in

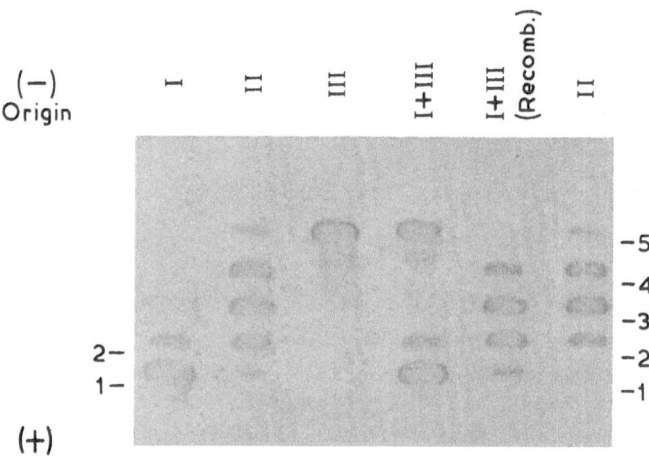

Fig. 40. Lactate dehydrogenase isoenzymes separated by starch-gel electrophoresis from extracts of wild pigeon hearts. I, showing the usual phenotype; II, the heterozygote showing all five isoenzymes; III, the rare phenotype, showing the predominance of an isoenzyme with the mobility of LD_5 but otherwise resembling LD_1; I + III is a mixture of extracts I and III; I + III (recomb.) is the same mixture subjected to dissociation and recombination of sub-units before electrophoresis. (Zinkham, Kupchyk, Blanco and Isensee, 1966.)

the LD_x components of testicular homogenates. Zinkham (1968) has suggested that the corresponding genotypes may be represented by *cc*, *cc'* and *c'c'*.

Zinkham and his colleagues have also observed another type of genetic polymorphism in the pigeon. In the course of a study of about 1,000 wild pigeons, one individual was found to exhibit a single zone of activity in the LD_5 region in all tissues other than the testis, which showed the normal isoenzyme complement. Studies of the catalytic properties and thermal inhibition, however, showed this isoenzyme to be a variant of the H_4 isoenzyme, hence in this particular bird the M sub-unit was absent and an H' variant took its place on electrophoresis. The bird thus appeared to be homozygous for the b'-gene. Heterozygotes were characterized by tissue isoenzyme patterns intermediate between those of usual and unusual homozygous forms (Fig. 40). So far no example of a pigeon homozygous for both the b' and c' variants has been observed (Zinkham, Kupchyk, Blanco and Isensee, 1966; Zinkham, 1968).

Diagnostic applications of lactate dehydrogenase isoenzymes
Serum lactate dehydrogenase determination was introduced into diagnostic enzymology in the years 1954–57, but it was soon noticed that elevations occur in a wide variety of pathological states, including cardiac, hepatic, malignant, haematological, muscular and renal diseases. This lack of specificity led to its temporary displacement as a diagnostic test by other enzyme procedures, but the discovery of its

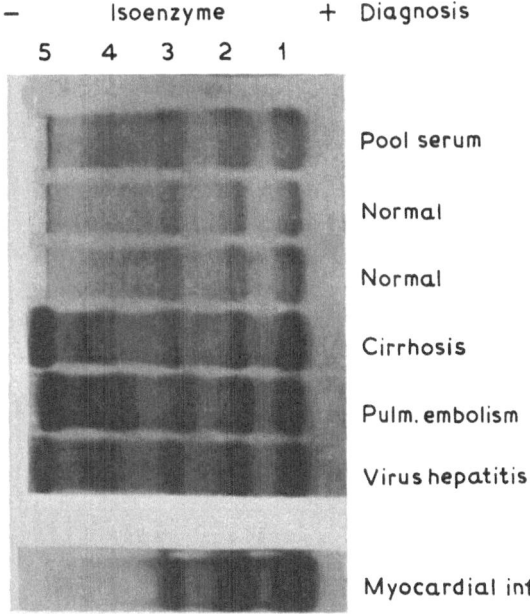

Fig. 41. Human serum lactate dehydrogenases from patients with myocardial infarction and liver diseases. The serum of the patient with pulmonary embolism also exhibits the presence of excess of LD_5 indicative of liver damage. The isoenzymes were separated by acrylamide-gel electrophoresis in tris buffer at pH 9·2.

heterogeneity soon led to its reinstatement. Much of the early work on isoenzymes was carried out in clinical laboratories, and consequently it was not long before variations in the isoenzymic composition of the serum enzyme found useful applications, since these changes reflect the pattern of diseased organs.

This is particularly the case in myocardial infarction when the serum contains excess of LD_1 and LD_2, which are characteristic of heart muscle, and in most cases of infective hepatitis, when the increase in the serum activity is largely due to an excess of LD_5. Examples of the serum enzyme patterns in patients with myocardial infarction and infective hepatitis are shown in Fig. 41. A variety of techniques has been used in clinical laboratories for the electrophoretic separation of isoenzymes (Wieme, 1959; Wróblewski and Gregory, 1960, 1961; Wróblewski, Ross and Gregory, 1960; Vesell and Bearn, 1961; Wieme and Van Maercke, 1961; Latner and Skillen, 1961; Van der Helm, Zondag and Klein, 1963; Batsakis, Preston and Briere, 1964; Fröhlich, 1965; Wieme, 1965; Wright, Cawley and Eberhardt, 1966; Bajolle, Deville and Borel, 1968; and others), while Hess has obtained similar results by chromatography on DEAE-cellulose (Hess, 1958; Hess and Walter, 1960a, 1960b, 1961).

Before discussing the clinical use of lactate dehydrogenase isoenzymes, it is pertinent to mention the recent observation of Cohen, Block and Djordjevich (1967) that there is a statistically significant sex difference between men and young women in their serum isoenzyme distribution, though not between the total serum enzyme activities of the two groups (Table 25). Young women, whether pregnant or not, appear to have high proportions of LD_1 and less of the other fractions in their serum than men of all ages and post-menopausal women. This difference

TABLE 25 *Distribution of lactate dehydrogenase isoenzymes in the serum of healthy adults of various ages*

(From Cohen, Block and Djordjevich, 1967)

Sex	Age (yr.)	LD isoenzymes (mean percentage of total)			
		LD_1	LD_2	LD_3	$LD_4 + LD_5$
Non-pregnant female	20–41	44	35	11	10
Pregnant female	19–41	48	35	10	7
Post-menopausal	44–78	37	43	13	7
Male	16–38	37	40	13	10
Male	41–75	36	43	12	9

seems to be related to the higher oestrogen circulation in young women, a suggestion supported by the finding of a similar change in men receiving oestrogen therapy. This factor must therefore be taken into account when applying serum isoenzyme tests for diagnostic purposes in young women. In this connection it is interesting to note that Clausen and Gerhardt (1963) had earlier reported that stilboestrol dipropionate selectively inhibits the slower isoenzymes, but on the other hand, Van der Heiden, Desplanque, Stoop and Wadman (1968) were unable to confirm a sex difference.

Another factor to be considered is the suggestion that the platelets are an important source of the serum enzyme in healthy individuals (Cohen and Larson, 1966). Nevertheless, the serum lactate dehydrogenase isoenzymes have proved valuable aids in diagnosis.

Myocardial infarction
A marked increase in LD_1 and LD_2, especially the former, is found 1–5 days after the onset of chest pain, though a change may often be observed within 6–12 hr. LD_1 and LD_2 reach peaks of activities about 48 hr. after an episode of myocardial infarction, but a relative excess usually persists for 1–3 weeks, even though the total serum lactate dehydrogenase activity may have returned to the normal range.

An extension of the infarct leads to a further release of these isoenzymes into the serum (Rettenbacker-Daubner and Reider, 1963; Cohen, Djordjevich and Ormiste, 1964). This sometimes presents problems, for renewed chest pain within a short period of the original episode may be due to a second infarction or to infection or to other causes. Since the electrocardiogram already shows changes due to the first infarction, isoenzyme studies are particularly useful in this situation.

The serum isoenzyme pattern may be altered when myocardial infarction is followed by congestive cardiac failure, since the resulting damage to the liver cells leads to the release of slow isoenzymes from this organ. In such cases an increase of LD_5 is superimposed upon the infarct pattern.

Experimental verification that the extent of the release of LD_1 is related to the size of the infarct has been obtained by Nachlas, Friedman and Cohen (1964), who produced graded infarcts in the dog by occlusion of the coronary artery.

Occasionally, however, slight elevation of the LD_5 fraction occurs in the absence of congestive failure about 6–10 days post-infarction, and it is uncertain whence this component arises. Latner and Skillen (1968) have pointed out that this band is not often seen in cases of congestive cardiac failure where there is no evidence of infarction, and suggest that it might be released from skeletal muscle as the result of disuse atrophy. However, since the appearance of the LD_5 band is not usually accompanied by an elevation of the serum creatine kinase, a very sensitive test for skeletal muscle damage, its origin cannot be clearly defined at present.

Patterns similar to those found in myocardial infarction also occur in other conditions in which there is damage to the heart muscle, e.g. viral myocarditis and active rheumatic carditis (Preston *et al.*, 1965; Bajolle *et al.*, 1968), but in angina pectoris, pericarditis and arrhythmias which do not involve myocardial necrosis the pattern is usually normal (Rettenbacher-Daubner and Reider, 1963; Cohen *et al.*, 1964; Bajolle *et al.*, 1968). While some investigators (Elliott, Jepson and Wilkinson, 1962; Cohen *et al.*, 1964) have observed normal patterns after pulmonary embolism, others (Van der Helm, Zondag, Hartog and Van der Kooi, 1962; Amelung, 1963; Mager, Blatt and Abelmann, 1966; Papadopoulos and Kintzios, 1967) have found abnormalities, usually involving an increase of LD_3. In the writer's experience, elevations of LD_1 and LD_2 are occasionally encountered after pulmonary embolism, but this is by no means a regular finding, and red-cell haemolysis may be a relevant factor.

An excess of LD_1 may also be encountered in certain non-cardiac conditions, including untreated megaloblastic anaemia, haemolytic anaemias and in progressive muscular dystrophy, but these do not usually present any problems of differential diagnosis with myocardial infarction.

Liver diseases

In liver disease generally the release of lactate dehydrogenase into the serum is somewhat variable, but high activities often occur in diseases such as infective hepatitis, infectious mononucleosis and toxic jaundice, in which hepatocellular damage is a predominant feature. In these a marked increase in the amount of LD_5 is found, but there are also indications of a change in the serum isoenzyme pattern towards LD_5, even in chronic liver diseases, such as obstructive jaundice and cirrhosis, even when the total serum lactate dehydrogenase remains within normal limits (Elliott and Wilkinson, 1963).

Lactate dehydrogenase isoenzymes are of value in the differential diagnosis of jaundice, for the LD_5 band is generally much more pronounced in hepatocellular disease than in biliary obstruction. This difference can be recognized more easily in the early stages of the disease, i.e.

within a day or two of the onset of jaundice. The isoenzyme pattern has also been used to differentiate hepatic from haemolytic jaundice (Woerner and Martin, 1961).

The patterns found in cirrhosis are somewhat variable and depend very largely upon the activity of the cirrhotic process and the degree of compensation. Thus in active, decompensated cases, LD_5 is a prominent feature of the electrophoretic pattern (Fig. 41). Bajolle *et al.* (1968) found elevation of this component in twenty-seven of thirty-one cirrhotics, but in some cases other fractions were also elevated due possibly to secondary effects on the circulatory system. The isoenzyme patterns of liver biopsy specimens from cirrhotic patients do not differ from those of normal subjects (Secchi, Mossa and Gallitelli, 1964).

Experimental liver damage, produced by the administration of carbon tetrachloride to rats and monkeys, causes an increase in the serum LD_5 within 24 hr. Prolonged administration, however, leads to a decrease of the serum LD_5 which is attributed to exhaustion of the capacity of the liver cells to produce the enzyme (Wieme and Van Maercke, 1961).

Malignant diseases
Very high total serum lactate dehydrogenase levels often occur in widely disseminated malignant states, especially lymphoma and melanomatosis, but in these conditions the isoenzyme distribution remains essentially normal with the principal activity in LD_2, LD_3 and LD_4, unless there is extensive liver involvement, when LD_5 is also usually increased.

The significance of the isoenzyme composition of malignant effusions has been studied by Richterich, Zuppinger and Rossi (1961), who found that the proportion of total LD migrating as a)'-globulin (LD_5) is greater than in the serum. Later the same group extended this observation by showing that in malignant effusions the)'-LD (LD_5) exhibits activity greater than 72 International units/litre and accounts for more than 30% of the total activity of the effusion. In transudates the isoenzyme composition is similar to that of the corresponding serum, while in inflammatory exudates there is a slight increase in the isoenzymes associated with the α_1- and α_2-globulins (LD_1 and LD_2) (Richterich, Locker, Zuppinger and Rossi, 1962).

The abnormal isoenzyme distribution is not always easily recognized in the serum, but a careful quantitative study by Wieme, Van Hove and Van der Straeten (1968) has recently shown that of 76 patients with carcinoma of the bronchus, 38 showed marked increases and 22 slight increases in the LD_3 fraction, although in some of these patients the total serum enzyme activity was within the normal range. In patients responding to cytostatic therapy the LD_3 component diminished, though there was sometimes a transient increase in this fraction and in the LD_5 soon after the commencement of treatment. Similar changes have been reported by others (Starkweather and Schoch, 1962; Bottomley, Locke and Ingram, 1964: Starkweather, Green, Spencer and Schoch, 1966).

An absolute increase in all fractions, with a significant relative increase in LD_3, has also been reported by Wright *et al.* (1966) in a variety of malignant diseases, including carcinoma of the breast, gall-bladder, thyroid, pancreas, bronchus, stomach and colon, and acute myeloid leukaemia. When the tumour had metastasized to the liver an increase in the serum LD_5 was also found. Similar changes have been described by Zondag and Klein (1968) in about 50% of a large series of patients with malignant lymphoma, leukaemia and metastatic carcinoma.

In acute leukaemia the circulating white cells show a marked reduction in LD_5 and a substantial increase in LD_1, LD_2 and LD_3 as compared with normal granulocytes. Similar though less pronounced changes are also observed in acute leukaemia in remission and in chronic myeloid leukaemia (Dioguardi, Agostoni and Fiorelli, 1962; Dioguardi and Agostoni, 1965). A similar shift has been observed in the bone marrows of patients with acute leukaemia (Starkweather, Spencer and Schoch, 1966).

The serum isoenzyme patterns found in leukaemia differ from those detected in most other forms of neoplastic disease in showing a shift towards the faster isoenzymes, but certain other examples have been reported. In seminoma and teratoma of the testis, Zondag and Klein (1968) found the increase to be mainly in the LD_1 and LD_2 fractions, whereas in other forms of carcinoma of the testis the largest relative increase is in the LD_3 component. An increase in the serum LD_1 has also been found in dysgerminoma of the ovary (Zondag, 1965).

Experimental studies in animals have shown similar variety in the patterns of serum isoenzymes observed, for while Cenciotti and Mariotti (1964) found LD_3 to be the isoenzyme most increased in tumour-bearing rats, infection of mice with the leukaemogenic Riley virus leads to an enormous increase in the serum LD_5 (Plagemann, Gregory, Swim and Chen, 1963). Subsequently Mahy and Rowson (1965) demonstrated that the clearance of LD_5 from the serum is strongly inhibited in mice infected with the Riley virus, whereas that of LD_1 is similar to that of control mice.

The distribution of lactate dehydrogenase isoenzymes in tumour tissues has attracted considerable attention, for numerous investigators have reported shifts towards the cationic isoenzymes in a variety of malignant tumours as compared with normal control or contiguous tissues (Pfleiderer and Wachsmuth, 1961; Goldman, Kaplan and Hall, 1964; Nissen and Bohn, 1965; Güttler, 1967; and others). An example is illustrated in Fig. 42. This is of great potential importance in view of the high glycolytic activity of tumour tissues first reported nearly half a century ago by Warburg and Minami (1923). The recent observations of Kaplan and his co-workers (see p. 144) suggesting that LD_5 (M_4) is adapted for anaerobic glycolysis, while LD_1 (H_4) is associated with

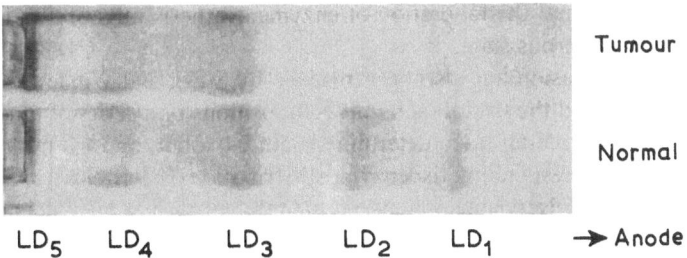

LD_5 LD_4 LD_3 LD_2 LD_1 → Anode

Fig. 42. Acrylamide-gel electrophoresis of lactate dehydrogenase isoenzymes from a lymph node tumour and a normal lymph node, showing the cationic shift. (Poznanska-Linde, Wilkinson and Withycombe, 1966.)

oxidative metabolism has revived interest in glycolytic processes in cancer tissues. Among the tissues in which such a shift has been detected are tumours of the prostate (Denis, Prout and Woolard, 1963; Van Camp, Denis and Van Sande, 1963); the breast (Richterich and Burger, 1963b; Barnett and Gibson, 1964; Stanislawski-Birencwajg and Loisillier, 1965); the brain (Gerhardt *et al.*, 1963); the cervix (Turner, 1964; Latner, Turner and Way, 1966; Okabe, Hayakawa, Hamada and Koike, 1968); the stomach (Yasin and Bergel, 1965; Leese, 1965; Baume, Builder, Fenton, Irving and Piper, 1966); the rectum and lymph nodes (Poznanska-Linde, Wilkinson and Withycombe, 1966); the colon (Langvad, 1968a), and the bronchus (Langvad, 1968b).

Gerhardt *et al.* (1963) have shown that in contrast to the findings in malignant tumours of the brain, the comparatively benign gliomas show a relative increase in the anionic isoenzymes.

Langvad (1968a) has recently measured the LD_2 and LD_4 contents of malignant tumours of the colon and of surrounding morphologically normal tissues, from which he has deduced the proportions of the M sub-unit present in each tissue. The shift towards the cathodic isoenzymes was demonstrated not only in the tumours but also in surrounding tumour-negative areas. Diffusion from the tumour was considered unlikely as a cause, since the fall in the ratio was not proportional to the distance of the sample site from the edge of the tumour and also because high ratios were sometimes observed in distinct areas remote from the tumour. Langvad therefore interprets his findings as evidence of a metabolic change in cells which may subsequently become neoplastic. Similar results were obtained in lung cancer, an observation which has led Langvad (1968b) to propose that the tumour develops in tissue which has already undergone a change to a more glycolytic pattern of metabolism, and that alteration of enzyme synthesis might be an index of a pre-cancerous state.

This conclusion has also been reached by Stagg and Whyley (1968) who compared the overall sub-unit compositions of a series of tumours of the female genital tract, determined with 2-oxobutyrate and pyruvate, with the corresponding isoenzyme distributions determined electrophoretically. Discrepancies between the two approaches might be due to the presence of two or more cell populations, e.g. malignant and non-malignant, malignant cells of more than one type or to a disturbance in the random association of sub-units leading to excessive formation of LD_5.

During chemical carcinogenesis a transient increase in the faster components has been observed. In experimental liver cancer induced in rats by the oral or parenteral administration of 3'-methyl-4-dimethylaminoazobenzene, Johnson and Kampschmidt (1965) observed all five isoenzymes in the liver during the precancerous stage, i.e. a shift towards the faster isoenzymes. An absolute increase in all five isoenzymes, but especially in LD_5, was also observed by Reeves (1967) during the pre-cancerous phase of the experimental induction of lung cancer in the rat by beryllium sulphate inhalation. The lung tumours showed a predominance of LD_5, but this appeared to be due to a diminution of the faster isoenzymes rather than to an increase in LD_5. It seems therefore that the shift towards the slower isoenzymes in neoplastic tissues may be due to accelerated synthesis of the M sub-unit, to suppression of H sub-unit synthesis, or to a combination of both factors.

There have been several reports of additional isoenzyme bands in the serum of cancer patients or in tumour tissues (Beautyman, 1962; Latner, 1964; Vesell, 1965b; Soetens et al., 1965). The writer and his colleagues recently had the opportunity to investigate the anomalous bands found in the serum of a patient with carcinoma of the oesophagus metastasizing to the liver (Fujimoto *et al.*, 1968). The serum lactate dehydrogenase was 3,700 I.U./l. (25°) and the '2-hydroxybutyrate dehydrogenase' (see p. 188) was similarly elevated (3,500 I.U./l. at 25°) indicating a marked excess of LD_1 and LD_2. This was confirmed by acrylamide-gel electrophoresis when additional bands between LD_1 and LD_2 and between LD_2 and LD_3 were also observed (Fig. 43).

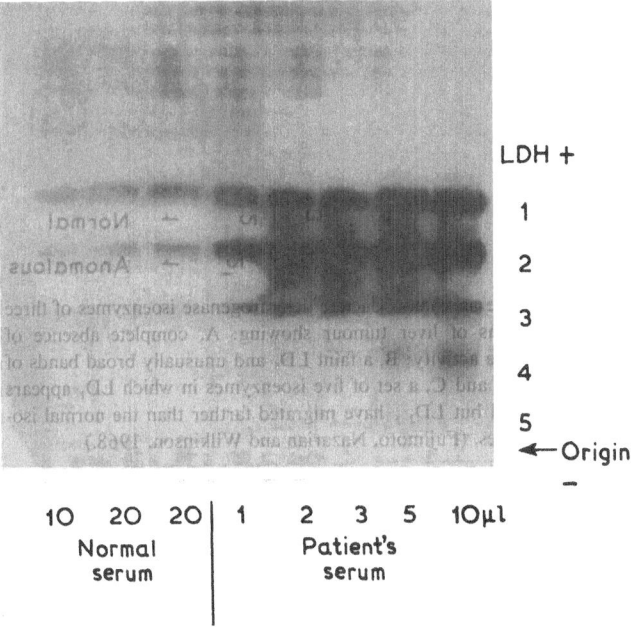

Fig. 43. Unusual lactate dehydrogenase isoenzyme pattern in the serum of a patient with a carcinoma of the oesophagus and metastases in the liver, showing the additional bands between LD_1 and LD_2 and between LD_2 and LD_3. (Fujimoto, Nazarian and Wilkinson, 1968.)

These appear to be due to polymorphism of the M sub-unit, since a portion of liver tumour removed at autopsy showed a pattern of five isoenzymes which, except for LD_1, differed from their normal counterparts in their electrophoretic mobilities (Fig. 44). The abnormal LD_5 had an appreciably greater mobility than normal LD_5 and seems to be composed of a different type of M sub-unit, since a complete set of regularly spaced hybrid isoenzymes with the normal H sub-unit was observed. The abnormal LD_5 also differed from normal LD$_5$ in its lability and in showing much greater affinity for 2-oxobutyrate relative to pyruvate. It is therefore suggested that the additional bands found in the serum could be H$_3$M′ and H$_2$M′$_2$ combinations. It seems therefore that the malignant change might in some cases in-

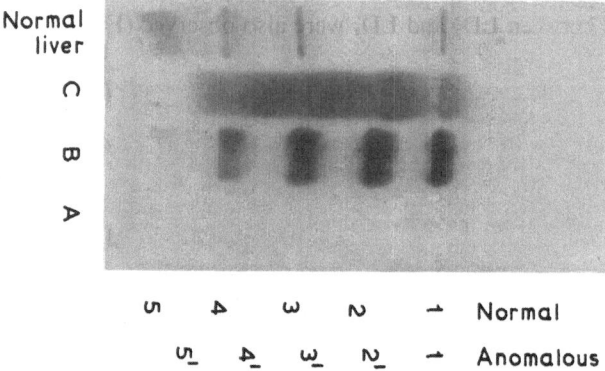

Fig. 44. The anomalous lactate dehydrogenase isoenzymes of three portions of liver tumour showing: A, complete absence of enzyme activity; B, a faint LD$_5$ and unusually broad bands of LD$_{1-4}$, and C, a set of five isoenzymes in which LD$_1$ appears normal but LD$_{2-5}$ have migrated farther than the normal isoenzymes. (Fujimoto, Nazarian and Wilkinson, 1968.)

volve an alteration in the templates on which the H and M sub-units are synthesized. If this occurs, entirely new sets of isoenzymes would result, since if only one sub-unit is involved, no fewer than fifteen isoenzymes might theoretically be produced (Fig. 45).

Neurological diseases
A marked increase in the serum LD$_3$ has recently been reported in children with Tay-Sachs disease by Saifer, Schneck, Perle and Volk (1969). These authors found that agar-gel electrophoresis of the serum isoenzymes was of value in differentiating primary ganglioside disorders such as Tay-Sachs disease and systemic late infantile amaurotic idiocy from other brain diseases involving storage defects of sphingomyelin (Niemann-Pick), sulphatides (metachromatic leucodystrophy) and mucopolysaccharides.

Diseases of muscle
The changes in the lactate dehydrogenase isoenzyme pattern in progressive muscular dystrophy has been discussed on pp. 167–169. Elevations

of LD_5 have been observed in the sera of patients with lesions of normal muscle, e.g. trichiniasis (Elliott and Wilkinson, 1963).

Diseases of joints
Increased amounts of LD_5 in the sera of patients with rheumatoid arthritis have been reported, but the total enzyme activity is usually normal (Wieme, 1963; Cohen, 1964). The synovial fluid of patients with inflammatory arthropathies usually exhibits increased LD_5 activity as compared with normal synovial fluid (Vesell, Osterland, Bearn and Kunkel, 1962; Greiling, Engels and Kisters, 1964).

Anaemia
The isoenzyme patterns found in megaloblastic and haemolytic anaemias will be discussed in the next section (p. 191).

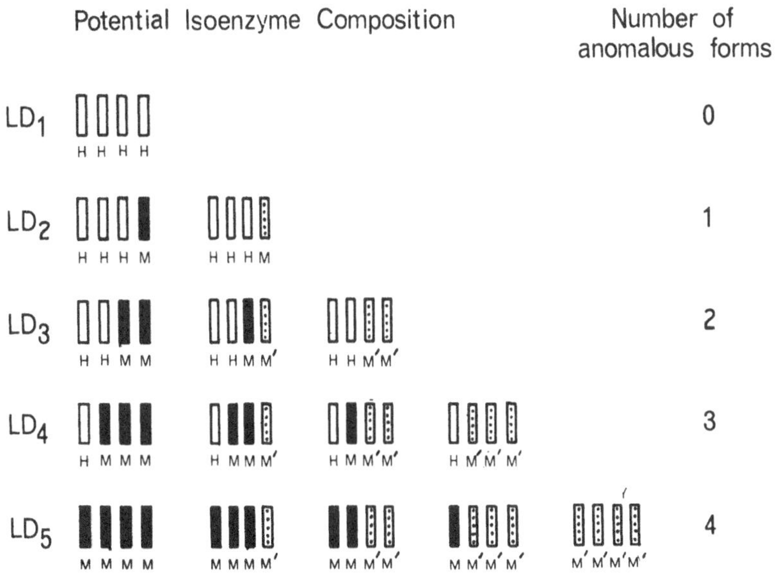

Fig. 45. Effects of a change in the composition of the monomeric sub-units from M to M' on the isoenzymic composition.

Pregnancy
The serum lactate dehydrogenase activity is frequently elevated in late pregnancy, especially during labour. The level in cord blood is usually higher than in the maternal serum, and it has been suggested that the placenta is the source (Little, 1959; Hagerman and Wellington, 1959; Linton and Miller, 1959; Lapan and Friedman, 1959). Meade and

Rosalki (1963) found a relative excess of LD_3 and LD_4 in the maternal serum during labour. These are the fractions they had previously found in placental extracts: the presence of LD_1 and LD_2 in the extracts is attributed to the red blood cells, which cannot easily be separated from this tissue (Meade and Rosalki, 1962). Hawkins and Whyley (1966), however, found LD_4 and LD_5 to be the principal isoenzymes in the placenta. They also found these components to be increased in the maternal serum during labour, but suggest that they may be derived from the uterine muscle, which is also rich in these fractions.

Geyer (1968) has shown that the myometrium of the gravid uterus contains a higher proportion of the slow isoenzymes as compared with that of the non-pregnant uterus (Table 10).

Renal diseases

All five isoenzymes occur in the kidney, with the anodic fractions LD_1 and LD_2 predominating in the cortex, while the medulla has a relatively larger complement of the slower components (Güttler and Clausen, 1965; Ringoir, 1967; Plummer and Leathwood, 1967). In acute tubular necrosis the LD_1 of the renal cortex is less prominent, a finding attributed to hypoxia, while at the same time the serum LD_5 is increased. The serum LD_5 is similarly increased in chronic pyelonephritis, and to a lesser extent in chronic glomerulonephritis (Ringoir, 1967).

The serum isoenzyme pattern has been extensively used in the monitoring of patients who have undergone renal homotransplantation as a means of detecting early signs of rejection of the homograft. An increase in the serum LD_5 has been interpreted by Prout, Macalalag and Hume (1964) as indicating impending rejection. Others, however, have found elevation of LD_5 in the sera of most patients who have received a transplant, and consider such a finding therefore to be of little value as an early diagnostic sign of rejection (Ringoir, 1967). In the terminal phase of rejection a rise in the serum LD_1 and LD_2 may be observed (Latner and Skillen, 1968). The rise in the LD_5 may be due to haemodialysis, which has been shown to cause an increase in the serum LD_5, attributed by Ringoir and Wieme (1965) to the release of LD_5 from diseased renal medullary cells, and by others to removal of inhibitors (Morgan, Morgan and Thomas, 1963; Emerson, Withycombe and Wilkinson, 1965).

Though it is generally agreed that most of the lactate dehydrogenase found in normal urines is in the LD_1 fraction (Kemp and Laursen, 1960; Macalalag and Prout, 1964; Guttler and Clausen, 1965; Gelderman, Gelboin and Peacock, 1965; Dubach, 1966), most investigators have reported an increase in LD_5 in a variety of renal diseases (Macalalag and Prout, 1964; Gelderman *et al.*, 1965; Dubach, 1966). This increase and the similar elevation in the serum are rather surprising in view of the preponderance of LD_1 and LD_2 in the kidney. Thiele and Mattenheimer (1968) have recently studied the isoenzyme patterns in micro-dissected portions of the human nephron, and have confirmed the predominance of the H sub-unit (Table 26). Only in the papilla was an excess of the M sub-unit observed. They point out that the finding of LD_1 in normal urine is consistent with the suggestion that it originates mainly in the renal cortex, and postulate that in disease states much of the urinary enzyme may be derived from the renal tubules (Mattenheimer, 1968).

TABLE 26 *Distribution of lactate dehydrogenase isoenzymes in the human nephron*

(From Thiele and Mattenheimer, 1968)

Unit of nephron	LD isoenzymes (mean percentage of total)				
	LD_1	LD_2	LD_3	LD_4	LD_5
Glomerulus	32	9	23	27	8
Convoluted tubules	57	31	10	2	< 1
Medullary rays	60	34	6	< 1	< 1
Outer medulla	46	30	18	4	2
Inner medulla	15	23	34	21	7
Papilla	8	17	35	29	11

Similar distribution in different parts of the nephron has also been detected by Nielsen, Kemp and Laursen (1968), who propose that in the initial phase of acute renal failure there might be an adaptive change leading to increased production of the M sub-unit. This it is suggested may be induced by a low oxygen tension. Thiele and Mattenheimer (1968), however, were unable to detect any close relation between the isoenzyme patterns of the various parts of the nephron and the type of metabolism. At present the origin of the increased urinary LD_5 remains unresolved, but the writer considers that hitherto insufficient attention has been paid to the effect on the isoenzyme distribution of other urinary constituents (Wilkinson, 1968) and of the low urinary pH (Jösch, Dubach and Strobel, 1967).

Somewhat different isoenzyme distributions have been found in the nephron of the rat. As compared with the human kidney, there is a greater proportion of LD_4 and LD_5 in all parts. A marked increase in these components is found in experimental ischaemia, but only minor changes occur in experimental hypertension (Emanuelli, Alessio, Fiorelli and Perpignano, 1967; Fiorelli, Emanuelli, Perpignano and Alessio, 1967).

Non-electrophoretic diagnostic techniques
Although some simplified techniques have been introduced, quantitative electrophoretic separation of isoenzymes is a tedious operation, and for routine purposes a number of simpler techniques has been devised with the object of obtaining an assessment of the relative proportions of the fast and slow isoenzymes in the serum (pp. 175–186). Among these are the serum 'α-hydroxybutyrate dehydrogenase' (SHBD) test (Rosalki and Wilkinson, 1960; Elliott and Wilkinson, 1961) and the relative heat-stability test (Wróblewski and Gregory, 1961), both of which have proved especially useful in myocardial infarction, but also have application in other conditions in which the serum lactate dehydrogenase is raised.

SERUM 2-HYDROXYBUTYRATE DEHYDROGENASE IN DIAGNOSIS. Since LD_1 and LD_2 exhibit much greater HBD activity (with 2-oxobutyrate as substrate) relative to their total activity (measured with pyruvate as substrate) than LD_4 and LD_5, it is probable that in pathological states in which excess of LD_1 and LD_2 are released into the serum, high SHBD activity will be found, whereas in conditions characterized by the release of LD_4 and LD_5 essentially normal SHBD levels would be expected. The value of SHBD determination in myocardial infarction and its low order of sensitivity in liver damage have now been confirmed by several groups of investigators (Elliott and Wilkinson, 1961; Konttinen, 1961; Pagliaro and Notarbartolo, 1961, 1962; Elliott, Jepson and Wilkinson, 1962; Hansson, Johansson and Sievers, 1962; Konttinen and Halonen, 1962; Rosalki, 1963; Bigazzi and Ciampi, 1963; Wilkinson and Rosalki, 1963; Preston, Batsakis and Briere, 1964; Rosalki and Wilkinson, 1964; Dubach and Margreth, 1965; Benson and Benedict, 1966).

The normal range for SHBD activity determined on 17 healthy subjects and 29 hospital patients was found by Elliott and Wilkinson (1961) to be 55–140 μmolar units per litre at 25°, and the results reported by other workers have been in close agreement with these figures. Bigazzi and Ciampi (1963) and Preston *et al.* (1964) have adopted 130 μmolar units and 120 μmolar units respectively as upper limits of the normal range.

In proven myocardial infarction most investigators have found the SHBD activity to be almost invariably raised (Table 27): false negative results are much less frequent than with other enzyme tests. SHBD elevation may be detected some 12 hr. after the onset of symptoms, and peak values are reached after 48–72 hr. Raised values persist for appreciably longer than those of other serum enzymes, and in a series of twenty consecutive patients Elliott and Wilkinson (1962) observed the

TABLE 27 *Serum HBD activities in normal subjects and patients with myocardial infarction*

	No. of individuals	Serum HBD activity (*µ*molar units per litre at 25)	
		Range	Mean $\pm$ S.D.
Normal subjects			
Elliott *et al.* (1962)	42	56–125	90 ╪ 19
Hansson *et al.* (1962)	22	68–111	95 ╪ 2
Konttinen and Halonen (1962)	108	37–105	74 ╪ 15
Pagliaro and Notarbartolo (1962)	30	48–77	64 ╪ 15
Rosalki (1963)	43	56–125	89 ╪ 15
Patients with myocardial infarction (highest observed values for each patient)			
Elliott *et al.* (1962)	122	163–1,100	--*
Hansson *et al.* (1962)	12	144–440	325 ╪ 27
Konttinen and Halonen (1962)	60	Up to 915	--
Pagliaro and Notarbartolo (1962)	30	Up to 720	--
Bigazzi and Ciampi (1963)	44	53–430	219 ╪ 104
Rosalki (1963)	70	—	520
Preston *et al.* (1964)	38	130–420	--

* Owing to the skew distribution of the enzyme activities in this group, expression of the results as a mean and standard deviation is of doubtful significance.

following mean periods during which elevated levels occur in the serum: glutamate-oxaloacetate transaminase (GOT, aspartate aminotransferase), 4·3 days $\pm$ S.D. 2·7 days; lactate dehydrogenase, 8·0 days $\pm$ 1·6 days; HBD, 13·3 days $\pm$ 3·3 days. Typical findings are illustrated in Fig. 46. Similar results have also been reported by Rosalki (1963) and by Preston *et al.* (1964).

Normal values are usually found in such diseases as angina of effort, coronary insufficiency, congestive cardiac failure and pulmonary infarction which might simulate myocardial infarction. In a total of 129 cases finally diagnosed as angina and coronary insufficiency reported by

Konttinen (1961), Elliott *et al.* (1962) and Rosalki (1963), only 10 marginally elevated values were observed, and in all these disorders the SHBD is less frequently raised than the SLD or SGOT. In an 'assessment of the HBD test Preston *et al.* (1964) conclude that it has an especial value in the differential diagnosis of coronary arterial disease, and that in several of their patients with myocardial infarction raised SHBD activities were observed before the characteristic electrocardiographic changes became apparent. It must be borne in mind, however, that it is a test for damage to the heart muscle, and consequently enzyme release occurs in myocarditis (Elliott *et al.*, 1962) and after cardiac surgery (Pyörälä, Gordin, Konttinen and Telivuo, 1963). Moreover, erythrocytes are rich in HBD activity, and visible haemolysis renders the serum sample unsuitable for this test.

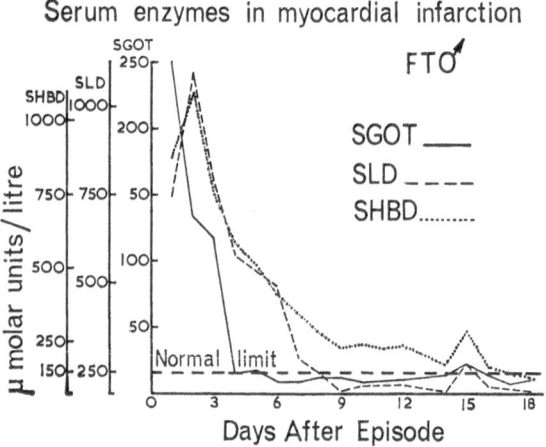

Fig. 46. Serial daily determinations of the serum aspartate transaminase (SGOT), lactate dehydrogenase (SLD) and 2-hydroxybutyrate dehydrogenase (SHBD) in a patient with acute myocardial infarction. (Elliott, Jepson and Wilkinson, 1962.)

Raised SHBD activities may also be found in hepatocellular diseases such as infective and toxic hepatitis and infectious mononucleosis. Elliott and Wilkinson (1963) found raised levels in 30 of 63 patients with acute liver disease, but in only 3 of 45 patients with chronic disorders (mainly obstructive jaundice and hepatic cirrhosis). That these elevations were due to the release of the slow isoenzymes LD_4 and LD_5 from the liver was shown by the finding of SHBD/SLD ratios below the normal range of 0·63–0·81. This is in marked contrast with the elevated ratios usually found after myocardial infarction.

In widely disseminated neoplastic diseases, such as lymphoma, melanomatosis and certain forms of leukaemia, high SHBD levels are often found, but the SHBD/SLD ratios are usually within the normal range unless there is liver involvement, when the ratio tends to be depressed. The only conditions, other than diseases of the myocardium, so far observed, in which raised serum activities are frequently accompanied by high ratios, are progressive muscular dystrophy of the pseudohypertrophic (Duchenne) type (p. 167) and megaloblastic anaemia due to vitamin B_{12} or folic acid deficiency (Elliott and Wilkinson, 1963; Fleming and Elliott, 1964) and occasionally in haemolytic anaemias (Emerson and Wilkinson, 1966).

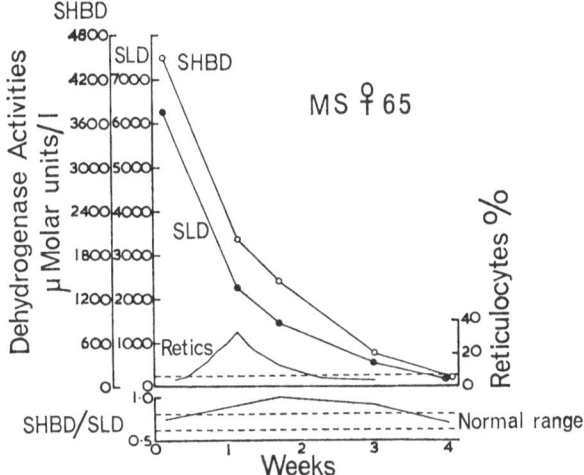

Fig. 47. Serial determination of the 2-hydroxybutyrate and lactate dehydrogenase activities in a patient with pernicious anaemia treated with vitamin B_{12}. The reticulocyte count may be used as an index of the patient's response to therapy. (Elliott and Wilkinson, 1963.)

Extremely high SHBD and SLD activities are encountered in untreated pernicious anaemia, and the electrophoretic patterns show a marked increase in LD_1 and LD_2. The enzyme levels and isoenzyme patterns rapidly return to normal when adequate vitamin B_{12} therapy is instituted (Fig. 47). The increase in the SHBD/SLD ratio which occurs when the reticulocyte response is maximal appears to be due to the more rapid clearance of the less-stable slow isoenzymes. The source of the exceptionally high serum enzyme activities has been studied by a number of investigators, and it is generally considered that the excess serum enzyme originates in the megaloblastic bone marrow (Gordin and Enari, 1959; Heller, Weinstein, West and Zimmerman, 1960; Grönvall, 1961). Isoenzyme studies have provided additional support for this view. Extramedullary sources of LD_1 and LD_2, e.g. heart, erythrocytes and kidney, are excluded by the absence of significant pathology in these tissues, and by the finding of normal aspartate transaminase activities in the serum. Extracts of the nucleated cells (buffy coat) of normal and megaloblastic bone marrows, however, show marked differences in their electrophoretic patterns and in their HBD/LD ratios. While normal marrow cells exhibit a preponderance of LD_4 and LD_5, the megaloblastic cells show LD_1 and LD_2 to be the most abundant isoenzymes (Yakulis, Gibson and Heller, 1962; Emerson, Withycombe and Wilkinson, 1967). Although the total enzyme activities vary widely, the HBD/LD ratios within each group are remarkably constant. Megaloblastic marrows have ratios significantly higher than normal marrows (Emerson *et al.*, 1967).

Studies of haem turnover rates suggest that erythropoiesis in the megaloblastic marrow proceeds at about three times the normal rate, but the delivery of viable erythrocytes into the circulating blood is not increased. This ineffective erythropoiesis indicates that about two-thirds of the developing erythrocytes are destroyed in the marrow before reaching maturity (Finch, Coleman, Motulsky, Donohue and Rieff, 1956). This substantial breakdown of cells rich in LD_1 and LD_2 could well account for the increase in the serum enzyme activity, a conclusion consistent with the normal serum enzyme values obtained in iron-deficiency anaemia and the only slightly elevated levels seem in haemolytic anaemias. In these forms of anaemia there is increased normal erythropoiesis, but no intramedullary cell destruction.

Bone marrow plasma has been found to exhibit substantially increased lactate dehydrogenase activity in pernicious anaemia, polycythaemia and certain other conditions. The increase occurs in all fractions, but especially in the slow isoenzymes, which suggest that it is due not only to intramedullary haemolysis but also to destruction of other marrow cells (Chury, Tovarék and Vojtková, 1968*a*, 1968*b*).

RELATIVE HEAT-STABILITY TEST. Attention has already been drawn (p. 161) to the relative lability of the slow isoenzyme (LD_5) when exposed to elevated temperatures, a property utilized by Wróblewski and Gregory (1961) to devise a simple test for clinical purposes. Samples of serum, to which NADH has been added, are heated at $57°$ or $65°$, after which the remaining enzyme activity is compared with that of an unheated sample. The control gives the total lactate dehydrogenase activity, while the difference between the activities of the control and the sample heated at $57°$ gives a measure of the heat-labile enzyme, principally LD_5. The activity of the heat-stable fraction (LD_1) is that of the sample heated at $65°$, while the difference between the activities of the two heated samples is an index of the total activity of the isoenzymes of intermediate heat-stability (LD_2, LD_3 and LD_4).

According to this test, normal sera contain 20–40% of the total activity as LD_1, 33–55% as LD_2, LD_3 and LD_4, and 10–30% as LD_5. In four patients with myocardial infarction the proportion of LD_1 ranged from 45 to 65% and in six cases of acute liver disease 33–85% appeared as the heat-labile LD_5. This procedure has been applied to the routine diagnosis of myocardial infarction by Takenaka, Gelderman and Brahen (1963), who obtained excellent correlation between the proportions of heat-stable lactate dehydrogenase in the serum, on the one hand, and the clinical and electrocardiographic findings, on the other.

Other investigators have employed similar techniques and have found the test to be particularly useful when cardiac infarction is suspected (Strandjord and Clayson, 1961; Dubach, 1961, 1963; Strandjord, Clayson and Freier, 1962; Wüst, Schön and Berg, 1962; Bell, 1963; Dubach and Orelli, 1963; Nutter, Trujillo and Evans, 1966). Latner and Skillen (1963) have used a 'heat-stability index', i.e. the ratio of the activity of serum heated at $60°$ for 1 hr. to that of unheated serum, and found that all but one sample of serum from fourteen patients with confirmed myocardial infarction gave values greater than $0·5$. Only one sample of serum from forty patients with other disorders gave such a value, while in normal subjects the index did not exceed $0·3$. Paunier and Rotthauwe (1963) have found the heat-stability test useful in the diagnosis of a variety of heart, liver and muscle diseases.

USE OF INHIBITORS. Urea and oxalate, which selectively inhibit lactate dehydrogenase isoenzymes (p. 151), have found a number of applications in diagnosis, especially in myocardial infarction. In this condition and in pernicious anaemia the serum lactate dehydrogenase was found to be inhibited by at least 68% by $0·2\text{mM}$-oxalate and less than 45% by 2M-urea, whereas in liver disease less than 50% inhibition is obtained with oxalate and more than 62% with urea (Emerson and Wilkinson, 1965). The test has proved useful in a number of problem cases in which myocardial infarction was suspected in the presence of pre-existing liver disease or congestive cardiac failure. Other investigators have used somewhat higher concentrations of urea in similar tests for the detection of LD_1 and LD_2 in myocardial infarction (Hardy, 1965; Welshman and Rixon, 1968).

The sensitivity of the urea-inhibition test has been improved by increasing the concentration of pyruvate. Under these conditions the inhibitory effect of urea on LD_1 and LD_2 was almost completely abolished, whereas LD_4 and LD_5 remain urea-sensitive. This modification is reported to be more discriminatory than existing tests in the diagnosis of myocardial infarction, pernicious anaemia and a variety of liver diseases (Konttinen and Lindy, 1967a, 1967b; Lindy and Konttinen, 1967b).

SOLVENT-PRECIPITATION TECHNIQUES. Diagnostic tests based upon the selective precipitation of the slow isoenzymes by the addition of acetone or chloroform have been referred to on p. 156, but their roles in clinical practice await further critical evaluation.

SELECTIVE ADSORPTION TECHNIQUES. The selective adsorption of LD_1 and LD_2 on DEAE-cellulose (p. 30) forms the basis of a simple test devised by Hess and Walter (1960b, 1961) for diagnostic purposes. Dialysed serum is treated with a suspension of DEAE-cellulose in phosphate buffer at pH 6·0, and after 10 min. the ion-exchange cellulose is removed by centrifugation. The lactate dehydrogenase activity of the supernatant is then compared with that of the untreated dialysed serum. In diseases such as myocardial infarction and haemolytic anaemia, which are characterized by the release of LD_1 and LD_2, less than 15% of the activity remains in the supernatant, whereas in virus hepatitis the serum enzyme consists largely of LD_4 and LD_5 and more than 70% of the original activity remains in the supernatant.

Summary

In view of the importance of lactate dehydrogenase isoenzymes, a brief summary of their main features is included here.

Human and animal tissue lactate dehydrogenases consist of five principal isoenzymes, and tissues can be differentiated into three groups according to whether the enzyme consists mainly of: (a) electrophoretically fast (anodic) components, LD_1 and LD_2, as in heart, kidney and erythrocytes; (b) slow isoenzymes, LD_4 and LD_5, as in liver and skeletal muscle; or (c) isoenzymes of intermediate mobility, as in lung, adrenal and thyroid.

Lactate dehydrogenase isoenzymes show a gradation in properties such as substrate affinity, stability at elevated temperatures, effects of inhibitors and reactions with coenzyme analogues. This phenomenon is readily explicable, since there is strong evidence showing that the enzyme molecule is a tetramer comprising monomers of two different types. It is considered that LD_1 consists of four identical B monomers, while LD_5 comprises four A sub-units, and LD_2, LD_3 and LD_4 are hybrid molecules containing monomers of both types. Dissociation of isoenzymes into monomers and subsequent reassociation may be effected by freezing overnight in $0 \cdot 5\text{M-NaCl}$. LD_1 and LD_5 exhibit a high degree of immunochemical specificity, and antibodies show little cross-reaction with the specific antigens, though they react with LD_2, LD_3 and LD_4.

The lactate dehydrogenase of human spermatozoa (LD_x) has an electrophoretic mobility intermediate between LD_3 and LD_4, but in most of its properties it resembles LD_1. Dissociation and reassociation studies indicate that LD_x is a tetramer comprising four identical sub-units of a type different from those found in the other isoenzymes.

The isoenzyme patterns in certain embryonic tissues often differ from those of the corresponding adult forms. There is a remarkable parallel between the isoenzyme patterns of human dystrophic muscle and of human foetal muscle which may be interpreted as indicating that failure of differentiation is an integral part of the dystrophic process.

The relative abundance of certain isoenzymes in the blood serum has found applications in diagnosis: an excess of LD_1 and LD_2 is characteristic of myocardial infarction, while a preponderance of LD_4 and LD_5 is found in hepatitis. Several simplified diagnostic techniques based upon differences in the properties of the isoenzymes have been introduced.

REFERENCES

ALLEN, J. M. (1961), *Ann. N.Y. Acad. Sci.* **94**, 937.

AMELUNG, D. I. (1963), *Dtsch. med. Wschr.* **88**, 1940.

APPELLA, E. and MARKERT, C. L. (1961), *Biochem. biophys. Res. Com.* **6**, 171.

AUERBACH, S. and BRINSTER, R. L. (1967), *Exptl. Cell Res.* **46**, 89.

BABSON, A. L. (1967), *Clin. chim. Acta* **16**, 121.

BAJOLLE, F., DEVILLE, A. and BOREL, J. P. (1968), *Rev. Franc. d'Etud. Clin. Biol.* **13**, 93.

BARNETT, H. and GIBSON, A. (1964), *J. clin. Path.* **17**, 201.

BATSAKIS, J. G., PRESTON, J. A. and BRIERE, R. O. (1964), *Milit. Med.* **129**, 1161.

BAUME, P. E., BUILDER, J. E., FENTON, B. H., IRVING, L. G. and PIPER, D. W. (1966), *Gastroent.* **50**, 781.

BAUR, E. W. and PATTIE, D. L. (1968), *Nature, Lond.* **218**, 341.

BEAUTYMAN, W. (1962), *Lancet* **ii**, 305.

BELL, R. L. (1963), *Amer. J. clin. Path.* **40**, 216.

BENITEZ, L. and FISCHER, R. (1965), *Nature, Lond.* **206**, 105.

BENSON, P. A. and BENEDICT, W. H. (1966), *Amer. J. clin. Path.* **45**, 760.

BIGAZZI, P. L. and CIAMPI, G. P. (1963), *Prog. Med.* **19**, 590.

BLANCHAER, M. C. (1962), *Pure appl. Chem.* **3**, 403.

BLANCO, A., RIFÉ, U. and LARSON, B. L. (1967), *Nature, Lond.* **214**, 1331.

BLANCO, A. and ZINKHAM, W. H. (1963), *Science,* **139**, 601.

BLANCO, A., ZINKHAM, W. H. and KUPCHYK, L. (1964), *J. exp. Zool.* **156**, 137.

BONAVITA, V. and GUANERI, R. (1962), *Biochim. biophys. Acta* **59**, 634.

BONAVITA, V., PONTE, F. and AMORE, G. (1962), *Nature, Lond.* **196**, 576.

BOTTOMLEY, R. R., LOCKE, S. J. and INGRAM, H. C. (1964), *Proc. Am. Assoc. Cancer Res.* **5**, 7.

BOYD, J. W. (1967), *Biochim. biophys. Acta* **146**, 590.

BOYER, S. H., FAINER, D. C. and WATSON-WILLIAMS, E. J. (1963), *Science,* **141**, 642.

BRODY, I. A. (1964a), *Nature, Lond.* **201**, 685.

BRODY, I. A. (1964b), *Neurology* **14**, 1091.

BRODY, I. A. (1965), *Nature, Lond.* **205**, 196.

BRODY, I. A. (1968), *Ann. N.Y. Acad. Sci.* **151**, 587.

CAHN, R. D., KAPLAN, N. O., LEVINE, L. and ZWILLING, E. (1962), *Science,* **136**, 962.

CENCIOTTI, L. and MARIOTTI, A. (1964), *Pathologica* **56**, 191.

CHILDS, V. A. and LEGATOR, M. S. (1965), *Life Sci.* **4**, 1643.

CHURY, Z., TOVARÉK, J. and VOJTKOVÁ, J. (1968a), *Fol. Haemat.* **89**, 15.

CHURY, Z., TOVARÉK, J. and VOJTKOVÁ, J. (1968b), *Z. klin. Chem.* **6**, 92.

CLAUSEN, J. and GERHARDT, W. (1963), *Acta Neurol. Scand.* **39**, 305.

COHEN, L., BLOCK, J. and DJORDJEVICH, J. (1967), *Proc. Soc. exp. Biol., N.Y.* **126**, 55.

COHEN, L., DJORDJEVICH, J. and ORMISTE, V. (1964), *J. Lab. clin. Med.* **64**, 355.

COHEN, L. and LARSON, L. (1966), *New Engl. J. Med.* **275**, 465.

COSTELLO, L. A. and KAPLAN, N. O. (1963), *Biochim. biophys. Acta* **73**, 658.

CRIDDLE, R. S., MCMURRAY, C. H. and GUTFREUND, H. (1968), *Nature, Lond.* **220**, 1091.

CROISILLE, Y. (1964), *Compt. rend. Acad. Sci.* **258**, 2214.

DAVIDSON, R. G., FILDES, R. A., GLEN-BOTT, A. M., HARRIS, H., ROBSON, E. B. and CLEGHORN, T. E. (1965), *Ann. Hum. Genet.* **29**, 5.

DAWSON, D. M., GOODFRIEND, T. L. and KAPLAN, N. O. (1964), *Science,* **143**, 929.

DENIS, L. J., PROUT, G. R. and WOOLARD, V. (1963), *Invest. urol.* **1**, 101.

DENNIS, D. and KAPLAN, N. O. (1960), *J. biol. Chem.* **235**, 810.

DIOGUARDI, N. and AGOSTONI, A. (1965), *Enzymol. biol. Clin.* **5**, 3.

DIOGUARDI, N., AGOSTONI, A. and FIORELLI, G. (1962), *Enzymol. biol. Clin.* **2**, 116.

DREYFUS, J.-C., DEMOS, J., SCHAPIRA, F. and SCHAPIRA, G. (1962), *Compt. rend. Acad. Sci.* **254**, 4384.

DUBACH, U. C. (1961), *Helv. med. Acta* **28**, 469.

DUBACH, U. C. (1963), *Diag. Terap.* **1**, 205.

DUBACH, U. C. (1966), *Helv. med. Acta* **33**, 139.

DUBACH, U. C. and MARGRETH, L. (1965), *Dtsch. med. Wschr.* **90**, 1429.

DUBACH, U. C. and VON ORELLI, A. (1963), *Helv. med. Acta* **30**, 685.

ELLIOTT, B. A., JEPSON, E. M. and WILKINSON, J. H. (1962), *Clin. Sci.* **23**, 205.

ELLIOTT, B. A. and WILKINSON, J. H. (1961), *Lancet* **i**, 698.

ELLIOTT, B. A. and WILKINSON, J. H. (1962), *Lancet* **ii**, 71.

ELLIOTT, B. A. and WILKINSON, J. H. (1963), *Clin. Sci.* **24**, 343.

EMANUELLI, G., ALESSIO, L., FIORELLI, G. and PERPIGNANO, G. (1967), *Boll. Soc. Ital. Biol. sper.* **44**, 130.

EMERSON, P. M. and WILKINSON, J. H. (1965), *J. clin. Path.* **18**, 803.

EMERSON, P. M. and WILKINSON, J. H. (1966), *Br. J. Haematol.* **13**, 656.

EMERSON, P. M., WILKINSON, J. H. and WITHYCOMBE, W. A. (1964), *Nature, Lond.* **202**, 1337.

EMERSON, P. M., WITHYCOMBE, W. A. and WILKINSON, J. H. (1965), *Lancet* **ii**, 571.

EMERSON, P. M., WITHYCOMBE, W. A. and WILKINSON, J. H. (1967), *Br. J. Haematol.* **13**, 656.

EMERY, A. E. H. (1964), *Nature, Lond.* **201**, 1044.

EMERY, A. E. H. (1967), *Biochem. J.* **105**, 599.

EMERY, A. E. H. (1968), *J. neurol. Sci.* **7**, 137.

FIELDHOUSE, B. and MASTERS, C. J. (1966), *Biochim. biophys. Acta* **118**, 538.

FIELDHOUSE, B. and MASTERS, C. J. (1968), *Biochim. biophys. Acta* **151**, 535.

FINCH, C. A., COLEMAN, D. H., MOTULSKY, A. G., DONOHUE, D. M. and RIEFF, R. H. (1956), *Blood* **11**, 807.

FINE, I. H., KAPLAN, N. O. and KUFTINEC, D. (1963), *Biochemistry* **2**, 116.

FINE, I. H., KAPLAN, N. O. and WHITE, S. (1962), *Federation Proc.* **21**, 409.

FIORELLI, G., EMANUELLI, G., PERPIGNANO, G. and ALESSIO, L. (1967), *Boll. Soc. Ital. Biol. sper.* **44**, 133.

FLEMING, A. F. and ELLIOTT, B. A. (1964), *Br. med. J.* **ii**, 1108.

FLEXNER, L. B., FLEXNER, J. B., ROBERTS, R. B. and DE LA HABA, G. (1960), *Develop. Biol.* **2**, 313.

FRITZ, P. J. (1965), *Science,* **150**, 364.

FRITZ, P. J. (1967), *Science,* **156**, 82.

FRITZ, P. J. and JACOBSON, K. B. (1963), *Science,* **140**, 64.

FRÖHLICH, C. (1965), *Z. klin. Chem.* **3**, 137.

FUJIMOTO, Y., NAZARIAN, I. and WILKINSON, J. H. (1968), *Enzymol. biol. clin.* **9**, 124.

FUJIMOTO, Y. and WILKINSON, J. H. (1970), *Biochem. biophys. Acta,* **206**, 38.

GELDERMAN, A. H., GELBOIN, H. V. and PEACOCK, A. C. (1965), *J. Lab. clin. Med.* **65**, 132.

GENIS-GALVEZ, J. M. and MAISEL, H. (1967), *Nature, Lond.* **213**, 283.

GERHARDT, W., CLAUSEN, J., CHRISTENSEN, E. and RIISHEDE, J. (1963), *Acta neurol. Scand.* **39**, 85.

GEYER, H. (1968), *Klin. Wschr.* **46**, 443, 446.

GIBSON, D. M., DAVISSON, E. O., BACHHAWAT, B. K., RAY, B. R. and VESTLING, C. S. (1953), *J. biol. Chem.* **203**, 397.

GOLDBERG, E. (1963), *Science,* **139**, 602.

GOLDBERG, E. (1965), *Science,* **148**, 391.

GOLDBERG, E. (1966), *Science,* **151**, 1091.

GOLDBERG, E. and CATHER, J. N. (1963), *J. cell. comp. Physiol.* **41**, 31.

GOLDBERG, E. and HAWTREY, C. (1967), *J. exp. Zool.* **164**, 309.

GOLDMAN, R. D., KAPLAN, N. O. and HALL, T. C. (1964), *Cancer Res.* **24**, 389.

GOODFRIEND, T. L., SOKOL, D. M. and KAPLAN, N. O. (1966), *J. molec. Biol.* **15**, 18.

GORDIN, R. and ENARI, T. M. (1959), *Acta Haemat., Basel* **21**, 360.

GRABOWSKI, C. T. (1961), *Science,* **134**, 1359.

GREEN, D. E. and BROSTEAUX, J. (1936), *Biochem. J.* **30**, 1489.

GREILING, H., ENGELS, G. and KISTERS, R. (1964), *Klin. Wschr.* **42**, 427.

GRÖNVALL, C. (1961), *Scand. J. clin. Lab. Invest.* **13**, 29.

GÜTTLER, F. (1967), *Enzymol. biol. clin.* **8**, 228.

GÜTTLER, F. and CLAUSEN, J. (1965), *Enzymol. biol. clin.* **5**, 55.

GÜTTLER, F. and CLAUSEN, J. (1969), *Biochem. J.* **114**, 839.

HAGERMAN, D. D. and WELLINGTON, F. M. (1959), *Amer. J. Obstet. Gynec.* **39**, 1043.

HANSSON, A., JOHANSSON, B. and SIEVERS, J. (1962), *Lancet* i, 167.

HARDY, S. M. (1965), *Nature, Lond.* **206**, 933.

HAWKINS, D. F. and WHYLEY, G. A. (1966), *Clin. chim. Acta* **13**, 713.

HAWTREY, C. and GOLDBERG, E. (1968), *Ann. N.Y. Acad. Sci.* **151**, 611.

HELLER, P., WEINSTEIN, H. G., WEST, M. and ZIMMERMAN, H. J. (1960), *J. Lab. clin. Med.* **55**, 425.

HELLUNG-LARSEN, P. and ANDERSEN, V. (1968), *Exptl. Cell Res.* **50**, 286.

HESS, B. (1958), *Ann. N.Y. Acad. Sci.* **75**, 292.

HESS, B. and WALTER, S. I. (1960a), *Verhand. dtsch. ges. inner. Med.* **66**, 639.

HESS, B. and WALTER, S. I. (1960b), *Klin. Wschr.* **38**, 1080.

HESS, B. and WALTER, S. I. (1961), *Klin. Wschr.* **39**, 213.

HEYCK, H., LAUDAHN, C. and LÜDERS, C. J. (1963), *Klin. Wschr.* **41**, 500.

HINKS, M. and MASTERS, C. J. (1966), *Biochim. biophys. Acta* **130**, 458.

HOCHACHKA, P. W. (1966), *Comp. Biochem. Physiol.* **18**, 261.

JOHNSON, H. L. and KAMPSCHMIDT, R. F. (1965), *Proc. Soc. exp. Biol., N.Y.* **120**, 557.

JOHNSTON, H. A., WILKINSON, J. H., WITHYCOMBE, W. A. and RAYMOND, S. (1966), *J. clin. Path.* **19**, 250.

JÖSCH, W., DUBACH, U. C. and STROBEL, M. (1967), *Experientia, Basel* **23**, 342.

KAPLAN, N. O. (1963), *Bact. Rev.* **27**, 155.

KAPLAN, N. O. and CAHN, R. D. (1962), *Proc. Natl. Acad. Sci., Wash.* **48**, 2123.

KAPLAN, N. O. and CIOTTI, M. M. (1961a), *Ann. N.Y. Acad. Sci.* **94**, 701.

KAPLAN, N. O. and CIOTTI, M. M. (1961b), *Biochim. biophys. Acta* **49**, 425.

KAPLAN, N. O., CIOTTI, M. M., HAMOLSKY, M. and BIEBER, R. E. (1960), *Science,* **131**, 392.

KAPLAN, N. O., EVERSE, J. and ADMIRAAL, J. (1968), *Ann. N.Y. Acad. Sci.* **151**, 400.

KEMP, E. and LAURSEN, T. (1960), *Scand. J. clin. Lab. Invest.* **12**, 463.

KONTTINEN, A. (1961), *Lancet* ii, 556.

KONTTINEN, A. and HALONEN, P. I. (1962), *Amer. J. Cardiol.* **10**, 525.

KONTTINEN, A. and LINDY, S. (1967a), *Clin. chim. Acta* **16**, 377.

KONTTINEN, A. and LINDY, S. (1967b) *Acta med. exp. Fenn.* **45**, 434.

KONTTINEN, A. and LINDY, S. (1967c), *Acta med. Scand.* **181**, 513.

KRAUS, A. P. and NEELY, C. L. (1964), *Science,* **145**, 595.

KREUTZER, H. H. and FENNIS, W. H. S. (1964), *Clin. chim. Acta* **9**, 64.

KREUTZER, H. J. H. and KREUTZER, H. H. (1965), *Clin. chim. Acta* **11**, 578.

KRIEG, A. F., ROSENBLUM, L. J. and HENRY, J. B. (1967), *Clin. Chem.* **13**, 196.

KUBOWITZ, F. and OTT, P. (1943), *Biochem. Z.* **314**, 94.

LANGVAD, E. (1968a), *Int. J. Cancer* **3**, 17.

LANGVAD, E. (1968b), *European J. Cancer* **4**, 107.

LAPAN, B. and FRIEDMAN, M. M. (1959), *J. Lab. clin. Med.* **54,** 417.

LATNER, A. L. (1964), *Proc. Assoc. clin. Biochem.* **3,** 120.

LATNER, A. L., SIDDIQUI, S. A. and SKILLEN, A. W. (1966), *Science,* **154,** 527.

LATNER, A. L. and SKILLEN, A. W. (1961), *Lancet* **ii,** 1286.

LATNER, A. L. and SKILLEN, A. W. (1963), *Proc. Assoc. clin. Biochem.* **2,** 100.

LATNER, A. L. and SKILLEN, A. W. (1964), *J. Embryol. exp. Morphol.* **12,** 501.

LATNER, A. L. and SKILLEN, A. W. (1968), *Isoenzymes in Biology and Medicine,* London and New York: Academic Press, p. 146.

LATNER, A. L. and TURNER, D. M. (1963), *Lancet* **i,** 1293.

LATNER, A. L., TURNER, D. M. and WAY, S. A. (1966), *Lancet* **ii,** 814.

LAURYSSENS, M. G., LAURYSSENS, M. J. and ZONDAG, H. A. (1964), *Clin. Chim. Acta* **9,** 276.

LEESE, C. L. (1965), *European J. Cancer* **1,** 211.

LINDSAY, D. T. (1963), *J. exp. Zool.* **152,** 75.

LINDY, S. and KONTTINEN, A. (1966), *Clin. chim. Acta* **14,** 615.

LINDY, S. and KONTTINEN, A. (1967a), *Clin. chim. Acta* **17,** 223.

LINDY, S. and KONTTINEN, A. (1967b), *Am. J. Cardiol.* **19,** 563.

LINTON, E. B. and MILLER, E. C. (1959), *Am. J. Obstet. Gynec.* **78,** 11.

LITTLE, W. A. (1959), *Obstet. and Gynec.* **13,** 152.

LÖWENTHAL, A., VAN SANDE, M. and KARCHER, D. (1961), *Ann. N.Y. Acad. Sci.* **94,** 988.

LUSH, I. E. and COWEY, C. B. (1968), *Biochem. J.* **110,** 33P.

MACALALAG, E. V. and PROUT, G. R. (1964), *J. Urol.* **92,** 416.

MAGER, M., BLATT, W. F. and ABELMANN, W. H. (1966), *Clin. chim. Acta* **14,** 689.

MALASKOVÁ, V. and HOLEYŠOVSKÁ, H. (1969), *Clin. chim. Acta* **24,** 39.

MAHY, B. W. J. and ROWSON, K. E. K. (1965), *Science,* **149,** 756.

MARKERT, C. L. (1962), *Hereditary, Developmental and Immunologic Aspects of Kidney Disease,* ed. METCOFF, J., Evanston: Northwestern Univ. Press, p. 54.

MARKERT, C. L. (1963b), *Cytodifferential and Macromolecular Synthesis,* New York: Academic Press, p. 65.

MARKERT, C. L. and FAULHABER, I. (1965), *J. exp. Zool.* **159,** 319.

MARKERT, C. L. and MØLLER, F. (1959), *Proc. Natl. Acad. Sci., Wash.* **45,** 753.

MASSARO, E. J. (1967), *Biochim. biophys. Acta* **147,** 45.

MATTENHEIMER, H. (1968), in *Enzymes in Urine and Kidney,* ed. DUBACH, U. C., Berne: Huber, p. 119.

MEADE, B. W. and ROSALKI, S. B. (1962), *Lancet* **i,** 1407.

MEADE, B. W. and ROSALKI, S. B. (1963), *J. Obst. Gyn. Brit. Cwlth.* **70,** 862.

MEISTER, A. (1950), *J. biol. Chem.* **194,** 117.

MEYERHOF, O. and LOHMANN, K. (1926), *Biochem. Z.* **171,** 421.

MORGAN, J. M., MORGAN, R. E. and THOMAS, G. E. (1963), *Metabolism* **12,** 1051.

MOYER, F. H., SPEAKER, C. B. and WRIGHT, D. A. (1968), *Ann. N.Y. Acad. Sci.* **151,** 650.

NACHLAS, M. M., FRIEDMAN, M. M. and COHEN, S. P. (1964), *Surgery, St. Louis* **55,** 700.

NANCE, W. E., CLAFLIN, A. and SMITHIES, O. (1963), *Science,* **142,** 1075.

NEAVES, W. B. and GERALD, P. S. (1968), *Science,* **160,** 1004.

NEBEL, E. J. and CONKLIN, J. L. (1964), *Proc. Soc. exp. Biol. N.Y.* **115,** 532.

NEMCHINSKAYA, V. L., GANELINA, L. SH. and BRAUN, A. D. (1968), *Nature, Lond.* **217,** 251.

NIELSEN, V. K., KEMP, E. and LAURSEN, T. (1968), *Acta med. Scand.* **184**, 109.

NISSELBAUM, J. S. and BODANSKY, O. (1961), *J. biol. Chem.* **236**, 323.

NISSEN, N. I. and BOHN, L. (1965), *European J. Cancer* **1**, 217.

NITOWSKY, H. M. and SODERMAN, D. D. (1964), *Exptl. Cell Res.* **33**, 562.

NOVOA, W. B., WINER, A. D., GLAID, A. J. and SCHWERT, G. W. (1959), *J. biol. Chem.* **234**, 1143.

NUTTER, D. D., TRUJILLO, N. P. and EVANS, J. M. (1966), *Am. Heart J.* **72**, 315.

NYGAARD, A. P. (1963), *The Enzymes*, ed. BOYER, P. D., LARDY, H. A. and MYRBÄCK, K., New York: Academic Press, Vol. 7, p. 557.

OKABE, K., HAYAKAWA, T., HAMADA, M. and KOIKE, M. (1968), *Biochemistry* **7**, 79.

ORLEANS-HARDING, J. G. and MAHLER, R. (1968), *Biochem. J.* **107**, 31P.

OTTOLENGHI, P. and DENSTEDT, O. F. (1958a), *Canad. J. Biochem. Physiol.* **36**, 1075.

OTTOLENGHI, P. and DENSTEDT, O. F. (1958b), *Canad. J. Biochem. Physiol.* **36**, 1085.

PAGLIARO, L. and NOTARBARTOLO, A. (1961), *Lancet* **ii**, 1261.

PAGLIARO, L. and NOTARBARTOLO, A. (1962), *Lancet* **i**, 1043.

PALMER, L. and KJELLBERG, B. (1967), *Experientia* **23**, 800.

PAPADOPOULOS, N. M. and KINTZIOS, J. A. (1967), *Am. J. clin. Path.* **47**, 96.

PAUNIER, L. and ROTTHAUWE, H. W. (1963), *Enzymol. biol. clin.* **3**, 87.

PEARSON, C. M., KAR, N. C., PETER, J. B. and MUNSAT, T. L. (1965), *Am. J. Med.* **39**, 91.

PESCE, A., FONDY, T. P., STOLZENBACH, F., COSTELLO, F. and KAPLAN, N. O. (1967), *J. biol. Chem.* **242**, 2151.

PFLEIDERER, G. and JECKEL, D. (1957), *Biochem. Z.* **329**, 370.

PFLEIDERER, G. and WACHSMUTH, E. D. (1961), *Biochem. Z.* **334**, 185.

PHILIP, J. and VESELL, E. S. (1962), *Proc. Soc. exp. Biol. N.Y.* **110**, 582.

PLAGEMANN, P. G. W., GREGORY, K. F., SWIM, H. E. and CHEN, K. K. W. (1963), *Canad. J. Microbiol.* **9**, 75.

PLAGEMANN, P. G. W., GREGORY, K. F. and WRÓBLEWSKI, F. (1960a), *J. biol. Chem.* **235**, 2282.

PLAGEMANN, P. G. W., GREGORY, K. F. and WRÓBLEWSKI, F. (1960b), *J. biol. Chem.* **235**, 2288.

PLAGEMANN, P. G. W., GREGORY, K. F. and WRÓBLEWSKI, F. (1961), *Biochem. Z.* **334**, 37.

PLUMMER, D. T., ELLIOTT, B. A., COOKE, K. B. and WILKINSON, J. H. (1963), *Biochem. J.* **87**, 416.

PLUMMER, D. T. and LEATHWOOD, P. D. (1967), *Biochem. J.* **103**, 172.

PLUMMER, D. T. and WILKINSON, J. H. (1961), *Biochem. J.* **81**, 38P.

PLUMMER, D. T. and WILKINSON, J. H. (1963), *Biochem. J.* **87**, 423.

PLUMMER, D. T., WILKINSON, J. H. and WITHYCOMBE, W. A. (1963), *Biochem. J.* **89**, 49P.

POZNANSKA-LINDE, H., WILKINSON, J. H. and WITHYCOMBE, W. A. (1966), *Nature, Lond.* **209**, 727.

PRESTON, J. A., BATSAKIS, J. G. and BRIERE, R. O. (1964), *Am. J. clin. Path.* **41**, 237.

PROUT, G. R., MACALALAG, E. V. and HUME, D. M. (1964), *Surgery, St. Louis* **56**, 283.

PYÖRÄLÄ, K., GORDIN, R., KONTTINEN, A. and TELIVUO, L. (1963), *Acta. med. Scand.* **174**, 361.

RAPOLA, J. and KOSKIMIES, O. (1967), *Science*, **157**, 1311.

RAUCH, N. (1968), *Ann. N.Y. Acad. Sci.* **151**, 672.

REEVES, A. L. (1967), *Cancer Res.* **27**, 1875.

RESSLER, N., COOK, U., OLIVERO, E. and JOSEPH, R. R. (1965), *Nature, Lond.* **206**, 828.

RESSLER, N., OLIVERO, E. and JOSEPH, R. R. (1965), *Nature, Lond.* **206**, 829.

RETTENBACHER-DAUBNER, H. and REIDER, H. (1963), *Wein. Klin. Wschr.* **75**, 833.

RICHTERICH, R. and BURGER, A. (1963a), *Helv. physiol. Acta* **21**, 59.

RICHTERICH, R. and BURGER, A. (1963b), *Enzymol. biol. clin.* **3**, 65.

RICHTERICH, R., BURGER, A. and WEBER, H. (1962), *Helv. physiol. Acta* **20**, C78.

RICHTERICH, R., GAUTIER, E., ZUPPINGER, K., EGLI, W. and ROSSI, E. (1961), *Klin. Wschr.* **39**, 346.

RICHTERICH, R., LOCKER, J., ZUPPINGER, K. and ROSSI, E. (1962), *Schweiz. med. Wschr.* **92**, 919.

RINGOIR, S. (1967), *LDH Isoënzymen bij nieraandoeningen experimentele en klinische studie*, Brussels: Arscia.

RINGOIR, S. and WIEME, R. J. (1965), *Lancet* **ii**, 906.

ROSA, J. and SCHAPIRA, F. (1964), *Nature, Lond.* **294**, 883.

ROSALKI, S. B. (1962), *Clin. chim. Acta* **8**, 415.

ROSALKI, S. B. (1963), *Brit. Heart J.* **26**, 795.

ROSALKI, S. B. and SINCLAIR, L. (1965), personal communication.

ROSALKI, S. B. and WILKINSON, J. H. (1960), *Nature, Lond.* **188**, 1110.

ROSALKI, S. B. and WILKINSON, J. H. (1964), *J. Amer. med. Assoc.* **189**, 61.

SAIFER, A., SCHNECK, L., PERLE, G. and VOLK, B. (1969), *Neurology, Minneapolis* **19**, 147.

SAYRE, F. W. and HILL, B. R. (1957), *Proc. Soc. exp. Biol., N.Y.* **96**, 695.

SCHAPIRA, F. and DEMOS, J. (1962), *Enzymol. biol. clin.* **2**, 45.

SCHAPIRA, F. and ROSA, J. (1967) *Rev. franç. d'Hémat.* **7**, 109.

SCHINDLER, R. and RICHTERICH, R. (1962), *Helv. physiol. Acta* **20**, C80.

SCHWERT, G. W., MILLER, B. R. and PEANASKY, R. J. (1967), *J. biol. Chem.* **242**, 3245.

SECCHI, G. C., MOSSA, R. and GALLITELLI, L. (1964), *Enzymol. biol clin.* **4**, 58.

SHAW, C. R. and BARTO, E. (1963), *Proc. Natl. Acad. Sci., Wash.* **50**, 211.

SHEPARD, T. H., GORDON, L. H. and WOLLENWEBER, J. E. (1965), *Nature, Lond.* **208**, 1107.

SHERMAN, I. W. (1962), *Trans. N.Y. Acad. Sci. Series 2*, **24**, 944.

SOETENS, A., KARCHER, D., VAN SANDE, M. and LÖWENTHAL, A. (1965), in *Enzymes in Clinical Chemistry*, ed. RUYSSEN, R. and VANDENDRIESSCHE, L., Amsterdam: Elsevier. p. 130.

STAGG, B. H. and WHYLEY, G. A. (1968), *Clin. chim. Acta* **22**, 521.

STAMBAUGH, R. and BUCKLEY, J. (1967), *J. biol. Chem.* **242**, 4053.

STAMBAUGH, R. and POST, D. (1966a), *Anal. Biochem.* **15**, 470.

STAMBAUGH, R. and POST, D. (1966b), *J. biol. Chem.* **241**, 1462.

STANISLAWSKI-BIRENCWAJG, M. and LOISILLIER, F. (1965), *European J. Cancer* **1**, 221.

STARKWEATHER, W. H., COUSINEAU, L., SCHOCH, H. K. and ZARAFONETIS, C. J. (1965), *Blood* **26**, 63.

STARKWEATHER, W. H., GREEN, R. A., SPENCER, H. H. and SCHOCH, H. K. (1966), *J. Lab. clin. Med.* **68**, 314.

STARKWEATHER, W. H. and SCHOCH, H. K. (1962), *Biochim. biophys. Acta* **62**, 440.
STARKWEATHER, W. H., SPENCER, H. H. and SCHOCH, H. K. (1966), *Blood* **28**, 860.
STRANDJORD, P. E. and CLAYSON, K. J. (1961), *J. Lab. clin. Med.* **58**, 962.
STRANDJORD, P. E., CLAYSON, K. J. and FREIER, E. F. (1962), *J. Amer. med. Assoc.* **182**, 1099.
SYNER, F. N. and GOODMAN, M. (1966), *Science*, **151**, 206.
TARMY, E. M. and KAPLAN, N. O. (1967a), *J. biol. Chem.* **243**, 2579.
TARMY, E. M. and KAPLAN, N. O. (1967b), *J. biol. Chem.* **243**, 2496.
THIELE, K. G. and MATTENHEIMER, H. (1968), *Z. klin. Chem.* **3**, 132.
TURNER, D. M. (1964), *Proc. Assoc. clin. Biochem.* **3**, 14.
VAN CAMP, K., DENIS, L. J. and VAN SANDE, M. (1963), in *Protides of the Biological Fluids*, ed. PEETERS, H., Amsterdam: Elsevier, Vol. 10, p. 45.
VAN DER HEIDEN, C., DESPLANQUE, J., STOOP, J. W. and WADMAN, S. K. (1968), *Clin. chim. Acta* **22**, 409.
VAN DER HELM, H. J., ZONDAG, H. A., HARTOG, A. P. and VAN DER KOOI, M. W. (1962), *Clin. chim. Acta* **7**, 540.
VAN DER HELM, H. J., ZONDAG, H. A. and KLEIN, F. (1963), *Clin. chim. Acta* **8**, 193.
VESELL, E. S. (1965a), *Science*, **150**, 1590.
VESELL, E. S. (1965b), *Science*, **148**, 1103.
VESELL, E. S. (1966), *Nature, Lond.* **210**, 421.
VESELL, E. S. and BEARN, A. G. (1957), *Proc. Soc. exp. Biol., N.Y.* **94**, 96.
VESELL, E. S. and BEARN, A. G. (1958a), *J. clin. Invest.* **37**, 672.
VESELL, E. S. and BEARN, A. G. (1958b), *Ann. N.Y. Acad. Sci.* **75**, 286.
VESELL, E. S. and BEARN, A. G. (1961), *J. clin. Invest.* **40**, 586.
VESELL, E. S. and BEARN, A. G. (1962a), *Proc. Soc. exp. Biol., N.Y.* **111**, 100.
VESELL, E. S. and BEARN, A. G. (1962b), *J. gen. Physiol.* **45**, 553.
VESELL, E. S., FRITZ, P. J. and WHITE, E. L. (1968), *Biochim. biophys. Acta* **159**, 236.
VESELL, E. S., OSTERLAND, K. C., BEARN, A. G. and KUNKEL, H. G. (1962), *J. clin. Invest.* **41**, 2012.
VESELL, E. S., PHILIP, J. and BEARN, A. G. (1962), *J. exp. Med.* **116**, 797.
VESELL, E. S. and YIELDING, K. L. (1966), *Proc. Natl. Acad. Sci., Wash.* **56**, 1317.
VESELL, E. S. and YIELDING, K. L. (1968), *Ann. N.Y. Acad. Sci.* **151**, 678.
WARBURG, O. and MINAMI, S. (1923), *Klin. Wschr.* **1**, 776.
WARBURTON, F. G. and SMITH, D. (1963), *Enzymologia* **26**, 125.
WARBURTON, F. G., SMITH, D. and LAING, G. S. (1963), *Nature, Lond.* **198**, 386.
WELSHMAN, S. G. and RIXON, E. C. (1968), *Clin. chim. Acta* **19**, 121.
WIELAND, T. and PFLEIDERER, G. (1957), *Biochem. Z.* **329**, 112.
WIELAND, T. and PFLEIDERER, G. (1961), *Ann. N.Y. Acad. Sci.* **94**, 691.
WIELAND, T., PFLEIDERER, G., HAUPT, I. and WÖRNER, W. (1959), *Biochem. Z.* **332**, 1.
WIELAND, T., PFLEIDERER, G. and ORTANDERL, F. (1959), *Biochem. Z.* **331**, 103.
WIEME, R. J. (1959), *Studies on Agar-Gel Electrophoresis*, Brussels: Arscia.
WIEME, R. J. (1963), *Nature, Lond.* **199**, 437.
WIEME, R. J. (1965). *Agar Gel Electrophoresis*, Amsterdam, Elsevier.
WIEME, R. J. and HERPOL, J. E. (1962), *Nature, Lond.* **194**, 287.
WIEME, R. J. and LAURYSSENS, M. J. (1962), *Lancet* i, 433.
WIEME, R. J., VAN HOVE, W. Z. and VAN DER STRAETEN, M. E. (1968), *Ann. N.Y. Acad. Sci.* **151**, 213.
WIEME, R. J. and VAN MAERCKE, Y. (1961), *Ann. N.Y. Acad. Sci.* **94**, 898.

WILKINSON, J. H. (1968), in *Enzymes in Urine and Kidney*, ed. DUBACH, U. C., Berne: Huber, p. 207.

WILKINSON, J. H., COOKE, K. B., ELLIOTT, B. A. and PLUMMER, D. T. (1961), *Biochem. J.* **80**, 29P.

WILKINSON, J. H. and ROSALKI, S. B. (1963), *Diag. Terap.* **1**, 309.

WILKINSON, J. H. and WITHYCOMBE, W. A. (1965), *Biochem. J.* **97**, 663.

WILSON, A. C., CAHN, R. D. and KAPLAN, N. O. (1963), *Nature, Lond.* **197**, 331.

WINER, A. D. and SCHWERT, G. W. (1958), *J. biol. Chem.* **231**, 1065.

WITHYCOMBE, W. A. (1965), Ph.D. Thesis, Univ. of London.

WITHYCOMBE, W. A., PLUMMER, D. T. and WILKINSON, J. H. (1965), *Biochem. J.* **94**, 384.

WITHYCOMBE, W. A. and WILKINSON, J. H. (1964), *Biochem. J.* **93**, 11P.

WOERNER, W. and MARTIN, H. (1961), *Klin. Wschr.* **39**, 368.

WRIGHT, E. J., CAWLEY, L. P. and EBERHARDT, L. (1966), *Am. J. clin. Path.* **45**, 737.

WRÓBLEWSKI, F. and GREGORY, K. F. (1960), *Proc. 4th. Int. Cong. clin. Chem. Edinburgh*, p. 62.

WRÓBLEWSKI, F. and GREGORY, K. F. (1961), *Ann. N.Y. Acad. Sci.* **94**, 912.

WRÓBLEWSKI, F., ROSS, C. and GREGORY, K. F. (1960), *New Engl. J. Med.* **263**, 531.

WUNTCH, T., CHEN, R. F. and VESELL, E. S. (1970), *Science* **167**, 63.

WÜST, H., SCHÖN, H. and BERG, G. (1962), *Klin. Wschr.* **40**, 1169.

YAKULIS, V. J., GIBSON, C. W. and HELLER, P. (1962), *Am. J. clin. Path.* **38**, 378.

YASIN, R. and BERGEL, F. (1965), *European J. Cancer* **1**, 203.

ZINKHAM, W. H. (1968), *Ann. N.Y. Acad. Sci.* **151**, 598.

ZINKHAM, W. H., BLANCO, A. and KUPCHYK, L. (1963), *Science*, **142**, 1303.

ZINKHAM, W. H., BLANCO, A. and KUPCHYK, L. (1964), *Science*, **144**, 1353.

ZINKHAM, W. H., KUPCHYK, L., BLANCO, A. and ISENSEE, H. (1965), *Nature, Lond.* **208**, 284.

ZINKHAM, W. H., KUPCHYK, L., BLANCO, A. and ISENSEE, H. (1966), *J. exp. Zool.* **162**, 45.

ZONDAG, H. A. (1965), in *Enzymes in Clinical Chemistry*, ed. RUYSSEN, R. and VANDENDRIESSCHE, L., Amsterdam: Elsevier, p. 120.

ZONDAG, H. A. and KLEIN, F. (1968), *Ann. N.Y. Acad. Sci.* **151**, 578.

CHAPTER 7

Multiple Molecular Forms of Other Oxidoreductases

Oxidoreductases other than lactate dehydrogenase have recently been shown to exist in several multimolecular forms, and in this chapter isoenzymes of the following will be considered: malate dehydrogenase, isocitrate dehydrogenase, glutamate dehydrogenase and alcohol ·dehydrogenase. Glucose 6-phosphate dehydrogenase, 6-phosphogluconate dehydrogenase and glyceraldehyde 3-phosphate dehydrogenase have been discussed in Chapter 5.

Malate dehydrogenase
Human serum malate dehydrogenase was separated by starch-block electrophoresis into three distinct fractions by Vesell and Bearn (1958), and several other investigators have since demonstrated its heterogeneity in tissues from various species of animals, plants and microorganisms.

Mitochondrial and supernatant malate dehydrogenases
The mitochondrial malate dehydrogenase of mammalian tissues has been shown to differ from the corresponding cell-sap enzyme in electrophoretic mobility (Wieland, Pfleiderer, Haupt and Wörner, 1959), kinetics (Delbrück, Zebe and Bücher, 1959; Delbrück, Schimassek, Bartsch and Bücher, 1959) and in its relative ability to utilize coenzyme analogues (Thorne, 1960). Grimm and Doherty (1961) have prepared highly purified specimens of both enzymes from ox heart by successive acetone precipitation, ammonium sulphate and ethanol fractionation, ion-exchange chromatography and finally starch-block electrophoresis. The cell-sap (soluble) enzyme migrates towards the anode when subjected to electrophoresis in citrate buffer at pH 6·25, whereas the mitochondrial enzyme moves towards the cathode. About 20–30% of the total activity of ox-heart homogenates occurs in the soluble fraction.

The purified mitochondrial enzyme consists of several electrophoretically distinct components. Two active fractions have been detected by the moving-boundary technique in phosphate buffer at pH 7·5, but ultracentrifuge studies indicated only a single symmetrical peak (Grimm and Doherty, 1961). The mitochondrial enzyme from pig heart has also been shown to be electrophoretically heterogeneous, four fractions being detected on starch gel, but the corresponding cell-sap enzyme exhibits only one band (Thorne, Grossman and Kaplan, 1963). Horse-heart and pig-heart mitochondrial enzymes have molecular weights of 50,000–60,000 and contain few aromatic amino-acid residues. The *N*-terminal alanine residue was the only one detected by Thorne (1962).

Ox-heart mitochondrial malate dehydrogenase is inhibited by *p*-chloromercuribenzoate, while the cell-sap enzyme is not affected (Wolfe and Neilands, 1956; Siegel and England, 1961).

Substrate analogue studies with the monofluoro- and difluoro-derivatives of oxaloacetate have provided a means not only for distinguishing between the cytoplasmic and mitochondrial forms of the enzyme but also for differentiating between the enzymes of different tissues. Kun and Volfin (1966) found the rat-liver cytoplasmic enzyme to utilize the fluoro-derivatives almost as effectively as oxaloacetate, while the corresponding kidney enzyme is appreciably less active with oxalo-monofluoroacetate, but readily reduced the difluoro-derivative. The liver and kidney mitochondrial forms reduce the difluoro-derivative at about 30% and 14% of the rate with oxaloacetate respectively, but both are almost completely inactive with the monofluoro-derivative.

TABLE 28 *Ratios of activities with coenzyme analogues of mitochon-drial and soluble malate dehydrogenase*

	APAD/NAD ratios		
	Mitochondrial enzyme	Soluble enzyme	References
Rat liver	1·05	0·83	Thorne (1960)
Ox heart	6·3	3·5	Grimm and Doherty (1961)

Both forms react with NAD and several of its analogues. Some typical figures showing the relative rates of reduction of APAD (Table 18) and NAD with malate as substrate are illustrated in Table 28. These observations have been greatly extended by Kaplan and Ciotti (1961), who observed that there is a considerable difference between the figures found for the soluble and mitochondrial enzymes of various rabbit tissues. The ratios for several pairs of coenzyme analogues for the soluble enzymes of various tissues, however, are in general very similar to each other. This is also the case with the mitochondrial enzymes, but an exceptional tissues is the heart, in which both soluble and mitochondrial enzymes appear to be distinct from those of other tissues.

Kaplan and Ciotti (1961) also report ratios of activities with various pairs of coenzyme analogues for the soluble enzymes of a number of tissues from the snail, the clam and the octopus. They suggest that malate dehydrogenase is differentiated to a greater extent in the tissues of invertebrates than of vertebrates.

There is some indication that the mitochondrial and cell-sap enzymes are immunochemically distinct, for antisera prepared in rabbits immunized with ox-heart cell-sap malate dehydrogenase give diffusion bands only with the specific antigen when tested by the agar-diffusion technique (Grimm and Doherty, 1961). The mitochondrial enzyme antiserum, however, gives diffusion bands with both enzymes, an observation which led these investigators to conclude that the response to this antiserum is probably the result of an immune reaction to non-enzyme protein.

Some mammalian malate dehydrogenase fractions are NADP-dependent. Various rat tissues, for example, show three bands of activity on cellulose acetate electrophoresis. Bands 1 and 3 are NAD-dependent and are inhibited by oxaloacetate, while band 2 requires NADP and is not sensitive to oxaloacetate (Sawaki, Morikawa and Yamada, 1965). Two electrophoretically separable forms of an NADP-dependent malate dehydrogenase have recently been detected in the supernatant fractions of various mouse and pig tissues (Henderson, 1968).

Although the catalytic, immunochemical, chromatographic and electrophoretic differences between the cell-sap and mitochondrial enzymes imply that the two forms are quite distinct, they appear to be structurally related at least in rat liver. The conversion of the cell-sap form of rat-liver malate dehydrogenase into the mitochondrial form by treatment with butanol suggests that the soluble enzyme might be a complex of the mitochondrial form with a lipoprotein (Sophianopoulos and Vestling, 1960).

The structure of malate dehydrogenase isoenzymes has been discussed in Chapter 4 (p. 82).

Biological significance of malate dehydrogenase isoenzymes
The Michaelis constants and turnover numbers of the mitochondrial and soluble forms of malate dehydrogenase (Table 29) show well-marked differences which suggest that the mitochondrial enzyme is better suited for the oxidation of malate, while the soluble enzyme is a more efficient catalyst for the reverse reaction. These observations

might well be related to their biological functions, since both forms are strongly inhibited by excess substrate, but whereas the mitochondrial enzyme is particularly sensitive to oxaloacetate, excess malate has a greater inhibitory effect upon the soluble enzyme. Kaplan (1961, 1963) has concluded therefore that these characteristics may prevent the reduction of oxaloacetate in mitochondria and the oxidation of malate in the soluble fraction.

TABLE 29 *Michaelis constants and turnover numbers for purified mitochondrial and soluble ox-heart malate dehydrogenases*

(Grimm and Doherty. 1961)

	Mitochondrial enzyme	Soluble enzyme
K_m (malate)	$9 \cdot 9 \times 10^{-4}$M	$5 \cdot 4 \times 10^{-4}$M
K_m (oxaloacetate)	$4 \cdot 0 \times 10^{-5}$M	$5 \cdot 1 \times 10^{-5}$M
Maximal turnover number (malate and NAD) (moles NAD reduced/min./mole enzyme)	35,000	20,000
Maximal turnover number (oxaloacetate and NADH) (moles NADH oxidized/min./mole enzyme)	59,000	72,000

These possibilities may play an important part in the energy relationships of mitochondria, and Kaplan envisages oxidation of soluble NADH by oxaloacetate to form malate, which enters the mitochondria, where it is again oxidized by the mitochondrial enzyme and mitochondrial bound NAD. The oxaloacetate produced might then be released into the cytoplasm, so enabling the cycle to be repeated. Oxidation of the mitochondrial NADH might be coupled with the formation of ATP, as shown in Fig. 48, in a scheme similar to that suggested for the soluble and mitochondrial forms of α-glycerol phosphate dehydrogenase (Bücher and Klingenburg, 1958; Sacktor, 1958).

Munkres and his co-workers have made a number of highly significant observations during their extensive studies on the mitochondrial malate dehydrogenase of *Neurospora*. This enzyme has a molecular weight of about 54,000 and consists of four sub-units. It reversibly dissociates into an $\alpha\alpha\alpha$-trimer and a β-monomer (Munkres, 1965a). In this species there is a remarkable similarity between malate dehydrogenase protein and that of aspartate aminotransferase, so much so that their primary and quaternary structures appear to be identical. They differ in conformation, probably owing to coenzyme binding (Munkres, 1965b).

Mutations of the enzyme not affecting the active centre produce abnormal conformations which lead to distortion of the active centre when the enzyme becomes associated with the mitochondrial membrane (Munkres and Woodward, 1966).

Up to five isoenzymes may be separated by acrylamide-gel electrophoresis, the major band being the slowest at pH 9. The various components have similar molecular weights, coenzyme binding, sub-unit types and catalytic properties, and appear to be conformational isoenzymes, since all enzyme preparations give the $\alpha\alpha\alpha$-trimer and the β-monomer in 3:1 ratios, and mutant forms always exhibit the same

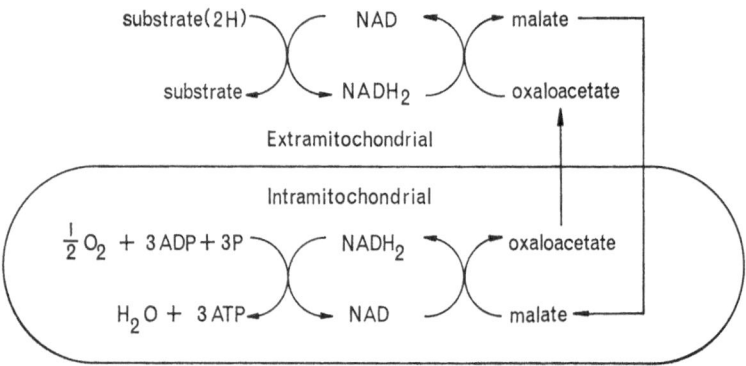

Fig. 48. Possible functions of mitochondrial and soluble malate dehydrogenase. (Kaplan, 1961.)

relative electrophoretic mobilities (Munkres, 1965*a*, 1968). Munkres (1968) has suggested that *Neurospora* malate dehydrogenase is an allosteric enzyme and that such allosteric effectors as pyridoxal phosphate, aspartate and mitochondrial structural protein produce marked conformational changes, such as to lead to the conformers occupying different sites on the mitochondrial membrane.

The results of these studies are clearly of considerable importance in correlating biochemical and structural concepts, but detailed discussion is beyond the scope of this monograph and the reader is referred to the review by Munkres (1968).

Malate dehydrogenase isoenzymes in mammalian tissues
Since Vesell and Bearn (1958) first established the heterogeneity of malate dehydrogenase in human serum and erythrocytes, the distribution of this enzyme in mammalian tissues has been investigated by

others. Various techniques have been employed, including the determination of enzyme activities in eluates after starch-block electrophoresis (Vesell and Bearn, 1958), 'enzymoelectrophoresis' (see Chapter 3) in agar gel (Löwenthal, Van Sande and Karcher, 1961; Kamarýt and Zázvorka, 1964) and tetrazolium staining procedures (Markert and Møller, 1959; Yakulis, Gibson and Heller, 1962). It seems that variations in the sensitivities of the techniques used are responsible for discrepancies in the number of isoenzymes reported, for while Yakulis *et al.* claim only four fractions, Löwenthal *et al.* and Kamarýt and Zázvorka detected as many as six zones of activity.

A series of human tissues was examined by Yakulis *et al.* (1962), who, in all cases except the erythrocytes, found identical patterns with malate dehydrogenase concentrated in two bands with mobilities similar to those of LD_3 and LD_4 (Fig. 9). Three bands, two of which were faintly staining, faster-moving anodic fractions, were found in haemolysates. These investigators were unable to detect any specific changes in the sera of patients with elevated serum malate dehydrogenase activities and conclude that determination of the isoenzyme pattern for this enzyme has little value in diagnosis at present.

A genetic variant of malate dehydrogenase has been observed in the erythrocytes of members of a Negro family studied by Davidson and Cortner (1967). After starch-gel electrophoresis the variant pattern consists of three major bands, the fastest of which has the same mobility as the major normal band, while the other two bands migrate more slowly. The two minor bands found in the usual pattern also appear in the variant. The supernatant leucocyte enzyme was found to exhibit the variant pattern, but the mitochondrial enzyme was normal.

Since Davidson and Cortner (1967) obtained a pattern of three bands after *in vitro* hybridization of purified samples of the normal supernatant enzyme and the slowest variant, it seems that the enzyme molecule is a dimer. Dissociation of a dimeric molecule into sub-units followed by random reassociation would be expected to produce three isoenzymes. The authors conclude that the variant pattern arose through mutation of an allele controlling the synthesis of one of the sub-units of the enzyme molecule.

MD_4, which has a mobility in agar gel similar to that of a slow β-globulin, was found by Kamarýt and Zázvorka (1964) to be the principal component of the malate dehydrogenase of homogenates of human heart and liver and of normal blood serum. Sera from patients with virus hepatitis give patterns resembling those of liver homogenates in containing increased amounts of the cathodic fraction, MD_6, but the patterns are not so clear-cut and distinctive as those of the LD isoenzymes discussed in Chapter 6.

A simple method for the separation of malate dehydrogenase isoenzymes by chromatography on DEAE-Sephadex has been devised by Schmidt, Schmidt and Möhr (1967), who used it to determine that the enzyme of rat liver consists of about 62% of the cytoplasmic form and 38% of the mitochondrial enzyme. The same technique has been applied to the study of the serum enzyme in patients with hepatitis. Both fractions are elevated during the acute phase of the illness, but during recovery the mitochondrial fraction remains abnormal for a much longer period. This is attributed to the relatively longer half-life of mitochondrial malate dehydrogenase in serum (Schmidt, Schmidt and Otto, 1967).

Although the malate dehydrogenase of the brain has received much attention, the results so far are rather inconclusive. A possible explanation of some of the anomalies has recently been suggested by Johnson (1962), who noted that 45% of the activity of rat-brain homogenates appeared in the soluble fraction, while 40% was associated with the mitochondria. About 20% of the particulate activity, however, may be released on treatment with water, and Johnson considers that this may be cytoplasmic enzyme in pieces of incompletely disintegrated axon occluded in the mitochondrial fraction. He concludes therefore that there is only one particulate and one soluble malate dehydrogenase in rat brain.

Löwenthal *et al.* (1961) separated the malate dehydrogenase isoenzymes of human and sheep brain and showed that the patterns, which exhibit six zones of activity, are very similar to those found in the corresponding cerebrospinal fluid and serum, though the relative electrophoretic mobilities of the sheep fractions differ considerably from those of their human counterparts. In both species the isoenzyme patterns obtained with white matter closely resemble those found with grey matter. In this connection it is interesting to note that Van der Helm (1962*a*) found no difference between the isoenzyme patterns of different parts of the human brain.

Both forms of the NADP-dependent malate dehydrogenase of mouse tissues migrate towards the cathode on starch-gel electrophoresis at pH 5. The faster-moving component is found in the liver, brain, lung, skeletal muscle and adipose tissue, while the slower form occurs in the heart. The kidney contains both isoenzymes. In the pig, however, the liver and brain forms migrate towards the anode at pH 5, while that from the heart is cathodic (Henderson, 1968).

Henderson (1968) has also reported the existence of genetically determined variants of the liver NADP-dependent enzyme, differing in their electrophoretic mobilities, in certain inbred strains of mice. Homozygotes for the 'fast' and 'slow' traits show single fast or slow bands respectively, but the heterozygote, produced by crossing the two strains, exhibits intermediate bands (Fig. 49).

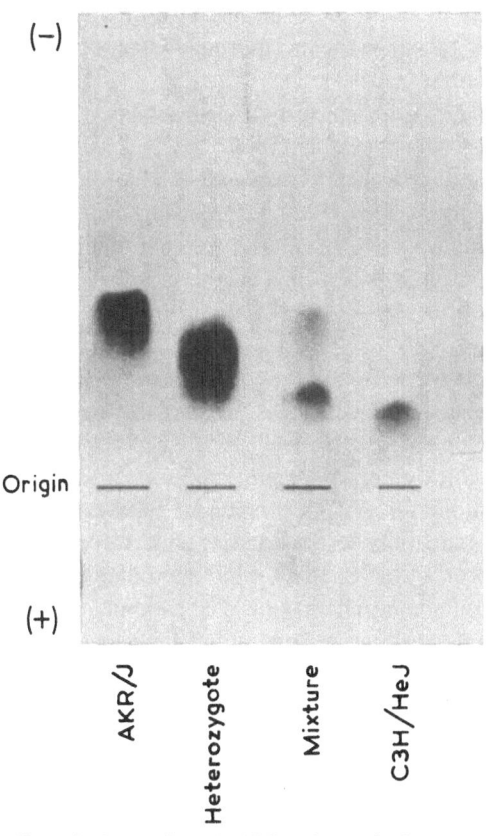

Fig. 49. Mouse-liver NADP-malate dehydrogenase isoenzymes separated by starch-gel electrophoresis showing two homozygous and one heterozygous phenotype. (Henderson, 1968.)

Malate dehydrogenase isoenzymes in insect development

Changes in the pattern of malate dehydrogenase isoenzymes occur during development in insects (Laufer, 1961). Blood from the diapausing pupa of *Cecropia* contains at least two isoenzymes, while that of the developing adult contains three or more; there is also a shift in the relative activities towards more acidic fractions. During the larval development of *Cynthia* there is a gradual increase in an electrophoretically fast anodic isoenzyme, while a fraction of intermediate mobility diminishes and is replaced by a slow anodic fraction shortly before pupation. During diapause enzyme activity falls to low levels, but subsequently there occurs an increase in the fast and slow isoenzymes which alone are found in the adult insect. Rather similar but less well-marked changes occur in the lactate and α-glycerol phosphate dehydrogenase isoenzymes.

Multiple forms of malate dehydrogenase also occur in the embryos of sea urchins, which exhibit three bands on acrylamide-gel electrophoresis. Unfertilized eggs, on the other hand, contain five fractions (Moore and Villee, 1963). The change occurs within about 90 min. of fertilization, and Villee (1968) has suggested that it may be related to changes in respiration and protein synthesis before the first cell division in the embryo.

Malate dehydrogenase isoenzymes of the chicken embryo
A total of six anionic isoenzymes and one cationic fraction has been observed by Conklin and Nebel (1965) when homogenates of various tissues of the chick embryo are subjected to starch-gel electrophoresis. The fastest (anionic) migrating component behaves like the supernatant enzyme in its sensitivity to urea and organic solvents and in its resistance to *p*-chloromercuribenzoate, while the other five anionic fractions resemble the mitochondrial enzyme. The cationic component, however, has properties quite different from either the supernatant or mitochondrial forms. The authors therefore conclude that there are three different types of avian malate dehydrogenase proteins, but it is not at present known why the slower anionic form splits into five bands on electrophoresis.

Isocitrate dehydrogenase
Isocitrate dehydrogenase, which catalyses the oxidation of isocitrate to α-oxoglutarate in the presence of manganous (Mn^{2+}) ions, occurs in two distinct forms, which for the purposes of this monograph may be regarded as two quite separate enzymes. The NAD-dependent enzyme found in yeast requires $5'$-adenosine monophosphate (AMP), whereas the NADP-dependent form which predominates in animal tissues does not (Kornberg and Pricer, 1951; Vignais and Vignais, 1961). There is considerable evidence to suggest that the two forms have different func-

tions (Kaplan, Swartz, Frech and Ciotti, 1956; Kaplan, 1963), but it is with the enzyme requiring NADP (L_s-isocitrate:NADP oxidoreductase, E.C. 1.1.1.41) that we are here concerned, for there have been several reports of its occurrence in multiple molecular forms.

TABLE 30 *Distribution of isocitrate dehydrogenase isoenzymes in extracts of rat tissues*

(Baron and Bell, 1962)

(0 = not seen, + = low activity, + + = high activity)

Tissue	ICD_1	ICD_2	ICD_3	ICD_4
Heart muscle	0 or +	+	0	+ +
Skeletal muscle	0 or +	+ +	0	+ +
Liver *	0	+ +	0 or + +	0
Kidney *	0	+ +	0 or + +	0
Spleen	+ +	+ +	0	0
Placenta	+	+ +	0	+
Brain	0	+ +	0	0
Testis †	0	+ +	0	+ +
Erythrocytes	0	+ +	0	0
Serum	0	+ +	0	0

* Some preparations of liver and kidney are found to contain both ICD_2 and ICD_3, while others contain only ICD_2.

† The ICD_4 of testis runs faster than the ICD_4 of other tissues.

Rat tissues contain three or four isoenzymes which may be separated by starch-gel electrophoresis and subsequently detected by staining with neotetrazolium (Tsao, 1960; Bell and Baron, 1962). Although better separation may be obtained at pH 8·6, considerable loss of activity occurs under these conditions, and recoveries are higher at pH 7·4. One of the fractions (ICD_4) moves towards the cathode at pH 8·6, while the fastest fraction (ICD_1) has a mobility similar to that of serum albumin (Baron and Bell, 1962). The distribution of the four fractions in rat tissues is summarized in Table 30. The liver enzyme appears to have a greater affinity for isocitrate than the heart enzyme, for Baron and Bell (1962) obtained values of 2×10^{-6}M and 5×10^{-6}M respectively for their approximate Michaelis constants.

There is some discrepancy in the numbers of isocitrate dehydrogenase isoenzymes found in human tissues, for while Baron and Bell (1962) found four bands in serum, only three were detected in tissue extracts. Campbell and Moss (1962), however, using different techniques, found only two peaks of activity in human heart and liver extracts after starch-gel electrophoresis at pH 6·2. These investigators eluted the enzyme and determined its activity by spectrofluorimetry. The fast and slow components from both tissues have similar mobilities, but the fast component preponderates in the liver, whereas heart extracts contain much larger amounts of the slow fraction. The slow fraction is much less stable than the fast isoenzyme, and incubation at $37°$ for 4 hr. before electrophoresis leads to a considerable increase in the ratio of fast to slow component in both tissue extracts. The Michaelis constant for the

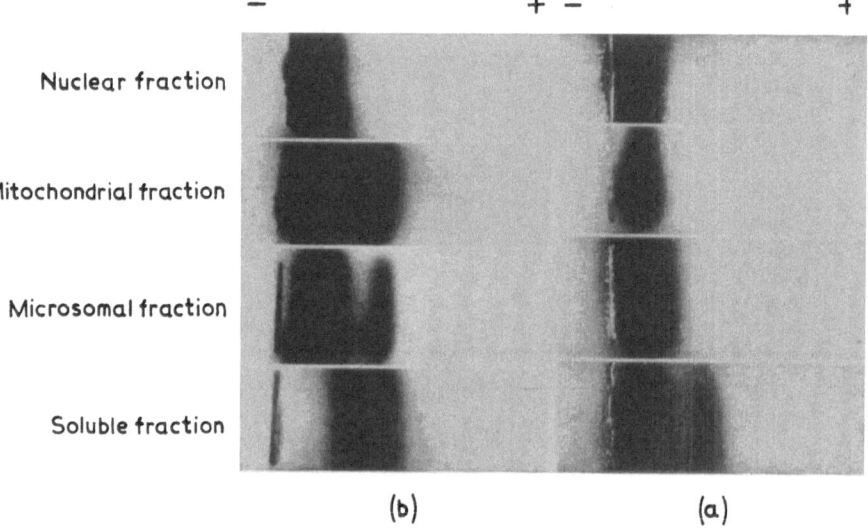

Nuclear fraction

Mitochondrial fraction

Microsomal fraction

Soluble fraction

(b)   **(a)**

Fig. 50. Subcellular distribution of isocitrate dehydrogenase isoenzymes in: (*a*) rat heart, and (*b*) rat liver. (Bell and Baron, 1968.)

slow band of heart was found by Campbell and Moss (1962) to be 3×10^{-4}M for isocitrate, while the K_m value for the corresponding band of liver was 1×10^{-3}M, an observation which according to these authors suggests that the slow bands of the two tissues may not be identical.

The supernatant and mitochondrial forms of isocitrate dehydrogenase have been shown to differ in their immunochemical properties (Lowenstein and Smith, 1962) and in their electrophoretic mobilities (Bell and Baron, 1964; Henderson, 1965). A recent study of carefully purified sub-cellular fractions of rat heart and rat liver by Bell and Baron (1968) has shown that the soluble fraction of the heart contains both the fast and slow isoenzymes, with the latter predominating. The heart particulate fractions, however, contain only the slow component. By contrast, the liver supernatant fraction contains only the fast isoenzyme, while the mitochondria and microsomes contain both bands and the liver nuclear fraction only the slow band (Fig. 50).

Henderson (1965, 1968) has described two allelic forms of the supernatant enzyme in the livers of inbred strains of mice. The single bands of activity exhibited by the homozygotes differ in their electrophoretic mobilities. The heterozygote contains three isoenzymes, two of which are identical with those of the homozygotes, while the third, which has a mobility intermediate between the other two forms, shows greatest activity. Henderson suggests therefore that in this set of genetic variants the enzyme molecule consists of two different types of sub-unit, the homozygous forms being represented by *aa* and *bb*, and the heterozygote by *ab*.

Heterogeneity of NADP-specific isocitrate dehydrogenase has also been recognized in bacteria. Reeves, Brehmeyer and Ajl (1968) described electrophoretically distinct forms of the enzyme in *E. coli* according to the conditions under which the cells are cultured. The enzyme from cell-free extracts of the same organism has now been resolved by CM-Sephadex chromatography into two well-separated peaks of activity, which differ markedly in their sensitivity to inhibition by 2M-urea.

Isocitrate dehydrogenase isoenzymes in heart and liver disease
Recognition that NADP-dependent isocitrate dehydrogenase occurs in several forms has provided an explanation for the oft-reported failure to observe an increase in the serum level after myocardial infarction, despite the fact that heart muscle is rich in this enzyme. Other enzymes present in heart muscle, e.g. lactate dehydrogenase and aspartate aminotransferase, are released into the serum, and increased activities persisting for several days are usually detectable after an infarct. By contrast, damage to liver cells which also contain appreciable amounts of these enzymes is reflected by an increase in the serum levels of all three.

Failure to detect raised isocitrate dehydrogenase activity in the serum after a clinical episode of myocardial infarction has been attributed to rapid clearance of the enzyme from the serum, since transient rises may be observed after experimental cardiac infarction in dogs, and normal levels are restored within a few hours, and since infarcted heart muscle contains less of the enzyme than normal myocardium (White, 1960). The mechanism for the removal of the enzyme is not known at present, but Strandjord, Thomas and White (1960) found that purified isocitrate dehydrogenase injected into dogs rapidly disappears from the circulation, while lactate dehydrogenase similarly administered is removed at a much slower rate. The excretory systems do not appear to be concerned, since in neither case is the disappearance rate affected by hepatectomy, nephrectomy or splenectomy.

The anomaly now appears to be related to the isoenzyme composition of the isocitrate dehydrogenase of the two organs, for the electrophoretically slow, unstable form predominates in the heart, while the liver enzyme consists primarily of the electrophoretically fast, stable form (Campbell and Moss, 1962). Thus, after myocardial infarction the unstable heart enzyme soon loses its activity, while in hepatitis the liver form retains its activity long enough for high serum levels to be detected.

Glutamate dehydrogenase

The glutamate dehydrogenases of human tissues may be resolved by agar-gel electrophoresis into five (or possibly six) separate isoenzymes, which may be detected by a tetrazolium-staining technique (Van der Helm, 1962b). Their relative mobilities and tissue distributions are listed in Table 31. Previous attempts employing starch-gel electrophoresis failed owing to the inability of the enzyme to penetrate the gel (Markert and Møller, 1959; Tsao, 1960).

TABLE 31 *Glutamate dehydrogenase isoenzymes of human tissues separated by agar-gel electrophoresis*

(Van der Helm, 1962b)

Fraction	1	2	3	4	5	6
Relative mobility	0·95	0·74	0·46	0·20	0·0	−0·12
Liver	+	+ + +	+	+	+ +	+
Heart	+ +	+ + +	+	+	+	
Kidney	+ + +	+ + +	+	+	+	
Spleen	+ + +	+ + +	+	+ +	+ +	

Glutamate dehydrogenase occurs in both NAD- and NADP-dependent forms in mammalian tissues and in yeast and *Neurospora* (Doherty, 1962; Samwal and Lata, 1962). At the present time it is uncertain how far these should be regarded as isoenzymes, and it may well be that the various forms are separate adaptive enzymes produced in response to changes in environment.

Glutamate dehydrogenase heterogeneity has also been recognized in plants, particularly legumes. Thurman, Palin and Laycock (1965) found up to seven bands of activity in the cotyledons, radicles and shoots of the broad bean (*Vicia Faba*) and the pea (*Pisum sativum*), while Grimes and Fottrell (1966) described three bands in the root nodules of *Pisum sativum*, clover (*Trifolium repens*) and lucerne (*Medicago sativa*).

Alcohol dehydrogenase

The first report of the heterogeneity of alcohol dehydrogenase was that of Watts and Donninger (1962), who found the crystalline yeast enzyme to contain no fewer than 18 electrophoretically distinct protein components, of which five exhibited enzyme activity. Only two fractions, however, were detectable when all metal ions were excluded from the preparation. Several distinct fractions have since been observed in alcohol dehydrogenases from other species, including insects, birds, mammals and plants.

Alcohol dehydrogenases of insects

The alcohol dehydrogenases of *Drosophilia melanogaster* have attracted much attention from the genetic and structural viewpoints, since homozygous inbred stocks were found to exhibit two different electrophoretic patterns. Johnson and Denniston (1964) found two isoenzymes in each of the homozygotes and five in heterozygotes produced by crossing. Other investigators have found sets of three isoenzymes in homozygous strains: a slow form in the Canton S and Oregon-RC inbred flies and a fast form in Swedish-6 and other wild-type flies. Hybrid flies displayed nine bands of activity on acrylamide-gel electrophoresis but only seven on agar gel (Grell, Jacobson and Murphy, 1965; Ursprung and Leone, 1965). It was originally considered that the multiplicity of bands observed in heterozygous insects was the result of new sub-unit combinations, but more recent work suggests that this mechanism is unlikely.

Grell, Jacobson and Murphy (1968) found that when purified preparations were subjected to acrylamide gel electrophoresis each homozygous strain exhibits five components, the slow isoenzymes being designated S1 to S5 and the fast isoenzymes F1 to F5. Although there was some overlap between S3 and F5 and between S1 and F3, incorporation of NAD into the gel altered the electrophoretic mobilities and enabled these pairs to be differentiated. The three-membered patterns of the crude extracts consisted of bands 1, 3 and 5, 2 and 4 not being detected. Bands 1, 3 and 5 of both fast and slow series were found to differ in their thermal stabilities.

Electrophoretic variants exhibiting chromosomal deficiencies including the alcohol dehydrogenase locus have been produced by X-irradiation, while injection with the chemical mutagen, ethyl methanesulphonate, led to the formation of new mutant electrophoretic patterns. Genetic studies with these mutants led Grell *et al.* (1968) to the conclusion that, in each homozygote, each isoenzyme is the product of a particular gene rather than the dimer of two sub-units. Ursprung and Carlin (1968) have demonstrated by *in vitro* studies involving dialysis against β-mercaptoethanol or treatment with NAD that the formation of additional bands is not consistent with the sub-unit hypothesis. They suggest that combination with the coenzyme plays a part in establishing the heterogeneity of alcohol dehydrogenase in this species. Strong support for this view is provided by the demonstration of Jacobson (1968) that the fast and slow isoenzymes are interconvertible. The heat-labile slowest form is converted into the heat-stable fastest form by treatment with NAD, while the reverse conversion is brought about by adsorption on DEAE-cellulose.

Avian alcohol dehydrogenases
In contrast to the single band of alcohol dehydrogenase activity found in the chicken, the Japanese quail, *Coturnix coturnix japonicus*, exhibits three distinct phenotypes, recognizable by starch-gel electrophoresis. Castro-Sierra and Ohno (1968) observed that in the quail each phenotype produces a single band (A, B or C), from which they conclude that in this species three alleles coexist at the alcohol dehydrogenase locus. In both the chicken and the quail the various isoenzymes are strongly positively charged and migrate towards the cathode at pH 8·7.

Three other phenotypes representing the heterozygous types, AB, AC and BC, were also recognized, but Castro-Sierra and Ohno (1968) were uncertain whether isoenzymes of intermediate electrophoretic mobility were produced in addition to the two parental bands which were already clearly demonstrated. Intermediate bands, however, were detected in hybrids between homozygous quails and the chicken.

Alcohol dehydrogenase in mammalian liver
In mammals alcohol dehydrogenase is almost exclusively confined to the liver, where its presence is something of a physiological curiosity, since its principal substrate, ethanol, is not normally ingested by animal species other than man. Reference has already been made (p. 50) to the probability that the 'nothing dehydrogenase' phenomenon is due to alcohol dehydrogenase (Shaw and Koen, 1965, 1967; Beutler, 1967).

Purified horse-liver alcohol dehydrogenase has been shown by McKinley-McKee and Moss (i965) to consist of four electrophoretically separable zones of activity, all of which migrate towards the cathode at pH 7·1. The most cationic band is numbered 1 but greatest activity is associated with band 2. Incorporation of NAD⁺ or NADH into the electrophoretic system causes the disappearance of band 2 and a substantial increase in the proportions of bands 3 and 4, but little alteration to band 1. NADH treatment also results in the production of a new band (5), which remains near the origin, a process which is accentuated by the presence of isobutyramide. The results are interpreted as indicating that the various bands represent enzyme–coenzyme complexes rather than true isoenzymes. The slow band 5 is regarded as a ternary enzyme–coenzyme–inhibitor complex. Band 1, which is relatively unaffected by the presence of coenzyme, disappears on treatment with phosphate and urea. It is suggested that this band may consist of an enzyme–coenzyme–buffer ion complex, though it is difficult to explain why combination with anions should produce a component which migrates towards the cathode at a faster rate.

The conclusions of McKinley-McKee and Moss (1965) are interesting in the light of more recent studies on the heterogeneity of the alcohol dehydrogenases of *Drosophila*, for which a similar explanation has been proposed (p. 217).

Although serum alcohol dehydrogenase determination has recently been recommended as a diagnostic test for hepatocellular necrosis and intrahepatic cholestasis (Mezey, Cherrick and Holt, 1968), human-liver alcohol dehydrogenase isoenzymes have so far received little attention. Von Wartburg, Papenberg and Aebi (1965), however, described their finding of an atypical alcohol dehydrogenase from specimens of two different human livers. Although the normal and atypical enzymes exhibit identical electrophoretic mobilities and similar chromatographic properties on CM-cellulose and DEAE-cellulose columns, they differ in several of their catalytic properties. With ethanol as substrate the atypical enzyme has a pH optimum at 8·5, whereas the corresponding figure for the normal enzyme is 10·8. Both enzymes have a pH optimum of 6·0–6·5 when acetaldehyde is used as substrate. The two enzymes also differ in their affinities for a series of short-chain aliphatic alcohols and in their sensitivities to a number of inhibitors. The normal enzyme is activated by thiourea, which inhibits the atypical form.

The same authors also report the isolation of two distinct fractions of alcohol dehydrogenase from the liver of the rhesus monkey by agar-gel electrophoresis and CM-cellulose chromatography. These fractions also differ in their substrate affinities and in their kinetic properties (Papenberg, von Wartburg and Aebi, 1965). These studies have recently been extended to other species by Moser, Papenberg and von Wartburg

(1968), who confirmed earlier findings of multiple bands of liver alcohol dehydrogenase in man, the horse and the rhesus monkey. They also described multiple components in the livers of cattle and chicken, but the liver alcohol dehydrogenase of the pig, rat and mouse appears to be homogeneous.

The occurrence of the atypical form of human-liver alcohol dehydrogenase has been confirmed by Edwards and Price Evans (1967), who found its properties to closely resemble those described by von Wartburg *et al.* (1965). Two of twenty-three patients exhibited the unusual form, but though each possessed about seven times the normal enzyme activity, their ethanol degradation rates were not strikingly different from those of patients with the typical form of the enzyme. Edwards and Price Evans (1967) therefore conclude that liver alcohol dehydrogenase is probably not a rate-limiting factor in human ethanol metabolism and that variations in alcohol tolerance do not appear to be related to the presence of the typical or atypical enzyme.

Alcohol dehydrogenase polymorphism in plants
An unusual form of genetic polymorphism of alcohol dehydrogenase has been reported in the kernel and scutellum of maize (Schwarz and Endo, 1966). Three alleles (F, S and C) are proposed to account for the occurrence of three different heterozygous phenotypes exhibiting three bands of enzyme activity on electrophoresis. Two single-banded homozygotes FF and SS have also been recognized. The presence of the C allele leads to the three heterozygous types: (*i*) CC, CS, SS; (*ii*) CC, CF, FF; and (*iii*) FF, FS, SS. The hybrid of CF and SS plants exhibits five bands, CC, CF, FF + CS, FS and SS. The C allele appears to consist of two components, a weak C(m) and a strong C(t), which specify enzymes with similar migration rates but different specific activities.

The existence of the homozygous 'fast' and 'slow' types of maize has been confirmed by Scandalios (1967), who also observed a faint faster band in each strain. In heterozygotes Scandalios found the major parental bands to be accompanied by a third band of intermediate mobility, but no such intermediate band occurred between the two parental minor fast-moving bands. He suggests that the latter may be monomers, since they do not hybridize, but agrees with Schwarz and Endo (1966) that the major bands are probably dimers.

Miscellaneous dehydrogenases
Kaplan and his co-workers have demonstrated by their abilities to utilize coenzyme analogues that several other dehydrogenases exhibit

heterogeneity (reviewed by Kaplan, 1963). Among these are the D-mannitol 1-phosphate dehydrogenases of various bacteria, especially strains of *B. subtilis* (S. Horwitz, quoted by Kaplan, 1963) and the glycol dehydrogenases of *Aerobacter aerogenes* (Lamborg and Kaplan, 1960). The lipoyl dehydrogenases of ox liver have been separated into six components by DEAE-Sephadex chromatography, and since the same fractions may also be detected when mild methods of isolation are used, they do not appear to be artifacts (Lusty, 1963).

The xanthine dehydrogenase of *Drosophila melanogaster* has been obtained in two chromatographically and electrophoretically distinguishable forms. Fresh homogenates contain a single component (XD-I), but partially purified extracts contain a second isoenzyme (XD-II). XD-I is converted into a form like XD-II by incubation with fly extracts, and it is suggested that the converter molecule is a large protein, possibly an esterase (Shinoda and Glassman, 1968).

REFERENCES

BARON, D. N. and BELL, J. L. (1962), *Proc. Assoc. clin. Biochem.* **2**, 8.
BELL, J. L. and BARON, D. N. (1962), *Biochem. J.* **82**, 5P.
BELL, J. L. and BARON, D. N. (1964), *Biochem. J.* **90**, 8P.
BELL, J. L. and BARON, D. N. (1968), *Enzymol. biol. clin.* **9**, 393.
BEUTLER, E. (1967), *Science*, **156**, 1516.
BÜCHER, T. and KLINGENBURG, M. (1958), *Angew. Chem.* **70**, 552.
CAMPBELL, D. M. and MOSS, D. W. (1962), *Proc. Assoc. clin. Biochem.* **2**, 10.
CASTRO-SIERRA, E. and OHNO, S. (1968), *Biochem. Genet.* **1**, 323.
CONKLIN, J. L. and NEBEL, E. J. (1965), *J. Histochem. Cytochem.* **13**, 510.
DAVIDSON, R. G. and CORTNER, J. A. (1967), *Nature, Lond.* **215**, 761.
DELBRÜCK, A., SCHIMASSEK, H., BARTSCH, K. and BÜCHER, T. (1959), *Biochem. Z.* **331**, 297.
DELBRÜCK, A., ZEBE, E. and BÜCHER, T. (1959), *Biochem. Z.* **331**, 273.
DOHERTY, M. D. (1962), *Federation Proc.* **21**, 57.
EDWARDS, J. A. and PRICE EVANS, D. A. (1967), *Clin. Pharmacol. Therap.* **8**, 824.
GRELL, E. B., JACOBSON, K. B. and MURPHY, J. B. (1965), *Science*, **149**, 80.
GRELL, E. B., JACOBSON, K. B. and MURPHY, J. B. (1968), *Ann. N.Y. Acad. Sci.* **151**, 441.
GRIMES, H. and FOTTRELL, P. F. (1966), *Nature, Lond.* **212**, 294.
GRIMM, F. C. and DOHERTY, D. G. (1961), *J. biol. Chem.* **236**, 1980.
HENDERSON, N. S. (1965), *J. exp. Zool.* **158**, 263.
HENDERSON, N. S. (1968), *Ann. N.Y. Acad. Sci.* **151**, 429.
JACOBSON, K. B. (1968), *Science*, **159**, 324.
JOHNSON, F. M. and DENNISTON, C. (1964), *Nature, Lond.* **304**, 906.
JOHNSON, M. K. (1962), *Biochem. J.* **84**, 25P.
KAMARÝT, J. and ZÁZVORKA, Z. (1964), *Clin. chim. Acta* **9**, 559.

KAPLAN, N. O. (1961), in *Mechanism of Action of Steroid Hormones*, ed. VILLEE, C. A. and ENGEL, A. A., Oxford: Pergamon Press, p. 247.

KAPLAN, N. O. (1963), *Bact. Rev.* **27**, 155.

KAPLAN, N. O. and CIOTTI, M. M. (1961), *Ann. N.Y. Acad. Sci.* **94**, 701.

KAPLAN, N. O., SWARTZ, M. N., FRECH, M. F. and CIOTTI, M. M. (1956), *Proc. Natl. Acad. Sci., Wash.* **42**, 481.

KUN, E. and VOLFIN, P. (1966), *Biochem. Biophys. Res. Commun.* **22**, 187.

LAMBORG, M. R. and KAPLAN, N. O. (1960), *Biochim. biophys. Acta* **38**, 272.

LAUFER, H. (1961), *Ann. N.Y. Acad. Sci.* **94**, 825.

LOWENSTEIN, J. M. and SMITH, S. R. (1962), *Biochim. biophys. Acta* **56**, 385.

LÖWENTHAL, A., VAN SANDE, M. and KARCHER, D. (1961), *Ann. N.Y. Acad. Sci.* **94**, 988.

LUSTY, C. J. (1963), *Federation Proc.* **22**, 241.

MCKINLEY-MCKEE, J. S. and MOSS, D. W. (1965), *Biochem. J.* **96**, 583.

MARKERT, C. L. and MØLLER, F. (1959), *Proc. Natl. Acad. Sci., Wash.* **45**, 753.

MEZEY, E., CHERRICK, G. R. and HOLT, P. R. (1968), *New Engl. J. Med.* **279**, 241.

MOORE, R. O. and VILLEE, C. A. (1963), *Science*, **142**, 389.

MOSER, K., PAPENBERG, J. and VON WARTBURG, J. P. (1968), *Enzymol. biol. clin.* **9**, 447.

MUNKRES, K. D. (1965a), *Biochemistry* **4**, 2180.

MUNKRES, K. D. (1965b), *Arch. Biochem. Biophys.* **112**, 347.

MUNKRES, K. D. (1968), *Ann. N.Y. Acad. Sci.* **151**, 294.

MUNKRES, K. D. and WOODWARD, D. O. (1966), *Proc. Natl. Acad. Sci., Wash.* **55**, 1217.

PAPENBERG, J., VON WARTBURG, J. P. and AEBI, H. (1965), *Biochem. Z.* **342**, 95.

REEVES, H. C., BREHMEYER, B. A. and AJL, S. J. (1968), *Science*, **162**, 359.

SACKTOR, B. (1958), *Proc. 4th. Int. Cong. Biochem. Vienna*, **12**, 138.

SAMWAL, B. D. and LATA, M. (1962), *Biochem. Biophys. Res. Comm.* **6**, 404.

SAWAKI, S., MORIKAWA, N. and YAMADA, K. (1965), *Nature, Lond.* **207**, 532.

SCANDALIOS, J. G. (1967), *Biochem. Genet.* **1**, 1.

SCHMIDT, E., SCHMIDT, F. W. and MÖHR, J. (1967), *Clin. chim. Acta* **15**, 337.

SCHMIDT, E., SCHMIDT, F. W. and OTTO, P. (1967), *Clin. chim. Acta* **15**, 283.

SCHWARZ, D. and ENDO, T. (1966), *Genetics* **53**, 709.

SHAW, C. R. and KOEN, A. L. (1965), *J. Histochem. Cytochem.* **13**, 431.

SHAW, C. R. and KOEN, A. L. (1967), *Science*, **156**, 1517.

SHINODA, T. and GLASSMAN, E. (1968), *Biochim. biophys. Acta* **160**, 178.

SIEGEL, L. and ENGLARD, S. (1961), *Federation Proc.* **20**, 239.

SOPHIANOPOULOS, A. J. and VESTLING, C. S. (1960), *Biochim. biophys. Acta* **45**, 400.

STRANDJORD, P. E., THOMAS, K. E. and WHITE, L. P. (1960), *J. clin. Invest.* **38**, 2111.

THORNE, C. J. R. (1960), *Biochim. biophys. Acta* **42**, 175.

THORNE, C. J. R. (1962), *Biochim. biophys. Acta* **59**, 624.

THORNE, C. J. R., GROSSMAN, L. I. and KAPLAN, N. O. (1963), *Biochim. biophys. Acta* **73**, 193.

THURMAN, D. A., PALIN, C. and LAYCOCK, M. V. (1965), *Nature, Lond.* **207**, 193.

TSAO, M. U. (1960), *Arch. Biochem.* **90**, 234.

URSPRUNG, H. and CARLIN, L. (1968), *Ann. N.Y. Acad. Sci.* **151**, 456.

URSPRUNG, H. and LEONE, J. (1965), *J. exp. Zool.* **160**, 147.

VAN DER HELM, H. J. (1962a), *J. Neurochem.* **9**, 325.

VAN DER HELM, H. J. (1962*b*), *Nature, Lond.* **194**, 773.

VESELL, E. S. and BEARN, A. G. (1958), *J. clin. Invest.* **37**, 672.

VIGNAIS, P. V. and VIGNAIS, P. M. (1961), *Biochim. biophys. Acta* **47**, 515.

VILLEE, C. A. (1968), *Ann. N.Y. Acad. Sci.* **151**, 222.

VON WARTBURG, J. P., PAPENBERG, J. and AEBI, H. (1965), *Canad. J. Biochem. Physiol.* **43**, 889.

WATTS, D. C. and DONNINGER, C. (1962), *Anal. Biochem.* **3**, 489.

WHITE, L. P. (1960), in *Enzymes in Health and Disease*, ed. GREENBERG, D. M. and HARPER, H. A., Springfield, Ill.: Thomas, p. 349.

WIELAND, T., PFLEIDERER, G., HAUPT, I. and WÖRNER, W. (1959), *Biochem. Z.* **331**, 103.

WOLFE, R. G. and NEILANDS, J. B. (1956), *J. biol. Chem.* **221**, 61.

YAKULIS, V. J., GIBSON, C. W. and HELLER, P. (1962), *Amer. J. clin. Path.* **38**, 378.

CHAPTER 8

Aminotransferase (Transaminase) Isoenzymes

Of the numerous enzymes which reversibly catalyse the transfer of an amino group from an amino acid to an α-keto acid, the two which have been most extensively studied are aspartate aminotransferase (formerly known as glutamate-oxaloacetate transaminase) and alanine aminotransferase (formerly known as glutamate-pyruvate transaminase). The former catalyses the reaction between aspartate and 2-oxoglutarate to form glutamate and oxaloacetate:

$$
\begin{array}{c}
\underset{\text{2-Oxoglutarate}}{\left.\begin{array}{c} \text{COOH} \\ | \\ \text{CH}_2 \\ | \\ \text{CH}_2 \\ | \\ \text{CO} \\ | \\ \text{COOH} \end{array}\right.} + \underset{\text{Aspartate}}{\left.\begin{array}{c} \text{COOH} \\ | \\ \text{CH}_2 \\ | \\ \text{CH·NH}_2 \\ | \\ \text{COOH} \end{array}\right.} \underset{\text{aminotransferase}}{\overset{\text{Aspartate}}{\rightleftharpoons}} \underset{\text{Glutamate}}{\left.\begin{array}{c} \text{COOH} \\ | \\ \text{CH}_2 \\ | \\ \text{CH}_2 \\ | \\ \text{CH·NH}_2 \\ | \\ \text{COOH} \end{array}\right.} + \underset{\text{Oxaloacetate}}{\left.\begin{array}{c} \text{COOH} \\ | \\ \text{CH}_2 \\ | \\ \text{CO} \\ | \\ \text{·COOH} \end{array}\right.}
\end{array}
$$

while alanine aminotransferase plays a similar part in the formation of glutamate and pyruvate from 2-oxoglutarate and alanine:

$$
\begin{array}{c}
\underset{\text{2-Oxoglutarate}}{\left.\begin{array}{c} \text{COOH} \\ | \\ \text{CH}_2 \\ | \\ \text{CH}_2 \\ | \\ \text{CO} \\ | \\ \text{COOH} \end{array}\right.} + \underset{\text{Alanine}}{\left.\begin{array}{c} \text{CH}_3 \\ | \\ \text{CH·NH}_2 \\ | \\ \text{COOH} \end{array}\right.} \underset{\text{aminotransferase}}{\overset{\text{Alanine}}{\rightleftharpoons}} \underset{\text{Glutamate}}{\left.\begin{array}{c} \text{COOH} \\ | \\ \text{CH}_2 \\ | \\ \text{CH}_2 \\ | \\ \text{CH·NH}_2 \\ | \\ \text{COOH} \end{array}\right.} + \underset{\text{Pyruvate}}{\left.\begin{array}{c} \text{CH}_3 \\ | \\ \text{CO} \\ | \\ \text{COOH} \end{array}\right.}
\end{array}
$$

These enzymes occur in most human and animal tissues, but their activities vary considerably. Heart, liver, skeletal muscle and kidney are particularly rich in aspartate aminotransferase, while liver and kidney are the richest sources of alanine aminotransferase. Since the tissue concentrations are several thousand times greater than the amounts in the blood serum, damage to a tissue usually leads to a marked increase in the serum activity, and these enzymes have been extensively used in diagnosis. Thus, after myocardial infarction there is a marked (but transient) increase in the serum aspartate aminotransferase, while in the early stages of infective hepatitis the activities of both aminotransferases are enormously increased.

Separation of aspartate aminotransferase isoenzymes
The serum aspartate aminotransferase has the same electrophoretic mobility after an episode of myocardial infarction as during infective hepatitis (Wróblewski, 1957; Pryse-Davies and Wilkinson, 1958;

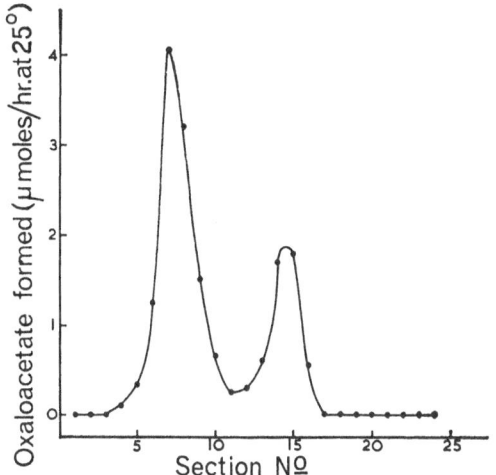

Fig. 51. Aspartate aminotransferase isoenzymes of rat liver separated by paper electrophoresis in phosphate buffer at pH 7·5. (Boyd, 1961.)

Ševela, 1958), an observation which suggests that human heart and human liver might contain the same enzyme. It migrates during paper electrophoresis as a slow α_2-globulin, whereas the serum alanine aminotransferase has the same mobility as a β-globulin. Although the serum aspartate aminotransferase appears to be substantially homogeneous, the purified heart and liver enzymes have been shown to consist of at least two distinct fractions separable by electrophoresis (Fig. 51) (Green, Leloir and Nocito, 1945; Jungner, 1957; Fleisher, Potter and Wakim, 1960; Boyd, 1961; Augustinsson and Erne, 1961; Borst and Peeters, 1961; Kormendy, Gantner and Hamm, 1965).

Fleisher *et al.* (1960) demonstrated that, when canine heart extracts were treated with carboxymethylcellulose in phosphate buffer of pH 7·0 and ionic strength 0·01, one isoenzyme was adsorbed while the other remained in solution. They confirmed these findings by paper electrophoresis in phosphate buffer of pH 7·4 containing 1% bovine albumin which was added to preserve enzyme activity. The paper was cut into strips, from each of which the enzyme was eluted and its activity determined by the Karmen (1955) spectrophotometric procedure. Similar results were obtained with human-, porcine- and canine-heart extracts, and in each species one isoenzyme migrated towards the anode and the other towards the cathode. While there is no significant species difference between the mobilities of the cathodic fractions, the mobilities of the anodic fractions varied appreciably: the human isoenzyme migrates farthest and is closely followed by the porcine fraction, whereas the canine isoenzyme remains near the origin.

The two isoenzymes have subsequently been purified from dog liver by chromatography, heat treatment and fractionation with ammonium sulphate (Fleisher, 1960). Separation may also be effected by agar-gel electrophoresis and also by zone electrophoresis against a sucrose-density gradient in a column electrophoresis apparatus (Boyd, 1961, 1962*a*). A chromatographic procedure has been described by Augustinsson and Erne (1961).

Boyd (1962*b*, 1966) has shown that the aspartate aminotransferase isoenzymes can readily be extracted from tissue homogenates with butan-1-ol and that they can subsequently be partly separated by ammonium sulphate fractionation. At 50–70% of saturation the precipitate consists predominantly of the anodic isoenzyme, while the cathodic isoenzyme is precipitated at higher concentrations. Further purification may be achieved by chromatography on DEAE-cellulose.

A very simple technique has been developed by Boyd (1962*a*) for the visualization of the bands containing aspartate aminotransferase activity after agar-gel electrophoresis by a modification of the Wieme (1959) technique. A second gel containing 2-oxoglutarate, aspartate, malate dehydrogenase and NADH is placed in contact with the gel containing the enzyme. The paired gels are incubated to permit enzyme action: the oxaloacetate formed is reduced by the malate dehydrogenase and the NADH is oxidized. When viewed under ultra-violet light the zones of enzyme activity appear as dark bands on the fluorescent gel. Permanent records may be prepared by ultra-violet photography. This technique has recently been applied by Boyde (1968*a*) to the semi-quantitative determination of the isoenzymes separated by acrylamide-gel electrophoresis.

The tetrazolium-staining procedure has been modified for the detection of aspartate aminotransferase bands after starch-gel electrophoresis (Boyde and Latner, 1961). A filter paper soaked in a solution of the substrates, NADH, pyridoxal phosphate and malate dehydrogenase is applied to the gel and incubated for up to 4 hr. at 36·5°. The gel is then immersed in a solution of tetrazolium MTT and N-methylphenazonium methosulphate, when the aminotransferase zones appear as pale areas on a dark background.

A much simpler staining technique (Schwartz, Nisselbaum and Bodansky, 1963; Decker and Rau, 1963) depends upon the production of a purple colour when oxaloacetate reacts with 6-benzamido-4-methoxy-m-toluidine-diazonium chloride (Azoene fast violet B), a reaction which forms the basis of a specific colorimetric method for the determination of serum aspartate aminotransferase activity (Babson, Shapiro, Williams and Phillips, 1962). An agar gel containing the dye and the substrate-buffer mixture is poured over the starch gel, and the paired gels are then incubated at 37° for 1 hr. Enzyme action occurs at the interface, and when the agar mixture is peeled off purple bands in the starch indicate the position of the aspartate aminotransferase isoenzymes.

Decker and Rau (1963) have reported the diazonium procedure to give better results than the tetrazolium technique, since blanks, prepared by omitting aspartate from the substrate mixture, were colourless in the former but exhibited the 'nothing dehydrogenase' phenomenon (see p. 50) in the latter. Furthermore, the diazonium method gives coloured bands on a colourless background, which these authors found preferable to the dark background obtained by the earlier technique.

When the activities of the enzyme fractions are compared it must be borne in mind that they are inhibited by oxaloacetate, one of the reaction products. Boyd (1961) accordingly recommends use of the Karmen (1955) method, since in this procedure oxaloacetate is removed as rapidly as it is formed, by reaction with malate dehydrogenase in the presence of NADH. The course of this reaction is followed spectrophotometrically at 340 nm. The colorimetric method of Reitman and Frankel (1957) and the spectrophotometric method of Cammarata and Cohen (1951) are less satisfactory, since they involve accumulation of oxaloacetate.

Properties of aspartate aminotransferase isoenzymes
An appreciable fraction of the total aspartate aminotransferase activity of rat-liver homogenates resides in the mitochondria (Hird and Rowsell,

1950; Müller and Leuthardt, 1950), and several investigators have reported that the activity of tissue homogenates increases on storage, presumably due to release of the enzyme from mitochondria (Schmidt, Schmidt and Wildhirt, 1958; Rosenthal, Thind and Conger, 1960). The various methods for disintegrating mitochondria, e.g. treatment in a Waring Blender or ultrasonic disintegration, all cause a considerable increase in the total activity, and it appears that the mitochondrial activity may be two to four times as great as that of the supernatant fraction (Eichel and Bukovsky, 1961; Boyd, 1961; Bhargava and Sreenivasan, 1965). Rosenthal *et al.* (1960) demonstrated that the mitochondrial and soluble aspartate aminotransferases of rat liver have properties very similar to those of the cationic and anionic forms, respectively, and Boyd (1961) confirmed that the anodic fraction (aspartate aminotransferase I) is derived from the supernatant fraction, while the cathodic fraction (aspartate aminotransferase II) originates in the mitochondria. Boyd (1962*b*, 1966) has since shown that butanol-treated extracts of rat-liver supernatant contain only the anodic isoenzyme, while similar extracts of mitochondria contain only the cathodic isoenzyme.

The two isoenzymes differ not only in their electrophoretic mobilities and chromatographic properties but also in their response to pH variation, in their immunochemical properties and in their substrate affinities (Nisselbaum and Bodansky, 1964; Wada and Morino, 1964). Fleisher *et al.* (1960) demonstrated a marked difference between the activities of canine-heart isoenzymes I and II when the pH is varied. When the pH is reduced below 7·4 the activity of isoenzyme I falls rapidly, and at pH 6·0 only about 40% of the initial activity remains, whereas no such fall occurs when isoenzyme II is similarly treated. Almost identical results have been observed by Boyd (1961), who used rat liver as the source of his isoenzymes (Fig. 52).

Fleisher *et al.* (1960) showed that the two isoenzymes exhibited a marked difference between their Michaelis (substrate) constants for aspartate at optimal concentrations of 2-oxoglutarate. At pH 6·0 canine-heart aspartate aminotransferase I has a K_m of $38·4 \times 10^{-3}$M, while at pH 7·4 the value is $11·9 \times 10^{-3}$M. Isoenzyme II from the same tissue, however, has K_m values of $2·48 \times 10^{-3}$M and $0·70 \times 10^{-3}$M at pH 6·0 and pH 7·4 respectively. Thus, the cationic isoenzyme has a much greater affinity for aspartate than has the anionic isoenzyme. However, the latter shows greater affinity for 2-oxoglutarate, but at pH 6·0 it is inhibited by higher concentrations of this substrate. Similar results are also obtained with rat-liver isoenzymes (Boyd, 1961).

A means for determining the relative proportions of the two isoenzymes in crude tissue extracts based upon the activities observed at different pH values has been devised by Fleisher *et al.* (1960). At pH 6·0 isoenzyme II can be determined almost exclusively by using high 2-oxoglutarate and low aspartate concentrations, while at pH 7·4, and with excess of both substrates, the sum of the activities of both isoenzymes is determined. The results obtained agree closely with those found by electrophoresis.

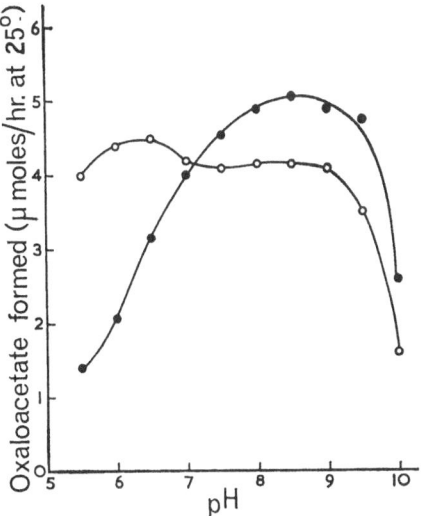

Fig. 52. The effect of pH changes on the activities of rat-liver aspartate aminotransferase isoenzymes. ●, isoenzyme I; ○, isoenzyme II. (Boyd, 1961.)

The properties of purified aspartate aminotransferase preparations with high specific activities obtained by a number of investigators (*vide, inter alia,* Green *et al.,* 1945; O'Kane and Gunsalus, 1947; Jenkins and Sizer, 1957; Jenkins, Yphantis and Sizer, 1959; Lis, 1958; reviewed by Velick and Vavra, 1962) are similar to those described by Fleisher *et al.* (1960) and Boyd (1961) for isoenzyme I. The highly purified ox-heart aspartate aminotransferase described by Marino, Greco, Scardi and Zito (1966) also appears to be the supernatant isoenzyme.

Several investigators have studied the properties of the purified supernatant and mitochondrial forms from various mammalian tissues, including rat liver (Hook and Vestling, 1962; Boyd, 1966; De Jiménez, Brunner and Soberón, 1967), ox-heart (Morino, Itoh and Wada, 1963), human heart and liver (Bodansky, Schwartz and Nisselbaum, 1966), and pig heart (Nisselbaum and Bodansky, 1966; Martinez-Carrion and Tiemeier, 1967). In general, these have confirmed and extended the results obtained with the relatively crude preparations available earlier. The cytoplasmic enzyme from chicken liver has been isolated and characterized (De Jiménez *et al.*, 1967; Bertland and Kaplan, 1968), but as yet little is known of the avian mitochondrial enzyme.

Martinez-Carrion and Tiemeier (1967) found the mitochondrial and supernatant pig-heart isoenzymes to consist of two similar polypeptide chains and to have similar molecular weights (about 100,000). However, the isoenzymes differ markedly in amino-acid composition and in the peptide maps obtained after tryptic digestion. The *N*-terminal amino-acid of the supernatant isoenzyme is alanine, while that of the mitochondrial enzyme is serine. It is remarkable that proteins of quite different structure should possess similar catalytic properties, especially since Munkres (1965) has demonstrated the apparent identity of aspartate apoaminotransferase and malate apodehydrogenase in *Neurospora* (p. 207).

The difference between the structures of the cytoplasmic and mitochondrial enzymes has been confirmed by antibody studies. Nisselbaum and Bodansky (1964) prepared antisera to the human-heart isoenzymes. These reacted with the homologous antigen, but no cross-reaction was observed. The ox-liver isoenzymes, administered to rabbits, produced specific antibodies, the presence of which was demonstrated by agar diffusion and tube precipitation (Morino, Itoh and Wada, 1963; Morino, Kagamiyama and Wada, 1964). The human serum aspartate aminotransferase and the purified cytoplasmic enzyme have both been shown to be sensitive to antisera to the pig-heart supernatant enzyme, while the purified human mitochondrial enzyme is unaffected (Lang and Massarat, 1965).

As expected, the supernatant and mitochondrial isoenzymes also differ in other properties. The mitochondrial component is appreciably more stable to heat (Wada and Morino, 1964; Nisselbaum and Bodansky, 1966), but in contrast to the supernatant enzyme, the particulate form is strongly inhibited by inorganic phosphate (Nisselbaum, 1968).

The two aspartate aminotransferase isoenzymes also differ from each other in the rates at which they disappear from blood (Wakim and Fleisher, 1963a; Fleisher and Wakim, 1963). After intravenous injection into dogs isoenzyme I does not disappear from the serum until after about 4–5 days, whereas isoenzyme II is almost completely eliminated in 6 hr. By contrast, alanine aminotransferase can be detected up to 19–25 days after injection.

Kinetic studies of their disappearance rates have shown that isoenzyme I resembles alanine aminotransferase in that the curves obtained are characteristic of those to be expected of a three-compartment system permitting reversible interchange between the compartments but only irreversible elimination from each. Isoenzyme II, however, does not fit this model scheme, for it disappears from serum at an accelerated rate after the first 2 hr., so that after 3 hr. only about 2% of the injected enzyme remains.

Both isoenzymes are rapidly transferred to the lymph, the activity of isoenzyme I reaching a maximum of 30% of the initial serum activity 1½–3 hr. after injection, while the corresponding figure for isoenzyme II is 10% about 45 min. after injection. Little of either enzyme is excreted in the urine.

Wakim and Fleisher (1963a) have also shown that the reticulo-endothelial system appears to play a major part in the removal of circulating enzymes from the serum. They injected into dogs suspensions of zymosan (an insoluble polysaccharide obtained from the cell walls of yeast), which is known to produce an initial marked inhibition of the reticulo-endothelial system during the first few hours after injection, but proliferation and increased activity of the reticulo-endothelial system subsequently extend over one or two days. When isoenzymes I and II are injected 2 hr. after zymosan their rates of disappearance are considerably retarded, whereas their removal is accelerated when they are administered 24 or 48 hr. later. This effect is somewhat greater with isoenzyme II than with isoenzyme I. Wakim and Fleisher (1963b) have extended their studies to the endogenous production of serum enzymes in response to zymosan injection, and have reported that approximately equal activities of isoenzymes I and II appear in the serum, although scarcely any isoenzyme II is detectable in serum collected before zymosan administration. These investigators suggest that the increase in total aminotransferase activity may be due to a combination of delayed clearance by the reticulo-endothelial system and to liberation of larger amounts of enzymes from the tissues.

Evidence has accrued to suggest that isoenzymes I and II may not be homogeneous, for Boyde and Latner (1961) have reported the presence of a second anodic band in the serum of a patient with myocardial infarction and also in that from a patient with carbon tetrachloride poisoning. The anodic isoenzyme occurs in multiple forms in various rat tissues: Decker and Rau (1963) have found that from liver, heart and muscle to consist of three distinct fractions, while that from kidney contains only two such bands. These investigators found that a commercial crystalline pig-heart aspartate aminotransferase contained four anodic fractions but no cathodic isoenzyme. The human-heart anodic isoenzyme contains three fractions, but that of the dog appears to be homogeneous. The significance of these findings has yet to be elucidated, but the authors suggest that the heterogeneity might be due to differences in the amino-acid composition of the protein fractions or to differences in other groups, e.g. sialic acid, associated with the enzyme proteins.

The complexity of human serum aspartate aminotransferase has been confirmed by Block, Carmichael and Jackson (1964), who subjected serum concentrates to starch-gel or starch-block electrophoresis. When a staining technique was employed only two fractions, both of which were anionic, could be detected, but eluates from either gels or blocks were found to contain three anionic (A, B, C) and two cationic components (D, E). The percentage distribution of the five isoenzymes is shown in Table 32. Fraction C was found to be the most stable to heat, while fraction A was the most labile.

TABLE 32 *Aspartate aminotransferase isoenzymes in concentrates of human sera*

(Block *et al.*, 1964)

	Isoenzymes				
	+ Anionic			Cationic −	
	A	B	C	D	E
Average percentage distribution	12·8	24·3	26·4	19·6	17·0

Evidence that the mitochondrial enzyme of rat liver consists of at least two components has recently been obtained by Bhargava and Sreenivasan (1968), who obtained one fraction by subjecting the mitochondrial pellet to ultrasonic disintegration and the other by subsequent treatment of the pellet with butanol. The two fractions were found to differ in their electrophoretic mobilities and in their chromatographic properties.

Another recent report is that of Gabrielli and Orfanos (1968*a*), who separated a total of five aspartate aminotransferase isoenzymes from rat serum and liver by starch-block electrophoresis. Two of these were anionic and appeared to be cytoplasmic, whereas the remaining three were cationic. When the animals were previously treated with carbon tetrachloride the serum pattern showed a marked increase in the middle cationic fraction and closely resembled that of a whole liver homogenate. The authors conclude that the increase in the middle cationic component is indicative of mitochondrial damage. They point out that with their technique only three components were detected in human serum.

These reports have been confirmed by Michuda and Martinez-Carrion (1969), who found that the pig-heart mitochondrial enzyme can be separated by column chromatography on CM-Sephadex into three fractions which differ in their electrophoretic mobilities on starch-gel at pH 8·6. The three components display similar reaction rates with a series of substrate analogues and contain the same proportions of pyridoxal phosphate, but differ in the relative amounts of 'catalytically active' and 'catalytically inactive' coenzyme. Optical rotatory dispersion measurements and immunological double diffusion studies on agar-gel suggest that the three enzymes are not conformational isomers.

Clinical applications of aspartate aminotransferase isoenzymes

Several investigators have reported the presence of the cationic (mitochondrial) enzyme in the serum of isolated patients with myocardial infarction (Fleisher *et al.,* 1960; Boyde and Latner, 1962; Zázvorka and Kamarýt, 1965), but recent studies suggest that it may be a regular finding in this disease. Massarat and Lang (1965) used a method depending upon the selective immuno-inhibition of the serum isoenzymes by specific antisera. In four patients studied serially, elevated levels of the mitochondrial enzyme were detected, especially on the first day after infarction, but the presence of this component could not be confirmed in six other patients from whom a single specimen only was collected two days or more after the episode.

Massarat and Lang's results correlate with the finding of Fleisher and Wakim (1963) that in dogs mitochondrial aspartate aminotransferase is rapidly cleared from the plasma, but are somewhat at variance with those reported by Boyde (1968*b*), who studied fifteen patients with myocardial infarction. Using acrylamide-gel electrophoresis with visual assessment of isoenzyme activities under ultra-violet light (Boyde, 1968*a*), he found not only marked elevation of the cationic isoenzyme in all but two cases but also that the cationic fraction returned to normal at a slower rate than the total serum enzyme activity. Boyde suggests that the prolonged elevation of the cationic isoenzyme in the serum might be reconciled with its rapid clearance if it were to be released gradually from damaged cells rather than as a single discharge.

Since he found the cationic isoenzyme to be elevated in eight cases when the total serum aspartate aminotransferase activity was normal, Boyde (1968*b*) further suggests that the mitochondrial component might be a more sensitive indicator of myocardial damage than the total enzyme activity. However, he used a rather insensitive colorimetric method to determine the latter, and it would be interesting to know how the results compare when a more precise spectrophotometric or fluorimetric method is used for the total enzyme activity.

Possibly the most suitable method yet devised for clinical purposes is that of Schmidt, Schmidt and Möhr (1967), who applied dialysed serum to a DEAE-Sephadex column. The mitochondrial enzyme is eluted with 0·008M-sodium phosphate buffer at pH 7·0, while the cytoplasmic enzyme remains on the column. The latter may be eluted with 0·2M-sodium phosphate buffer, pH 7·0, containing 0·2M-sodium chloride. This procedure appears to offer several advantages: it can be used on a relatively small scale with limited elution volumes; separation is clearcut; and it offers a means for the precise determination of the two fractions.

It has recently been applied to the study of the isoenzymes of malate dehydrogenase and aspartate aminotransferase in the sera of patients with liver diseases. While mitochondrial aspartate aminotransferase could not be detected in the sera of healthy individuals, high levels were observed in acute and chronic hepatitis and in active cirrhosis. During the recovery phase of acute hepatitis the mitochondrial enzyme disappeared from the serum at a faster rate than the cytoplasmic enzyme (Schmidt, Schmidt and Otto, 1967).

Comparable results with a similar method have been reported by Gabrielli and Orfanos (1968*b*), who found two cationic components not only in the sera of patients with chronic alcoholism and cirrhosis but also in the sera of normal subjects. One of the cationic fractions, however, was not detectable in sera from patients with virus hepatitis nor in human-liver extracts.

The release of enzymes into the circulation after experimental partial hepatectomy in the rat has been attributed by Bengmark, Ekholm and Olsson (1967) at least partly to increased synthesis in the surviving liver cells, since both supernatant and mitochondrial fractions of aspartate aminotransferase expressed per mg. liver protein show post-operative increases in parallel with that found in the serum enzyme.

Detection of the mitochondrial component in normal serum seems to depend upon the technique employed, for Schwartz and Bodansky

(1966) found only the anionic isoenzyme on starch-block electrophoresis. Of 37 patients with neoplastic disease, 20 showed the single-banded normal serum pattern, while the sera of 11 were found to contain both anionic and cationic isoenzymes. The other six showed the normal band accompanied by a faster fraction which migrated with the albumin. Four members of this group also exhibited the cationic band, the appearance of which generally coincided with an acute phase, usually involving the liver.

The results so far reported in muscular dystrophy have been rather conflicting, for while Kar and Pearson (1964) failed to detect the mitochondrial isoenzyme, and Mannucci, Ideo, Cao and Macciotta (1965) found small amounts in one of two cases, Boyde (1968b) found marked elevation of this component in each of six cases, five of which were of the Duchenne type and the sixth of the limb-girdle type.

Although mature rabbit erythrocytes have been found to contain only the anionic form of aspartate aminotransferase, reticulocytosis induced by bleeding or treatment with acetylphenylhydrazine leads to a substantial increase in the overall red-cell enzyme activity and to the appearance of the cationic component. The latter appears to originate in the mitochondrial fraction of the reticulocytes (Nisselbaum and Bodansky, 1965). Similarly in normal human erythrocytes only the anionic isoenzyme is detectable (Nisselbaum, 1965; Mannucci and Dioguardi, 1966), while in patients with reticulocytosis both anionic and cationic isoenzymes can be detected in the reticulocyte-rich red cells (Mannucci and Dioguardi, 1966).

Aspartate aminotransferase isoenzymes in developing tissues
Few reports have appeared so far concerning changes in aspartate aminotransferase isoenzymes in developing tissues. Waksman and Rendon (1968) recently studied the development of rat heart, kidney and brain at various ages from birth to 20 days, by separating the anionic and cationic isoenzymes with Cellogel electrophoresis. Markedly different development patterns were found in the three tissues: in the heart both fractions increase in the same proportions throughout post-natal development, while in the kidney there is a slight parallel increase in both fractions which begins about the 12th day. Thus, in these tissues the isoenzyme ratio remains virtually constant. In the brain the mitochondrial fraction remains very low, exhibiting only a slight increase, while the cytoplasmic isoenzyme increases about ten-fold. Hence there is a marked increase in the ratio of supernatant isoenzyme to mitochondrial isoenzyme in this tissue.

Alanine aminotransferase

Alanine aminotransferase occurs predominantly in the supernatant fraction of animal tissues, and many investigators have reported that this enzyme forms only one band when preparations are subjected to electrophoresis (Fleisher, 1960; Boyd, 1962a). The rat-liver enzyme has been extensively studied by Segal and his co-workers, who found enzyme preparations from normal and corticoid-treated rats to be indistinguishable in pH optima and kinetic, immunological and other properties (Segal, Beattie and Hopper, 1962; Segal, Kim and Hopper, 1965).

However, Ziegenbein (1966) has recently reported that under certain conditions rat-heart cytoplasmic alanine aminotransferase is adsorbed on DEAE-cellulose, whereas the mitochondrial fraction from the same tissue passes through the column. Both fractions could be determined spectrophotometrically, but the mitochondrial fraction showed no activity measurable by the Reitman–Frankel colorimetric method, presumably, Ziegenbein suggests, because of its sensitivity to inhibition by accumulated pyruvate.

Further evidence of the heterogeneity of this enzyme is provided by the occurrence of two components from rat-liver alanine aminotransferase, separable on acrylamide-gel electrophoresis (Gatehouse, Hopper, Schatz and Segal, 1967).

REFERENCES

AUGUSTINSSON, K. B. and ERNE, K. (1961). *Experientia* **17**, 396.

BABSON. A. L., SHAPIRO, P. O., WILLIAMS, P. A. R. and PHILLIPS, G. E. (1962). *Clin. chim. Acta* **7**, 199.

BENGMARK, S., EKHOLM, R. and OLSSON, R. (1967), *Acta Hepato-Splenologica* **14**, 80.

BERTLAND, L. H. and KAPLAN, N. O. (1968). *Biochemistry* **7**, 134.

BHARGAVA, M. M. and SREENIVASAN, A. (1965). *Enzymologia* **29**, 65.

BHARGAVA, M. M. and SREENIVASAN, A. (1968). *Biochem. J.* **108**, 619.

BLOCK, W. D., CARMICHAEL, R. and JACKSON, C. E. (1964), *Proc. Soc. exp. Biol. N.Y.* **115**, 941.

BODANSKY, O., SCHWARTZ, M. K. and NISSELBAUM, J. S. (1966), *Adv. Enzyme Reg.* **4**, 299.

BORST. P. and PEETERS, E. M. (1961), *Biochim. biophys. Acta* **54**, 188.

BOYD, J. W. (1961). *Biochem. J.* **81**, 434.

BOYD, J. W. (1962a). *Clin. chem. Acta* **7**, 424.

BOYD, J. W. (1962b). *Biochem. J.* **84**, 14P.

BOYD, J. W. (1966). *Biochim. biophys. Acta* **113**, 302.

BOYDE, T. R. C. (1968a). *Z. klin. Chem.* **6**, 56.

BOYDE, T. R. C. (1968b), *Enzymol. biol. clin.* **9**, 385.

BOYDE, T. R. C. and LATNER, A. L. (1962), *Biochem. J.* **82**, 51P.

CAMMARATA, P. S. and COHEN, P. P. (1951), *J. biol. Chem.* **193**, 53.

DECKER, L. E. and RAU, E. M. (1963), *Proc. Soc. exp. Biol. N.Y.* **112**, 144.

DE JIMÉNEZ, E. S., BRUNNER, A. L. and SOBERÓN, G. (1967), *Arch. Biochem. Biophys.* **120**, 175.

EICHEL, H. J. and BUKOVSKY, J. (1961), *Nature, Lond.* **191**, 243.

FLEISHER, G. A. (1960), *Federation Proc.* **19**, 6.

FLEISHER, G. A., POTTER, C. S. and WAKIM, K. G. (1960), *Proc. Soc. exp. Biol. N.Y.* **103**, 229.

FLEISHER, G. A. and WAKIM, K. G. (1963), *J. Lab. clin. Med.* **61**, 98.

GABRIELLI, E. R. and ORFANOS, A. (1968a), *Proc. Soc. Exp. Biol., N.Y.* **127**, 766.

GABRIELLI, E. R. and ORFANOS, A. (1968b), *Proc. Soc. exp. Biol., N.Y.* **128**, 803.

GATEHOUSE, P. W., HOPPER, S., SCHATZ, L. and SEGAL, H. L. (1967), *J. biol. Chem.* **242**, 2319.

GREEN, D. E., LELOIR, L. F. and NOCITO, V. (1945), *J. biol. Chem.* **161**, 559.

HIRD, F. J. R. and ROWSELL, E. V. (1950), *Nature, Lond.* **166**, 517.

HOOK, R. H. and VESTLING, C. S. (1962), *Biochim. biophys. Acta* **65**, 358.

JENKINS, W. T. and SIZER, I. W. (1957), *J. Amer. chem. Soc.* **79**, 2655.

JENKINS, W. T., YPHANTIS, D. A. and SIZER, I. W. (1959), *J. biol. Chem.* **234**, 51.

JUNGNER, G. (1957), *Scand. J. clin. Lab. Invest.* **10**, supp. 31, 280.

KAR, N. C. and PEARSON, C. M. (1964), *Proc. Soc. exp. Biol., N.Y.* **116**, 733.

KARMEN, A. (1955), *J. clin. Invest.* **34**, 131.

KORMENDY, K., GANTNER, G. and HAMM, R. (1965), *Biochem. Z.* **342**, 31.

LANG, N. and MASSARAT, S. (1965), *Klin. Wschr.* **43**, 597.

LIS, H. (1958), *Biochim. biophys. Acta* **28**, 191.

MANNUCCI, P. M. and DIOGUARDI, N. (1966), *Clin. Chim. Acta* **14**, 215.

MANNUCCI, P. M., IDEO, G., CAO, A. and MACCIOTTA, A. (1965), *Rass. Med. Sarda* **68**, 287.

MARINO, G., GRECO, A. M., SCARDI, V. and ZITO, R. (1966), *Biochem. J.* **99**, 589.

MARTINEZ-CARRION, M. and TIEMEIER, D. (1967), *Biochemistry* **6**, 1715.

MASSARAT, S. and LANG, N. (1965), *Klin. Wschr.* **43**, 602.

MICHUDA, C. M. and MARTINEZ-CARRION, M. (1969), *Biochemistry* **8**, 1095.

MORINO, Y., ITOH, H. and WADA, H. (1963), *Biochem. biophys. Res. Commun.* **13**, 348.

MORINO, Y., KAGAMIYAMA, H. and WADA, H. (1964), *J. biol. Chem.* **239**, PC943.

MÜLLER, A. F. and LEUTHARDT, F. (1950), *Helv. chim. Acta* **33**, 268.

MUNKRES, K. D. (1965), *Arch. Biochem. Biophys.* **112**, 347.

NISSELBAUM, J. S. (1965), *Federation Proc.* **24**, 356.

NISSELBAUM, J. S. (1968), *Anal. Biochem.* **23**, 173.

NISSELBAUM, J. S. and BODANSKY, O. (1964), *J. biol. Chem.* **239**, 4232.

NISSELBAUM, J. S. and BODANSKY, O. (1965), *Science,* **149**, 195.

NISSELBAUM, J. S. and BODANSKY, O. (1966), *J. biol. Chem.* **241**, 2661.

O'KANE, D. E. and GUNSALUS, I. C. (1947), *J. biol. Chem.* **170**, 425.

PRYSE-DAVIES, J. and WILKINSON, J. H. (1958), *Lancet* i, 1249.

REITMAN, S. and FRANKEL, S. (1957), *Amer. J. clin. Path.* **28**, 56.

ROSENTHAL, O., THIND, S. K. and CONGER, N. (1960), *Abs. 138th Meeting Amer. Chem. Soc.* New York, 10c.

SCHMIDT, E., SCHMIDT, F. W. and MÖHR, J. (1967), *Clin. Chim. Acta* **15**, 337.

SCHMIDT, E., SCHMIDT, F. W. and OTTO, P. (1967), *Clin. Chim. Acta* **15**, 283.

SCHMIDT, E., SCHMIDT, F. W. and WILDHIRT, E. (1958), *Klin. Wschr.* **36**, 172.

SCHWARTZ, M. K. and BODANSKY, O. (1966), *Amer. J. Med.* **40**, 231.

SCHWARTZ, M. K., NISSELBAUM, J. S. and BODANSKY, O. (1963), *Amer. J. clin. Path.* **40**, 103.

SEGAL, H. L., BEATTIE, D. S. and HOPPER, S. (1962), *J. biol. Chem.* **237**, 1914.

SEGAL, H. L., KIM, Y. S. and HOPPER, S. (1965), *Adv. Enzyme Reg.* **3**, 29.

ŠEVELA, M. (1958), *Nature, Lond.* **181**, 915.

VELICK, S. F. and VAVRA, J. (1962), in *The Enzymes*, 2nd edn., ed. BOYER, P. D., LARDY, H. A. and MYRBÄCK, K., New York: Acad. Press., Vol. 6, p. 219.

WADA, H. and MORINO, Y. (1964), *Vitamins & Hormones* **22**, 411.

WAKIM, K. G. and FLEISHER, G. A. (1963a), *J. Lab. clin. Med.* **61**, 86.

WAKIM, K. G. and FLEISHER, G. A. (1963b), *J. Lab. clin. Med.* **61**, 107.

WAKSMAN, A. and RENDON, A. (1968), *Arch. Biochem. Biophys.* **123**, 201.

WIEME, R. J. (1959), *Clin. chim. Acta* **4**, 317.

WRÓBLEWSKI, F. (1957), in *Hepatitis Frontiers* (Ford Foundation Symposium, ed. HARTMAN, F. W., LOGRIPPO, G. A., MATEER, J. G. and BARRON, J.), Boston, Mass.: Little Brown, p. 447.

ZÁZVORKA, A. and KAMARÝT, J. (1965), *Čas. Lěk. čec.* **104**, 970.

ZIEGENBEIN, R. (1966), *Nature, Lond.* **212**, 935.

CHAPTER 9

Phosphatase Isoenzymes

Alkaline phosphatase

Alkaline phosphatase has been extensively used in diagnosis during the past three decades, and there has been much discussion concerning the origin of the serum enzyme in hepatobiliary disease. It was formerly considered to originate mainly in the osteoblasts of bone and to be excreted in the bile (Gutman, 1959). Many other tissues including the liver, intestinal mucosa, kidney and placenta have been shown to contain the enzyme (Bodansky, 1937, 1948; Armstrong and Banting, 1935; Cloetens, 1939; Sherlock and Walshe, 1947; Burke, 1950; Ross, Iber and Harvey, 1956; Ahmed and King, 1959; Grossberg, Harris and Schlamowitz, 1961; Hodson, Latner and Raine, 1962; Fishman, Green and Inglis, 1962; Moss and King, 1962; and others, reviewed by Posen, 1967), and there is now strong evidence that other tissues, especially the liver, contribute to the serum alkaline phosphatase activity. A number of chromatographic and electrophoretic techniques have been applied in attempts to identify the tissue of origin of the serum enzyme activity, and these led to the recognition of its heterogeneity.

These procedures have frequently proved helpful in resolving the question of the source of the serum enzyme in several clinical contexts, e.g. in malignant disease when a raised serum alkaline phosphatase is sometimes the first indication of metastatic invasion of bone or liver, or in the assessment of suspected liver disease in childhood, adolescence or pregnancy when the serum activity is physiologically elevated above normal adult levels.

As mentioned in Chapter 1, there is some discussion concerning the true nature of the components of human and animal serum and tissue alkaline phosphatases, but according to a recent recommendation of the Enzyme Commission, they should be regarded as isoenzymes (Webb, 1964). The methods used for their identification and separation are applicable to isoenzymes generally, and an account is therefore included in this monograph.

Distribution of alkaline phosphatase isoenzymes

PAPER ELECTROPHORESIS. The numbers and intensities of alkaline phosphatase bands vary considerably according to the techniques employed for their separation and even to the methods of extraction from tissues. Baker and Pellegrino (1954) and Eisfeld and Koch (1954) found alkaline phosphatase to migrate principally in association with the α_2-globulin fraction when serum is subjected to paper electrophoresis, and this has been confirmed by Wolfson (1957). Taleisnik, Paglini and Zeitune (1955) showed that the serum alkaline phosphatase of rats could be separated by paper electrophoresis into two fractions migrating as α_2- and β-globulins. An additional band appeared in the α_1-globulin fraction after electrophoresis of the sera of rats in which the bile duct had been ligated.

STARCH-BLOCK ELECTROPHORESIS. Quantitative studies with the aid of starch-block electrophoresis also indicate that the major zone of the serum alkaline phosphatase activity migrates with the α_2-globulins (Rosenberg, 1959), but Keiding (1959) succeeded in resolving this into an 'α_2-component' migrating near the α_2-globulin peak and a 'β-component' located between the α_2-globulin and the β-globulin. Both workers found a minor band in the α_1-globulin region in the sera of patients with liver diseases. Since the β-fraction is considerably increased during childhood and is particularly prominent in the sera of patients with Paget's disease or bone metastases, it would appear to be very largely derived from bone (Keiding, 1959). The α_1-component, on the other hand, appears to originate in the liver, and the finding that an appreciable fraction of the alkaline phosphatase of the bile migrates as an α_1-globulin is consistent with this conclusion (Rosenberg, 1959). Further support for this view is provided by Cooke and Zilva (1961), who reported a case in which extensive liver metastases but no other secondaries were found at autopsy. After starch-block electrophoresis about 50% of the serum alkaline phosphatase activity was found in the α_1-peak.

More recently, Keiding (1964) has extended his studies to lymph and intestinal fluid, in which he has demonstrated the presence of a β_2-fraction with a mobility somewhat slower than the β-globulin. There is a close correlation between the lipid content of the lymph and its β_2-phosphatase activity. After the ingestion of cream there is an increase in the β_2-fraction of lymph, and it is suggested that alkaline phosphatase is secreted by the intestinal mucosa into the intestinal lumen, where it becomes bound to lipid. No β_2-phosphatase is found in the portal blood, which suggests that this fraction is not reabsorbed directly from the intestine. Little of the β_2-fraction is detectable in bile, and Keiding postulates that during passage through the liver it might be transformed into the α_1-fraction.

Results similar to those observed during starch-block electrophoresis of human serum alkaline phosphatase by Keiding (1959) have been reported by Nordentoft-Jensen (1964), who used Pevikon C-870 (Stockholms superfosfat fabrik A.B., Stockholm) as supporting medium.

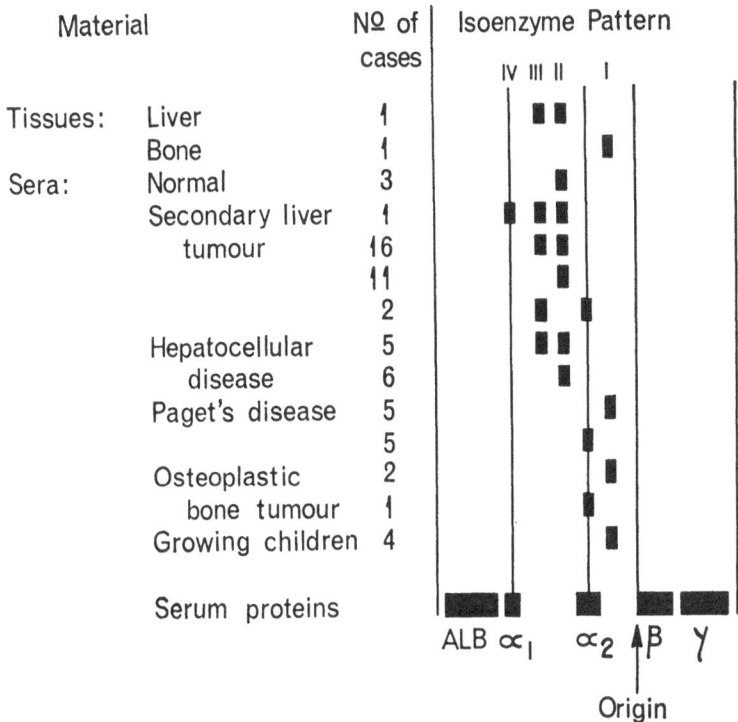

Fig. 53. The alkaline phosphatase isoenzymes of human bone and liver and of normal and pathological sera, separated by agar-gel electrophoresis. (From Haije and de Jong, 1963.)

This is a co-polymer of polyvinyl chloride and polyvinyl acetate, introduced as a medium for electrophoresis by Müller-Eberhard (1960). Nordentoft-Jensen (1964) showed that the ratio of the serum β_1- and α_2-alkaline phosphatase fractions was in the range of 0·7–1·3 in normal adults. Higher values were found in children and in patients with bone diseases, while patients with liver diseases give lower ratios. When the total serum alkaline phosphatase activity was raised the results were of diagnostic significance.

AGAR-GEL ELECTROPHORESIS. Haije and de Jong (1963) have applied agar-gel electrophoresis to the study of the isoenzyme patterns of the alkaline phosphatases of human sera and tissues. The results (Fig. 53) resemble those obtained by starch-block electrophoresis in showing three main bands of activity: I slightly slower and II slightly faster than the α_2-globulins, while III has a mobility between that of the α_1- and α_2-globulins. In one serum a fourth band (IV) is associated with the α_1-globulins. The single band appearing between I and II is attributed to

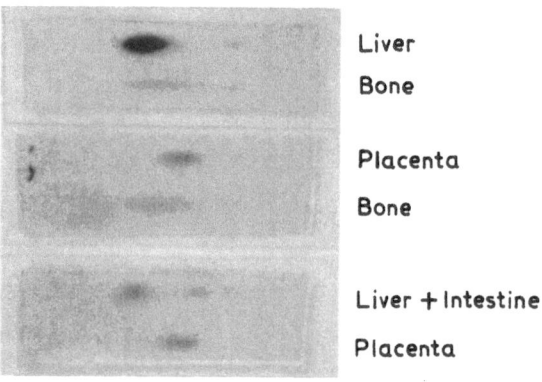

Liver

Bone

Placenta

Bone

Liver + Intestine

Placenta

Fig. 54. Agar-gel electrophoresis of human-tissue alkaline phosphatases.

superposition of these two bands, since it frequently occurs in patients with diseases involving both bone and liver, and bands in this region may be obtained by the electrophoresis of mixtures of serum containing band I only with serum containing band II only.

There is general agreement between the results of Haije and de Jong (1963) and those obtained by starch-block electrophoresis, in that band I appears to be associated with bone diseases, while the faster fractions II and III predominate in liver disease.

In the writer's laboratory agar-gel electrophoresis has proved a useful technique for the differentiation of liver and bone as the source of an elevated serum alkaline phosphatase, provided that suitable markers were available. Some typical results are illustrated in Fig. 54. The human liver isoenzymes sometimes appeared as two bands, but usually showed only a single band, which migrated slightly faster than the bone enzyme. The placental component appeared midway between the bone band and the intestinal isoenzyme, which remained near the origin (Metz and Wilkinson, unpublished observations).

Agar-gel electrophoresis has also been used in the study of alkaline phosphatase synthesis after ligation of the bile duct in the rat (Börnig, Štěpán, Horn, Giertler, Thiele and Večerek, 1967).

STARCH-GEL ELECTROPHORESIS. The improved resolution of protein fractions obtainable by starch-gel electrophoresis extends to the phosphatase components (Markert and Møller, 1959). Estborn (1959) showed that the principal alkaline phosphatase in the serum of a patient with obstructive jaundice due to cholelithiasis migrated as a distinct band between the β- and haptoglobin fractions. In certain other sera this band is accompanied by a weak band of activity occurring in the prealbumin (acid α_1) fraction. This component is also found in bile, but the principal phosphatase of this fluid migrates somewhat more slowly than the main serum component. Estborn concludes that the serum alkaline phosphatase migrates separately from the main serum protein fractions and that the association with the α_1- and α_2-globulins reported by investigators using starch grains as support media is coincidental. Hodson, Latner and Raine (1962) came to the same conclusion as a result of their observation that the positions of the alkaline phosphatase bands relative to the major serum proteins may be varied by using a discontinuous buffer system.

Numerous investigators (Kowlessar, Pert, Haeffner and Sleisenger, 1959; Boyer, 1961; Latner, 1961; Paul and Fottrell, 1961; Kowlessar, Haeffner and Riley, 1961; Moss, Campbell, Anagnostou-Kakaras and King, 1961a, 1961b; Hodson et al., 1962; Latner and Skillen, 1962; Moss, 1962a; Moss and King, 1962; Chiandussi, Greene and Sherlock, 1962; Cunningham and Rimer, 1963; Taswell and Jeffers, 1963; Warnock, 1966) have now separated human and animal serum and tissue alkaline phosphatase components by starch-gel electrophoresis, and though the patterns obtained differ in detail, certain general conclusions may be drawn. The principal alkaline phosphatase band of liver extracts migrates close to the transferrin (β-globulin) fraction of serum, but minor bands in the β-lipoprotein and slow α_2-globulin zones are usually detected. There may also be activity near the point of insertion. The main band of kidney extracts occurs in the haptoglobin zone, while minor bands may be seen in the β-lipoprotein and transferrin (β-globulin) regions. Most of the intestinal alkaline phosphatase also moves rather more slowly than the main band of the liver enzyme, but slower fractions in the β-lipoprotein region and near the origin may be detected. Bone alkaline phosphatase activity is usually found close to the transferrin (β-globulin) region, but some investigators (e.g. Moss, 1962a) have reported the presence of slower bands. Some typical findings are illustrated in Fig. 55.

Evidence that the slow-moving bands of serum originate in the liver has been provided by elution of the enzyme after starch-gel electrophoresis. The slow band of alkaline phosphatase also exhibits considerable 5′-nucleotidase activity, whereas the fraction associated with the β-

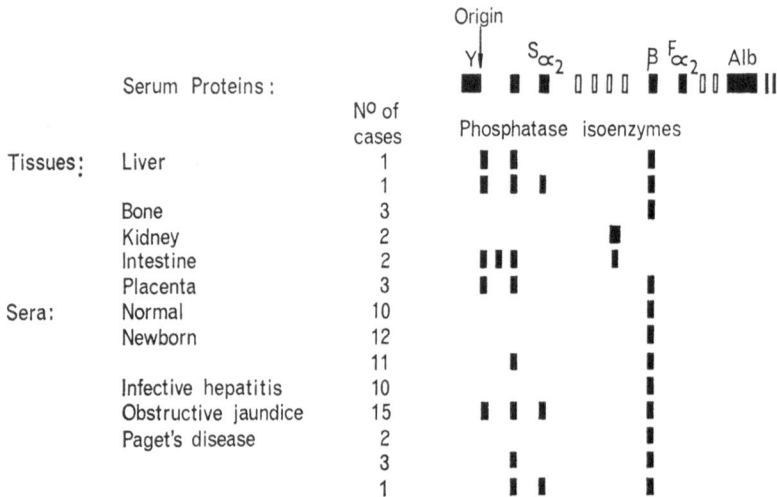

Fig. 55. The alkaline phosphatase isoenzymes of human tissues and of normal and pathological sera, separated by starch-gel electrophoresis. (From Chiandussi, Greene and Sherlock, 1962.)

globulins is devoid of such activity (Kowlessar *et al.*, 1959, 1961). Since elevation of the serum 5′-nucleotidase is highly specific for liver diseases (Dixon and Purdom, 1954), these observations are consistent with the view that the liver is the source of the slow components of the serum alkaline phosphatase.

The degree of resolution obtained by different investigators varies considerably, and Boyer (1961) reports the occurrence of no fewer than sixteen bands of activity in human serum, though not all could be detected in the serum of a single individual. For convenience this author has classified the bands into six zones of activity ($ABCDEF$), the most rapid of which migrates immediately behind transferrin C, while the D zone has a mobility similar to that of the haptoglobin 1–1 band. The fast zones *A* and *B* and the slow *D* components occur only in the sera of pregnant women (see p. 259) and appear to be derived from the placenta, since they occur after placental extracts are subjected to electrophoresis, and since their activities are destroyed by rabbit anti-human placental alkaline phosphatase. The *C* components appear to be derived from bone, since they are prominent in the sera of growing children and are inactivated by rabbit anti-human bone alkaline phosphatase, but not by the placental anti-enzyme. Moreover, the principal isoenzyme of purified bone and milk alkaline phosphatase migrates in the *C* zone. The *F* zone is shown by two-dimensional electrophoresis (paper–starch gel) to contain several active fractions which have mobilities corresponding to those of α_1, α_2 or α_2-β-globulins of human sera when separated by paper electrophoresis.

Of 120 patients with a variety of diseases and having raised serum alkaline phosphatase activities, only two exhibited bands in other than the *C* and *F* zones, and Boyer therefore concludes that simple starch-gel electrophoresis is not likely to be useful in diagnosis. Other workers, however, suggest that there are clinical problems in which study of the serum alkaline phosphatase isoenzymes might provide information of diagnostic significance. Since the principal fractions of liver and bone alkaline phosphatase occur in the β-globulin region, it is clear that a band in this zone might originate in either tissue (Chiandussi *et al.*, 1962; Moss, 1962*a*). Nevertheless, the presence and distribution of the minor bands appear to be significant. Chiandussi *et al.* (1962) observed bands in the γ-globulin, β-lipoprotein and slow α_2-globulin regions in the sera of patients with biliary obstruction due to a variety of causes. Since a similar pattern also occurs in the bile, this observation is consistent with the view that in obstructive jaundice there is regurgitation of bile into the blood. Hodson *et al.* (1962) point out that serum isoenzyme separation might aid in the differential diagnosis of jaundice, for in cases of obstructive jaundice they noted a doublet of phosphatase activity near the origin which is less prominent in sera from patients with hepatocellular disease.

Starch-gel electrophoresis has been employed in attempts to resolve the controversy concerning the origin of the alkaline phosphatase of normal human serum. Schlamowitz and Bodansky (1959) reported that immunological studies with anti-human bone phosphatase (see p. 262) indicate that about half the normal serum enzyme originates in bone, but the authors point out that their anti-serum cross-reacted to some extent with phosphatases from other tissues. On electrophoresis most of the activity occurs close to the β-globulin region, a fraction attributed by Kowlessar et al. (1959) to bone phosphatase and by Hodson et al. (1962) to the liver enzyme. Chiandussi et al. (1962), however, conclude that it could be of either hepatic or osseous origin.

Considerable evidence has now accrued to confirm that the liver is the main source of the normal serum enzyme (Hodson et al., 1962; Cunningham and Rimer, 1962). Keiding (1966) found the normal serum enzyme to behave like the liver enzyme in its resistance to inhibition by L-phenylalanine, though Fishman, Inglis and Krant (1965), using a somewhat different technique for measuring L-phenylalanine inhibition (see p. 255), had earlier suggested that normal serum contains a substantial proportion of the intestinal component. Warnock (1966) used a combination of starch-gel electrophoresis, heat inactivation (p. 258) and L-phenylalanine inhibition, and confirmed that the serum enzyme of normal adults is largely derived from the liver, but in children it appears to be of mixed bone and liver origin. An additional component behaving like the intestinal enzyme was occasionally detected in the sera of both children and adults. A similar conclusion was reached by Yong (1966) on the basis of agar-gel electrophoresis.

The presence of the bone enzyme in the sera of children is well established (Taswell and Jeffers, 1963; Beckman and Grivea, 1965; Warnock, 1966; Metz and Wilkinson, unpublished observations), and is due to increased osteoblastic activity associated with bone growth during childhood.

Starch-gel electrophoresis has also been applied in the study of the alkaline phosphatases of other body fluids. The isoenzyme patterns of bile differ from those of the corresponding serum, though preparative ultracentrifugation indicates that their molecular weights are similar (Pope and Cooperband, 1966).

Urinary alkaline phosphatase migrates on starch gel at a much faster rate than that of kidney, liver, bone, small intestine or serum. It is considered that the fast-moving urine component might consist of the kidney enzyme which has been altered after its release into the urine, since in acute tubular necrosis there is a considerable increase in urinary phosphatase excretion (Butterworth, Pitkänen and Moss, 1963; Moss, 1964a; Butterworth, Moss, Pitkänen and Pringle, 1964).

Among the alkaline phosphatases of other species investigated by starch-gel electrophoresis are those of *Drosophila melanogaster* (Beckman and Johnson, 1964; Schneiderman, 1967; Wallis and Fox, 1968), sheep (Rendel and Stormont, 1964; Rasmusen, 1965), cattle (Gahne, 1963) and the domestic fowl (Law and Munro, 1965). In some of these species enzyme heterogeneity appears to be under genetic control, an aspect discussed later in this chapter (p. 264).

Few comparative electrophoretic studies of the alkaline phosphatases of different species have so far been reported, but Paul and Fottrell (1961) found wide variation in mobilities of the liver isoenzymes in starch gel. In the rat and the mouse there are two fast-moving bands, whereas in the guinea-pig, frog and pigeon a single, slower, more diffuse band can be seen. A single fast band occurs in the liver of the perch.

ACRYLAMIDE GEL ELECTROPHORESIS. Among the applications of acrylamide gel electrophoresis to the study of alkaline phosphatase heterogeneity are those of Epstein, Wolf, Horwitz and Zak (1967), Sussman, Small and Cotlove (1967), Smith and Moss (1968), Dunne, Fennelly and McGeeney (1968), and Kaschnitz, Patch and Peterlik (1968). So far relatively little experience has been gained with this medium, but Smith and Moss found it to be a useful preparative tool for the quantitative separation of the human tissue isoenzymes.

Rat liver, kidney and bone phosphatases have been identified in the serum by acrylamide disc electrophoresis (Akhtar, Hansen and Kaercher, 1968). An indigogenic technique, depending upon the hydrolysis of 5-bromo-4-chloroindolyl 3-phosphate, has been devised for the location of alkaline phosphatase activity after disc electrophoresis (Epstein *et al.*, 1967).

ELECTROPHORESIS ON SEPHADEX G 200. Sephadex G 200 has been used as a medium for the gel electrophoresis of human serum and tissue alkaline phosphatases. Either thick or thin layers may be used; the latter may be dried to produce permanent records (Fishman and Ghosh, 1967; Inglis, Ghosh and Fishman, 1967).

CELLULOSE ACETATE ELECTROPHORESIS. Cellulose acetate electrophoresis has also been applied to the separation of alkaline phosphatase isoenzymes, and the results obtained suggest that this convenient medium might well be of value in diagnosis (Korner, 1962; Kitchener, Neale, Posen and Brudenell-Woods, 1965; Posen, Neale, Birkett and Brudenell-Woods, 1967; Romel, LaMancusa and Dufrene, 1968). Cellulose acetate gel (Cellogel) seems to give the best resolution, but other forms of cellulose acetate have also been used with satisfactory results.

As with other electrophoretic media, α-(or β-) naphthyl phosphate has been employed as a chromogenic substrate, the liberated naphthol being coupled with a suitable diazo reagent (Azoene fast violet B or Fast blue RR) to indicate the sites of phosphatase activity (Barka, 1961; Kitchener *et al.*, 1965). Very satisfactory results have been obtained in the writer's laboratory with this technique, which appears to be rather more sensitive than the agar-gel method described above.

More recently phenolphthalein monophosphate has been recommended as a suitable substrate, as this substance offers a means for the quantitative determination of the various enzyme fractions. For this purpose the bands are cut out, individually eluted, treated with alkali and determined colorimetrically. Alternatively, the strips may be cleared with a solution of sodium hydroxide in glycerol and mounted between microscope slides for subsequent scanning in a suitable densitometer (Romel *et al.*, 1968).

ION-EXCHANGE CHROMATOGRAPHY. The application of more refined techniques has shown that the human serum enzyme also consists of several distinct components. Chromatography on DEAE-cellulose columns enabled Fahey, McCoy and Goulian (1958) to detect two peaks of activity, both of which were eluted long before the single peak of acid phosphatase activity. Similar methods have also been used for the separation of various human-tissue alkaline phosphatase isoenzymes (Grossberg *et al.*, 1961), those of sheep brain (Saraswathi and Bachhawat, 1966) and the genetically determined mouse duodenal isoenzymes (Moog, Vire and Grey, 1966; Nayudu and Moog, 1967). DEAE-cellulose column chromatography has also been used to separate the alkaline phosphatase of *Escherichia coli* into three distinct components (Lazdunski and Lazdunski, 1967).

An interesting variant of this technique using chromatography on DEAE-paper has been employed by Hosoyama, Fukuda and Shimadate (1967) to separate five components from rat tissues (Fig. 56).

Several investigators have used DEAE-Sephadex column chromatography for the study of tissue alkaline phosphatases. For example, two fractions have been separated from the human small intestine (Moss, 1964*b*) and four from human kidney (Butterworth and Moss, 1966), while one major fraction and a number of minor components have recently been isolated from porcine kidney (Butterworth, 1968*a*).

Properties of alkaline phosphatases

Alkaline phosphatase is relatively non-specific in its substrate requirements, and several substrates have been proposed for the determination of its activity, e.g. β-glycerophosphate (Bodansky, 1933; Shinowara, Jones and Reinhart, 1942), phenyl phosphate (King and Armstrong, 1934; King and Wootton, 1956), *p*-nitrophenyl phosphate (Ohmori, 1937; King and Delory, 1939), phenolphthalein diphosphate (King, 1938; Klein, Read and Babson, 1960) and phenolphthalein monophosphate (Babson, 1965; Babson, Greeley, Coleman and Phillips, 1966; Wilkinson and Vodden, 1966).

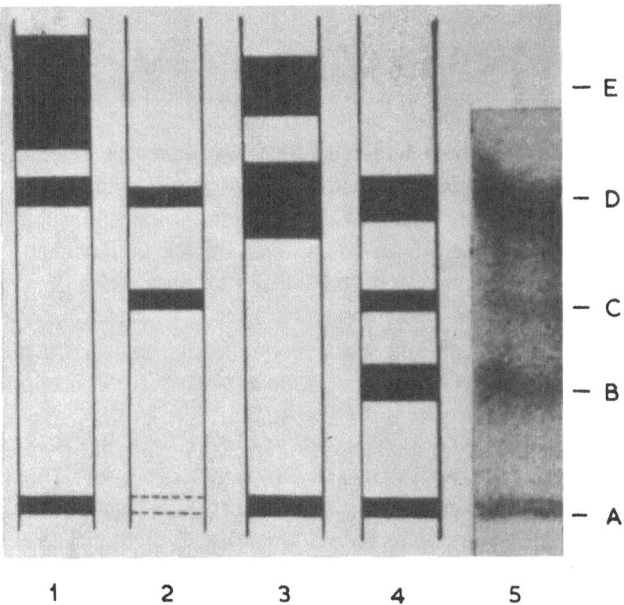

Fig. 56. Rat-tissue alkaline phosphatases separated by chromatography on DEAE-paper. 1, small intestine; 2, liver; 3, kidney; 4, testis; 5, photograph of testis chromatogram. (Hosoyama, Fukuda and Shimadate, 1967.)

Determination of their relative activities with a number of substrates may enable the alkaline phosphatases of different tissues to be distinguished (Table 33), but there is no evidence to suggest that such techniques can be used to differentiate between the various fractions separable by chromatography or electrophoresis. Nucleotide phosphates, such as adenosine 5'-phosphate, appear to be exceptional, since Kowlessar *et al.* (1961) have shown that only certain fractions of alkaline phosphatase exhibit ability to hydrolyse this substrate (5'-nucleotidase activity). Similar observations have also been made by Sandler and Bourne (1961, 1963).

TABLE 33 Relative activities of human tissue alkaline phosphatases with various ester phosphate substrates

(From Nisselbaum, Schlamowitz and Bodansky, 1961)

Substrate	Relative activity of enzyme from				
	Intestine	Bone	Liver	Spleen	Kidney
β-Glycerophosphate	100	100	100	100	100
p-Nitrophenyl phosphate	45	103	83	100	99
Phenolphthalein phosphate	47	71	82	71	69
Adenosine 5′-phosphate	131	66	65	57	53
Phosphoglyceric acid	48	23	34	30	31

The nucleoside phosphatases of the Golgi body have been separated by acrylamide-gel electrophoresis into two groups which differ according to their substrate specificities for thiamine pyrophosphate, on the one hand, and cytidine, guanosine and inosine diphosphates, on the other. Neither reacts with adenosine diphosphate (Allen, 1963; Allen and Hyncik, 1963).

5′-Nucleotidase has long been recommended as a useful means for differentiating between liver and bone as the source of a raised serum alkaline phosphatase activity. Elevated activities occur in hepatobiliary diseases, but the serum nucleotidase remains normal in disease of bone (Dixon and Purdom, 1954; Young, 1958).

Eaton and Moss (1967c) have compared the relative activities of liver and intestinal alkaline phosphatases with the three adenosine phosphates. The liver enzyme has a marked preference for AMP, especially in the presence of added Mg^{2+}, though the K_m values for AMP, ADP and ATP were similar (about 2 mM). The intestinal enzyme hydrolysed the three substrates at approximately equal rates, but in the presence of added Mg^{2+}, the activities with ADP and ATP were much reduced, while that with AMP was unchanged.

SUBSTRATE AFFINITIES. The introduction of a highly sensitive spectrofluorimetric procedure for the determination of alkaline phosphatase, in which the rate of the hydrolysis of β-naphthyl phosphate is followed (Moss, 1960a; 1960b), has enabled the enzyme kinetics of alkaline phosphatases from various tissues, partly purified by starch-gel electrophoresis, to be determined. Moss et al. (1961a) evaluated the Michaelis constants for the principal bone, liver, intestine and kidney alkaline phosphatases, all of which migrate at a similar rate slightly behind the transferrin C fraction of serum globulin. During this study they confirmed the observation of Motzok (1959) that the pH optimum changes with alterations in substrate concentration. The effects on the K_m values of partial purification by electrophoresis were well marked, especially in the case of the liver and kidney enzymes. The results are summarized in Table 34.

TABLE 34 *Michaelis constants* (K_m) *for β-naphthyl phosphate of electrophoretically purified human tissue alkaline phosphatases* (*major bands*)

(Moss, Campbell, Anagnostou-Kakaras and King, 1961a)

Tissue of origin	K_m (mM)
Bone	0·110
Liver	0·067
Intestine	0·090
Kidney	0·103

These studies have been extended to the alkaline phosphatases of pathological sera by Moss *et al.* (1961b), some of whose results are listed in Table 35. K_m values similar to that of the bone enzyme occur in

TABLE 35 *Michaelis constant* (K_m) *for β-naphthyl phosphate of the principal alkaline phosphatases electrophoretically separated from human pathological sera*

(From Moss, Campbell, Anagnostou-Kakaras and King, 1961b)

Group	Case	Diagnosis	Serum alkaline phosphatase (King-Armstrong units)	K_m for β naphthyl phosphate (mM)
I (Serum enzyme resembles that of bone)	1	Paget's disease	53	0·11
	2	Cancer of breast with bone secondaries	57	0·11
	3	Cancer of pancreas with biliary obstruction	96	0·094
II (Serum enzyme resembles that of liver)	4	Cancer of pancreas with secondaries	200	0·067
			310	0·061
	5	Biliary cirrhosis	97	0·071
III (Serum enzyme resembles mixtures of bone and liver enzymes)	3	After operation for relief of obstruction	56	0·083
	6	Lymphosarcoma	160	0·081

the sera of patients with Paget's disease or with bony metastases, while the K_m values found with the serum enzyme of patients with biliary cirrhosis or with hepatic metastases resemble that of the liver enzyme. In a third group the serum enzyme appears to consist of a mixture of the bone and liver enzymes. It is of interest that in the early stages of a case (no. 3) of extrahepatic obstructive jaundice the serum enzyme had a K_m value similar to that of the bone enzyme, while in later stages of the disease there was a change towards the K_m value of the liver phosphatase.

Moss and King (1962) have since demonstrated that the K_m values for the minor bands of phosphatase activity of bone, liver, intestine and kidney are the same as those of the major bands of the respective tissues. Furthermore, the various fractions of the same tissue show similar pH optima, similar rates of thermal denaturation and similar activation by Mg^{2+} ions. These results are interpreted as indicating that each organ may elaborate a single alkaline phosphatase which might subsequently undergo slight structural changes to give multiple bands on electrophoresis (Moss, 1962a). This explanation appears probable, but it would be interesting to learn whether the same conclusion is reached when an alternative substrate is used.

Such changes might occur during extraction or as the result of post-mortem autolysis or, alternatively, the minor bands might result from the binding of the phosphatase molecule with other proteins. The occurrence of a band of phosphatase activity in the β-lipoprotein region in each tissue extract studied is consistent with this view, and Moss (1962b) has obtained further evidence from a comparative study of autolysed and butanol-extracted preparations of human liver. Most activity was found near the transferrin C globulin zone (band 1) in both preparations, but there was a considerable difference between the two extracts in the positions and relative intensities of the minor bands. The butanol-extracted preparation exhibited much greater activity in the β-lipoprotein region (band 4) than did the autolysed specimen, while the latter contained more activity in band 2, which had a mobility only slightly less than band 1. Both preparations contained trace amounts in the slow α_2-globulin region (band 3). Moss suggests that butanol may extract a phosphatase–lipoprotein complex which migrates as a β-lipoprotein (band 4). The relative absence of band 4 in the autolysed preparation may be attributed to the breakdown of such a complex during autolysis. The chromatographic behaviour of butanol-extracted and autolysed preparations of hog kidney might be explained on the same basis. While butanol-extracted preparations of fresh material exhibit a single phosphatase component on DEAE-cellulose chromatography, the enzyme from autolysates is eluted in two fractions which, however, appear to have almost identical relative activities with a number of substrates (Nisselbaum *et al.*, 1961).

It is interesting to note in this connection that similar complex formation with lipids has been postulated by Keiding (1964) to account for the presence of a β_2-alkaline phosphatase in human lymph (see p. 240).

Although there is strong evidence in support of the lipoprotein complex theory, this explanation may not apply in all cases, and it is difficult to reconcile it with observations on the intestinal phosphatases. The single broad band found by Moss and King (1962) has been resolved into three components. Using the butanol-extraction technique, Moss (1963) demonstrated the presence of two electrophoretically faster components by DEAE-cellulose column chromatography and starch-gel electrophoresis. Prolonged storage of the butanol extracts or prolonged autolysis of the tissue brought about an increase in the mobility of the fastest component, but the changes in relative intensities were rather slight. Moss concludes from studies with Sephadex G-200 that the resolution of intestinal alkaline phosphatase during starch-gel electrophoresis is due to differences in charge rather than in molecular size.

Moss (1964*b*) has extended his study of the two main human intestinal phosphatase fractions separated by DEAE-Sephadex column chromatography, and has shown that, though they are electrophoretically distinct and differ in thermal stabilities, they have similar Michaelis constants for phenyl phosphate and similar inhibition constants for competitive inhibition by inorganic phosphate. Both are similarly inhibited by L-phenylalanine (p. 255). It seems, therefore, that the difference between the fractions does not involve the enzymically active centre, and may possibly be due to degradation (Moss, personal communication).

This view is supported by Robinson and Pierce (1964), who demonstrated that the anodic migration of three of the four human alkaline phosphatase isoenzymes separated by starch-gel electrophoresis is reduced by previous incubation with neuraminidase. This result is interpreted as indicating that in three of the isoenzymes there is a terminal neuraminic (sialic) acid residue, which is not present in the fourth. It appears likely, therefore, that alkaline phosphatase isoenzymes may differ from each other in their carbohydrate components (see p. 259).

Although early work (Folley and Kay, 1936; Morton, 1955) suggested that the enzymes hydrolysing orthophosphates and pyrophosphate were distinct from each other, evidence has recently accumulated indicating that human-tissue alkaline phosphatases also possess pyro-phosphatase activity (Cox and Griffin, 1965; Fernley and Walker, 1966; Moss, Eaton, Smith and Whitby, 1967; Eaton and Moss, 1967*b*). Moss *et al.* (1967) compared the alkaline phosphatase (determined with *p*-nitrophenyl phosphate) and pyrophosphatase activities of the human liver and intestinal enzymes at various stages during purification, and found an exact coincidence of both types of activity during gel-filtration, ion-exchange chromatography and starch-gel electrophoresis. Moreover, the ratio of the two activities remained constant during inactivation of each preparation at 55°. Similar results have since been obtained with bone alkaline phosphatase (Eaton and Moss, 1968*a*). Close correlation was also observed between the alkaline phosphatase and pyrophosphatase activities of sera from patients with bone and hepatobiliary diseases (Eaton and Moss, 1967*a*).

Comparison with the pyrophosphatase activity may be of value in differentiating certain tissue alkaline phosphatases, for Eaton and Moss (1967*a*) found the ratio of *ortho*phosphatase activity to *pyro*-phosphatase for the intestinal enzyme to differ markedly from those of the liver and bone enzymes. Eaton and Moss (1968*b*) have extended their studies by comparing the kinetics for the hydrolysis of both sub-strates with purified liver, bone and intestinal phosphatases. In each case the K_m values for both reactions were similar at pH 7–9, but above pH 9 there is marked increase in the K_m for orthophosphate while that for pyrophosphate remains virtually unchanged. Discontinuities in the pK_m–pH curves for pyrophosphatase activity at pH 7·5, 8·1 and 8·4 for the liver, bone and intestinal enzymes respectively are attributed to changes in enzyme rather than substrate ionization. The authors deduce that at physiological pH values the enzymes are likely to be as effective inorganic pyrophosphatases as they are orthophosphatases, a conclusion of interest in view of the suggestion of Russell (1965) that bone alkaline phosphatase may act as a pyrophosphatase in calcification.

EFFECTS OF INHIBITORS. Early work with enzyme inhibitors added little to our knowledge of the various alkaline phosphatases. Differences in the degree of inhibition by amino acids (Bodansky, 1936, 1948; Thannhauser, Reichel and Grattan, 1937; Abul-Fadl and King, 1949), bile salts (Bodansky, 1937), formalin (Abul-Fadl and King, 1949) and iodoacetate (Gryden, Friedenwald and Carlson, 1955) have been reported, but Ahmed, Abul-Fadl and King (1959) suggest that most of these effects are relatively non-specific. Similarly, the inhibitory effect of

Zn^{2+} ions appears to be exerted irrespective of the tissue of origin of the enzyme.

Mg^{2+} (or Co^{2+}) ions appear to be essential activators of alkaline phosphatase (Ahmed *et al.*, 1959), and consequently, any substance such as ethylenediamine tetra-acetic acid (EDTA) which removes Mg^{2+} ions inactivates the enzyme. Rat-liver and rat-intestinal glucose 6-phosphatase, however, have recently been shown to be unaffected by 0.01M-EDTA at pH 6·5, but to be completely inhibited at pH 8·6, whereas the corresponding β-glycerophosphatases are strongly inhibited at both pH values (Di Bella, Richetta and Pichierri, 1963). Similar effects are observed with 0.06M-cyanide.

EDTA appears to have an effect on alkaline phosphatase other than the chelation of Mg^{2+} ions, and it has been suggested that it may form a complex with the enzyme molecule (Wallach and Ko, 1964). Placental alkaline phosphatase differs from that of other tissues in being resistant to inactivation by EDTA (Kitchener *et al.*, 1965; Conyers, Birkett, Neale and Posen, 1966). Two types of rat-liver alkaline phosphatase have been separated from the particulate and supernatant fractions by chromatography on DEAE-cellulose. Both are inhibited by EDTA, but only the supernatant enzyme is sensitive to Mg^{2+}. This component is strongly inhibited by fluoride, whereas the particulate enzyme is not affected by fluoride (Kaneko, 1968).

Some of the early reports have been re-assessed by Fishman, Green and Inglis (1962), who observed some organ-specific inactivation of rat alkaline phosphatases. These investigators found that certain amino acids, e.g. phenylalanine and leucine, specifically inhibit intestinal alkaline phosphatases, and confirmed the observation of Bodansky (1937) that taurocholate inhibits the liver, bone, lung and kidney enzymes but not the intestinal phosphatase.

The specific effect of L-phenylalanine on the intestinal enzyme has been confirmed by other investigators (Moss, 1965; Newton, 1966; Warnock, 1966), and this inhibitor has since found a number of applications in the study of alkaline phosphatase isoenzymes (reviewed by Fishman and Ghosh, 1967). It is interesting to note that Eaton and Moss (1967*b*) found both the alkaline phosphatase and pyrophosphatase activities of human intestine to be inhibited by L-phenylalanine, whereas both activities of a liver preparation were unaffected by this reagent. This is consistent with the view (p. 253) that both the phosphatase and pyrophosphatase activities are properties of the same enzyme protein.

Fishman and Ghosh (1967) recommend the use of D-phenylalanine in the control determination to eliminate the possibility of non-specific inhibition or activation. Although the intestinal component is maximally inhibited with 0·007M-L-phenylalanine, these authors found a concentration of 0·005M to have the greatest differential effect. The placental enzyme has also been shown to be equally sensitive to L-phenylalanine (Fishman, Inglis, Sarke and Ghosh, 1966).

The mechanism of the stereospecific inhibition has recently been investigated by Ghosh and Fishman (1966, 1968), who found it to be pH-dependent and to be of the 'uncompetitive' type. The small entropy changes associated with L-phenylalanine inhibition, the finding of less than one L-phenylalanine molecule combined with one enzyme molecule and the limited effect of conformational changes induced by urea or heat on the extent of L-phenylalanine inhibition are consistent with a homosteric rather than an allosteric mechanism.

A histological study of the L-phenylalanine-sensitive alkaline phosphatase of rat intestine by Watanabe and Fishman (1964) showed that it is localized in the striated border of the epithelial cells, an observation which led Fishman and Ghosh (1967) to speculate upon the possibility that L-phenylalanine sensitivity might be organelle- or site-specific rather than organ-specific.

Another inhibitor which has recently found application in phosphatase studies is urea (Butterworth and Moss, 1966; Birkett, Conyers, Neale, Posen and Brudenell-Woods, 1967; Bahr and Wilkinson, 1967; Horne, Cornish and Posen, 1968). At low concentrations (*ca.* 0·5M) it appears to have rather variable effects, and in the writer's laboratory some liver extracts and occasional sera from patients with liver disease have shown marked activation, particularly when phenyl phosphate has been used as substrate. This action is as yet unexplained, but it seems to be of little diagnostic value. At concentrations greater than 2M, urea exerts a time-dependent, temperature-dependent and pH-dependent effect which may be of some value in differentiating tissue alkaline phosphatases. Some examples are illustrated in Fig. 57. Placental phosphatase is the most resistant to inhibition by urea, but the intestinal enzyme is also quite insensitive, whereas the kidney, liver and bone enzymes are strongly inhibited.

Exposure to acid conditions at 0° inactivates various human tissue alkaline phosphatases, which differ in their sensitivities. Thus incubation of the bone enzyme at pH 3·5 for 20 min. causes almost complete inactivation, whereas the liver enzyme retains about 50% of its activity under such conditions, when phosphatase activity is subsequently determined at pH 9·9 (Eaton and Moss, 1968b). The intestinal enzyme, however, is little affected by acid treatment (Butterworth and Moss, 1967; Scutt and Moss, 1967).

The pig-kidney enzyme is also inactivated by exposure to acid conditions, but is reactivated on subsequent incubation for 30 min. at 30° and pH 7·6. Reactivation is complete when the acid incubation is performed at pH 3·5, but is less so when the enzyme is exposed to lower pH values.

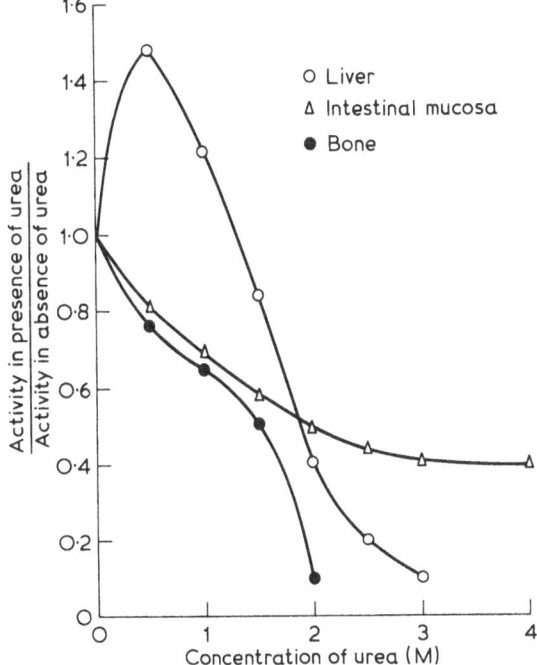

Fig. 57. Effect of urea on some human-tissue alkaline phosphatases with phenyl phosphate as substrate in bicarbonate–carbonate buffer at pH 10·0. (Bahr and Wilkinson, 1967.)

Very little reactivation occurs when the enzyme is inactivated at pH values below 3·3. These effects are attributed to a reversible conformational change which becomes permanent on prolonged acid treatment (Butterworth, 1968*b*).

Arsenate behaves as a competitive inhibitor of human liver and intestinal orthophosphatase and pyrophosphatase activity, and provides further confirmation that both types of activity are associated with the same enzyme molecule in each case (Eaton and Moss, 1967*b*).

It is convenient at this stage to refer to the precipitation of alkaline phosphatases by 20% ethanol at 0°, a procedure which Peacock, Reed and Highsmith (1963) suggest may aid differentiation of the bone and liver enzymes in pathological sera. Less than one Bessey–Lowry–Brock (1946) unit is found in the precipitate produced by the sera of patients with bone disease, while that from the sera of patients with liver involvement may contain up to 3 units or more. The authors obtained evidence suggesting that the hypoalbuminaemia which often occurs in liver disease is not responsible for the precipitation of the liver enzyme. The technique, however, appears to be of limited value when there is simultaneous bone and liver disease.

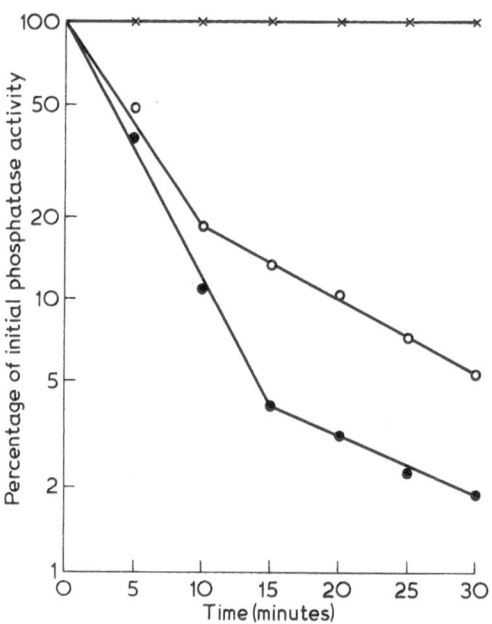

Fig. 58. Effect of heating at 56° on the alkaline phosphatase activity of O, normal serum; ●, serum from a patient with bone disease (Paget's); and x, placental homogenate. (Neale, Clubb, Hotchkis and Posen, 1965.)

HEAT INACTIVATION. Human placental alkaline phosphatase differs from the enzyme from other tissues in being remarkably stable to heat. In the presence of Mg^{2+} ions, Neale, Clubb, Hotchkis and Posen (1965) found the placental enzyme to withstand heating at 70° for 30 min. without loss of activity, while alkaline phosphatases from other human tissues are almost completely inactivated when heated at 56° for the same period (Fig. 58). Bone phosphatase is more sensitive to heat inactivation than the liver enzyme, and this technique has been suggested as an aid in the identification of the tissue of origin of a raised serum alkaline phosphatase (Posen, Neale and Clubb, 1965; Kerkhoff, 1968).

The heat stability of the placental enzyme has had some important clinical applications. The elevated serum alkaline phosphatase observed during the third trimester of pregnancy has been shown to contain an appreciable heat-stable component which is therefore considered to be of placental origin (Fig. 59) (McMaster, Tennant, Clubb, Neale and

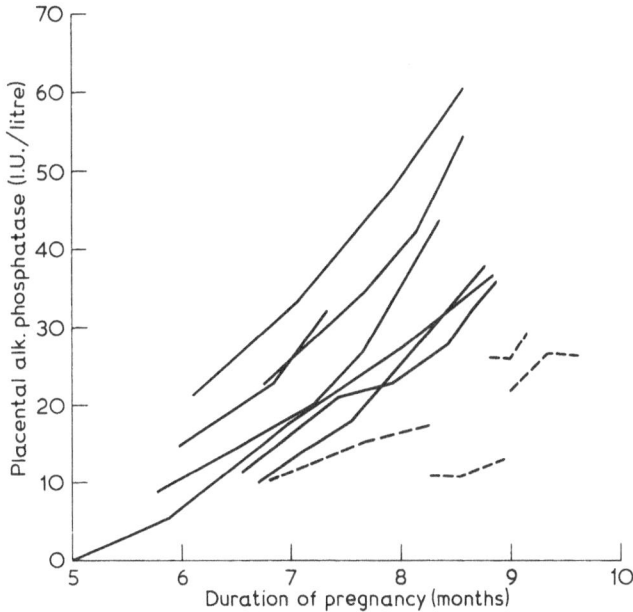

Fig. 59. Heat-stable (placental) alkaline phosphatase in pregnancy serum, ___, normal pregnancies; ---, complicated pregnancies.

Posen, 1964), a conclusion consistent with the results of electrophoretic studies (Chiandussi *et al.*, 1962; Latner, 1965) and immunological investigations (Birkett, Done, Neale and Posen, 1966; Sussman, Bowman and Lewis, 1968). The foetal serum enzyme, however, does not contain any significant amount of the heat-stable component and does not appear to be derived from the placenta or the maternal circulation (Kitchener *et al.*, 1965).

These results have been confirmed by others (Curzen and Morris, 1966; Messer, 1967; Shaper and Patel, 1969), and it seems that the heat-stable fraction of the serum alkaline phosphatase may be of value as a test of placental function. Messer (1967) has shown that placental dysfunction is associated with a fall in the level of this component in the serum which correlates well with urinary oestriol determination.

The ready identification of human placental alkaline phosphatase by means of its heat stability enabled Clubb, Neale and Posen (1965) to study its clearance from the circulation after intravenous infusion into human subjects. The enzyme was found to disappear from the circulation according to a multiphasic exponential equation similar to that describing the clearance of injected plasma proteins, at rates independent of the amount of endogenous phosphatase administered and of its concentration in the plasma.

The clearance rates in patients with biliary obstruction were similar to those found in normal subjects, a highly significant observation in view of the prolonged controversy concerning the possible biliary excretion of the enzyme. For more than thirty years since Armstrong, King and Harris (1934) demonstrated a marked rise in the serum alkaline phosphatase to follow ligation of the bile duct in the dog, the view has been widely held that the serum enzyme originates principally in bone and is excreted in the bile. This retention theory (reviewed by Gutman, 1959) is not consistent with the results of Posen and his colleagues, who question whether there is any need to postulate the excretion of enzymes.

An example of an unusual form of heat-stable alkaline phosphatase has recently been detected by Fishman, Inglis, Stolbach and Krant (1968) in a male patient with a bronchogenic carcinoma with multiple secondaries in the adrenals, spleen, kidneys and brain, but not in the liver or bone. The serum and tumour alkaline phosphatase closely resembled the placental enzyme not only in heat stability but also in electrophoretic mobility, sensitivity to L-phenylalanine and the effect of neuraminidase.

In this connection it is also of interest to note that Warnock and Reisman (1969) have recently reported the occurrence of a fast-moving variant alkaline phosphatase in liver tumours. The variant differs from the normal liver enzyme in being more sensitive to inhibition by L-phenylalanine and in being less retarded during electrophoresis after treatment with neuraminidase.

EFFECT OF NEURAMINIDASE. As mentioned on p. 77, treatment with neuraminidase reduces the electrophoretic mobility of the faster human serum and tissue alkaline phosphatase, while that of the slowest component is unaffected (Robinson and Pierce, 1964). This suggests that all but the slowest isoenzyme possess a terminal neuraminic (sialic) acid group, removal of which reduces the negative charge on the enzyme molecule. Other investigators have confirmed this observation (Fig. 16, p. 78). Posen (1967) showed that bone phosphatase, which normally travels with the α_2-globulins on cellulose acetate electrophoresis, remains with the γ-globulins after incubation with neuraminidase, while Moss *et al.* (1966, 1967) found that after such treatment the orthophosphatase and pyrophosphatase zones of human liver extracts were retarded to the same extent on starch-gel electrophoresis. Both groups, however, reported that the mobilities of human and rat intestinal phosphatase are unaffected by treatment with neuraminidase. Fishman and his co-workers have demonstrated that placental phosphatase, the electrophoretic mobility of which is also retarded by neuraminidase, contains at least 1·5% of sialic acid, most of which can be liberated by enzymic hydrolysis (Fishman and Ghosh, 1967; Ghosh, Goldman and Fishman, 1967).

The effects of neuraminidase on the electrophoretic mobilities of the serum alkaline phosphatases from two patients with cirrhosis have been described by Fishman, Inglis and Ghosh (1968). One patient, in whom 40% of the serum enzyme was of intestinal origin, exhibited two bands on electrophoresis, but after treatment with neuraminidase only a single slow band could be detected. About 90% of the second patient's serum alkaline phosphatase was intestinal in nature, and the single banded electrophoretic pattern was unchanged after incubation with neuraminidase.

The two fractions of sheep-brain alkaline phosphatase isolated by DEAE-cellulose chromatography differ in their *N*-acetylneuraminic acid contents, since after treatment with neuraminidase the elution pattern of enzyme II shifts towards that of enzyme I. The kinetic properties of enzyme II, however, are unaffected by neuraminidase treatment and remain distinct from those of enzyme I (Saraswathi and Bachhawat, 1968).

SUMMARY OF CATALYTIC PROPERTIES OF TISSUE ALKALINE PHOSPHATASES. A list of some of the properties which have been found useful for differentiating human tissue alkaline phosphatases is given in Table 36. As will be seen, they fall into three classes, the placental enzyme, the intestinal enzyme and other tissue enzymes. While the first two are relatively easily distinguished, the enzymes in the third group have very similar properties.

TABLE 36 *Catalytic properties useful as aids in the differentiation of human tissue alkaline phosphatases*

	Tissue phosphatase				
	Bone	Liver	Kidney	Placenta	Intestine
Heating at 56° 30 min.	Inactivated	Inactivated	Inactivated	Stable	Inactivated
15 min.	Almost completely inactivated	Partly inactivated	—	Stable	Partly inactivated
0·005M-L-Phenyl-alanine	Stable	Stable	Stable	Inhibited	Inhibited
EDTA	Inactivated	Inactivated	Inactivated	Stable	Inactivated
Urea 0·5M	Slight inhibition	No effect. occasional activation	Slight inhibition	No effect	No effect
2M	Inhibition almost complete	70% inhibition	80% inhibition	25% inhibition	40% inhibition
Anionic electrophoretic mobility (order on cellulose acetate)	2	1	5	4	3
Effect of Neuraminidase on electrophoretic mobility	Retarded	Retarded	Retarded	Retarded	Unaffected

IMMUNOCHEMICAL REACTIONS. The application of immunological techniques to the study of the tissue specificities is largely due to Schlamowitz. The sera of rabbits immunized with purified dog intestinal alkaline phosphatase precipitate almost all the phosphatase activity of the antigenic protein preparation without impairing its activity, an observation which suggests that the catalytic site of the enzyme is not involved in the antigen–antibody reaction. Although this antiserum precipitates the homologous enzyme, it exhibits a high degree of species and tissue specificity, for it exerts little or no effect upon rat, rabbit or bovine intestinal alkaline phosphatase, nor upon the dog-liver or dog-kidney enzymes (Schlamowitz, 1954a, 1954b).

Similar techniques have been applied by Schlamowitz and Bodansky (1959) to human tissue alkaline phosphatases. Antisera to human intestinal and human bone phosphatases almost completely precipitate the homologous enzymes, but show scarcely any cross reaction. These authors found that the anti-human intestinal phosphatase does not precipitate human kidney or liver phosphatases, but their bone antiserum exhibited considerable action against these enzymes. These cross reactions, it is suggested, may have been due to the presence of enzymes from other tissues in the antigenic preparation, since the authors were unable to obtain a sufficient quantity from normal bone and used the enzyme from the tumour mass of an osteogenic sarcoma.

These antisera have been used in an attempt to determine the tissues of origin of raised serum alkaline phosphatase activities in patients with liver and bone metastases (Table 37). It appears that about a quarter of

TABLE 37 *The precipitation of human serum alkaline phosphatase by anti-intestinal phosphatase and anti-bone phosphatase sera*

(Schlamowitz and Bodansky, 1959)

Diagnosis	Serum alkaline phosphatase[+] (K.-A. units)	Phosphatase precipitated by:			'Bone' phosphatase in serum*	'Bone'/Intestinal phosphatase
		Anti-intestine	Anti-bone	Anti-intestine + Anti-bone		
		%	%	%	%	
Normal	8·4	29	67	69	40	1·4
Normal	4·7	13	60	61	48	3·8
Normal	8·4	13	72	72	59	4·6
Cancer of:						
Kidney; liver metastases	50	11	87	86	75	6·9
Breast; liver and skeletal metastases	118	8	92	91	83	10
Prostate; skeletal metastases	122	0·5	94	94	93	190

* Differences between figures in columns 3 and 5.
+ K.-A., King-Armstrong.

the activity in normal sera is derived from the intestine and about half from bone, but in the cancer patients the intestinal contribution falls to one-tenth or less, while the 'bone' fraction increases to three-quarters or more. Schlamowitz and Bodansky point out, however, that they were unable to assess the effect of the bone phosphatase antiserum on the possible presence of the liver and kidney enzymes in the serum.

Reference has already been made (p. 245) to the work of Boyer (1961), who used rabbit anti-human placental and anti-human bone alkaline phosphatases for the identification of fractions separated by starch-gel electrophoresis. Anti-human placental alkaline phosphatase has been employed by Birkett *et al.* (1966) and Sussman, Bowman and

Lewis (1968) to establish the placental origin of the pregnancy serum enzyme (p. 259), and Posen *et al.* (1967) have used the anti-human intestinal anti-enzyme to confirm the presence of intestinal alkaline phosphatase in normal human serum.

Sussman, Small and Cotlove (1968) have described the preparation of specific antisera to human liver and placental alkaline phosphatases by inoculating sheep with highly purified human enzymes demonstrated to be free from non-enzyme protein. The antisera almost completely inhibited their respective homologous antigens, but did not react with alkaline phosphatase from bone, intestine, kidney or neutrophils, from which the authors deduced that there are at least three antigenic forms of the enzyme: one from liver, one from placenta and one or more from other tissues.

These results suggest that specific antisera might eventually provide a reliable means of differentiating between the liver and bone phosphatases in pathological human sera. However, Sussman (1968, 1970) has recently reported the surprising finding that the serum alkaline phosphatase in patients with osteomalacia or Paget's disease proved immunochemically to be of hepatic origin. This suggests that synthesis of hepatic alkaline phosphatase is stimulated in these diseases or, alternatively, that enzymes derived from a particular tissue (in these cases, bone) might undergo modification in the liver. If either suggestion proves to be correct, identification of a tissue phosphatase in the serum might prove to be illusory as a diagnostic aid.

Immunochemical techniques have also been employed in the study of phosphorylating enzymes (Henion and Sutherland, 1957). The homologous enzyme is strongly inhibited by antisera prepared in the chicken to dog-heart and dog-liver phosphorylases, but phosphorylases from other dog tissues and from the heart and liver of other species are inhibited only slightly.

Genetic studies

The main band of alkaline phosphatase found in normal serum is probably derived from the liver, but a second slow-moving form was first reported by Hodson *et al.* (1962). This appears to be of intestinal origin and has now been shown to be under genetic control (Cunningham and Rimer, 1963). In a twin study Arfors, Beckman and Lundin (1963a, 1963b) demonstrated a highly significant correlation between the presence of the slow component and the ABO blood group, most individuals exhibiting the second phosphatase isoenzyme being of group O. In an extensive population study Beckman (1964) found that all individuals with the slow phosphatase component in their sera secreted the ABH substance in their saliva, but a previously suggested association with the Lewis blood group substances could not be confirmed.

These observations have been confirmed by several investigators (Schreffler, 1965; Evans, 1965; Bamford, Harris, Luffman, Robson and Cleghorn, 1965; Robinson, Levene, Blumberg and Pierce, 1967). Shreffler (1965) found that the slow isoenzyme cannot be detected in persons with blood group A or in non-secretors of ABH. Robinson *et al.* (1967) demonstrated the frequency of the slow band to be about 70% in North American Indians, appreciably higher than in other populations so far studied.

The intestinal origin of the slow band was confirmed in a series of cirrhotic patients by Stolbach, Krant, Inglis and Fishman (1967), who found the ratio of the serum L-phenylalanine-sensitive alkaline phosphatase to the total serum enzyme to be significantly higher in patients of group O than in those of other blood groups.

Genetic control of the alkaline phosphatase of pregnancy sera was first observed by Boyer (1961), whose results have been referred to on p. 245. Of the six zones detected by starch-gel electrophoresis, zones A, B and D were found only in pregnancy sera, the comparatively rare zone D in Negroes only. Women in the last trimester of pregnancy may be classified according to whether or not zone A phosphatases are present. A much higher proportion (0·536) of U.S. white women exhibit zone A phosphatases in their sera than U.S. Negro women (0·250) or Nigerian women (0·122). Beckman and Grivea (1965) found the zone A component in the sera of 45% of Swedish women, an incidence not very different from that found by Boyer for white U.S. women.

Boyer (1961) found three main patterns, A, AB and B, in placental extracts. Zones A and B appear to be the corresponding bands found in the serum, while pattern AB contains both A and B components together with a band of intermediate mobility. Robson and Harris (1965), however, in a study of over 700 placentae, detected six distinct phenotypes described as S, FS, SI, I, FI and F (Fig. 60), by electrophoresis at pH 6·0 and 8·6. They suggest that Boyer's zone A included the F, FI and I phenotypes, which are indistinguishable at pH 8·6, while the AB type included FS and SI. Robson and Harris proposed that the six phenotypes are determined by three autosomal allelic genes, and tested their hypothesis by studying a series of 380 placentae from 190 dizygotic twin pairs, when close agreement between the observed and expected distributions was observed. They also obtained evidence that the placental phosphatase phenotype is dependent upon the genotype of the foetus: in 78 cases the twins had different phosphatase patterns, thus excluding the maternal genotype as a determinant factor.

Beckman and his colleagues have studied placental phosphatase polymorphism in several other populations, e.g. Sweden, Greece and Nigeria (Beckman, Bjorling and Christodoulou, 1966; Beckman, Beckman,

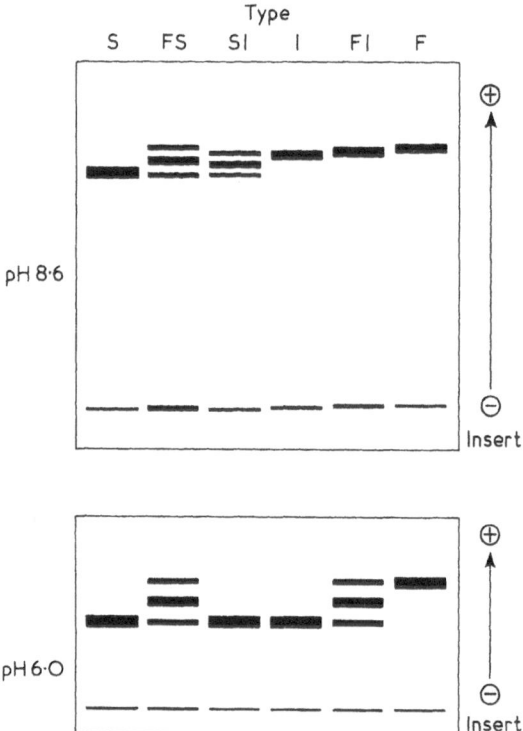

Fig. 60. Placental alkaline phosphatase phenotypes detected by starch-gel electrophoresis at pH 8·6 and 6·0. (Robson and Harris, 1965.)

Christodoulou and Ifekwunigwe, 1967), and Iceland (Beckman and Johannssen, 1967), during the course of which several rare phenotypes were encountered. One of the anomalies was a deficiency of placental alkaline phosphatase in a twin-pair found to have dysostosis craniofacialis (Crouzon), which the authors suggest might be causally related to the enzyme deficiency.

The genetic variation of human alkaline phosphatases has recently been reviewed by Beckman (1966). Genetic variation of alkaline phosphatase has also been observed in other species. Two phenotypes, one of which exhibits a fast-moving serum component, have been described in cattle (Rendel and Gahne, 1963; Gahne, 1963), while a similar variant in sheep (B phosphatase) has been shown to be associated with the presence of a blood group substance O in adult ewes (Rendel and Stormont, 1964; Rendel, Aalund, Freedland and Møller, 1964; Rasmusen, 1965).

Duodenal alkaline phosphatase also appears to be genetically determined in the mouse. Nayudu and Moog (1967) crossed two inbred strains of Swiss mice, one with three to four times the intestinal phosphatase activity of the other and studied the frequency distribution of the high-activity character in the F_1 generation, and in the F_2 generation produced by sib-mating the F_1 hybrids. They concluded that the enzyme activity is under polygenic control, but the isoenzyme patterns were qualitatively indistinguishable, though the proportions of the two forms varied.

Two genetically determined zones of alkaline phosphatase activity have been detected in fowl plasma by starch-gel electrophoresis. In a study of two generations the faster-moving isoenzyme appeared to be determined by an autosomal gene, dominant to its allele governing the slower component (Law and Munro, 1965; Wilcox, 1966). In a later report Law (1967) demonstrated the presence of two electrophoretic variants of leucine aminopeptidase (p. 303) associated with the alkaline phosphatase variants. A given individual displayed either the fast or the slow forms of both enzymes. Since in each case the fast forms were convertible to the slow forms by treatment with neuraminidase, it seems that the occurrence of the two forms may be due to the presence or absence of a gene which controls the coupling of sialic acid to the enzyme molecules.

Three variants of alkaline phosphatase have been observed in *Drosophila melanogaster* showing electrophoretically fast, slow and intermediate bands of activity controlled by a pair of codominant allelic genes (Beckman and Johnson, 1964; Johnson, 1966). Schneiderman, Young and Childs (1966), however, demonstrated as many as seven different components in the larvae, pupae and adult forms of this insect. Schneiderman (1967) has recently shown that the pupal phosphatase is the larval skin phosphatase modified by larval gut enzymes, and points out that the relationships of the alkaline phosphatases in different tissues of this species are as yet ill-defined. Nevertheless, Wallis and Fox (1968) reported further evidence indicating genetic control of alkaline phosphatase variation in *Drosophila*.

Alkaline phosphatase isoenzymes in bacteria
Much work has been performed on the multiple forms of the alkaline phosphatase of *Escherichia coli*. This heterogeneity was first recognized by Bach, Signer, Levinthal and Sizer (1961), who observed different isoenzyme patterns in mutant forms of the bacterium. Both the normal and mutant isoenzymes appear to be dimers which dissociate into monomers on reduction with thioglycollate in 6M-urea, but recombine during air oxidation in the presence of mercaptoethanol. Mixtures of monomers recombine to form hybrids which differ from both parental forms (Levinthal, Signer and Fetherolf, 1962). The inactive monomers from a given isoenzyme appear to be identical (Rothman and Byrne, 1963).

Schlesinger and Levinthal (1963) found that dissociation into monomers could also be induced by treatment at $0°$ with acetate buffer at pH 5, and they used this technique and thioglycollate reduction to prepare partly active, artificial hybrid isoenzymes from normal *E. coli* phosphatase and inactive, but antigenically related, proteins from a phosphatase-free mutant. These partly active hybrids appear to contain one sub-unit from each strain (Fan, Schlesinger, Torriani, Barrett and Levinthal, 1966). Schlesinger (1967) reported that the monomers of *E. coli* phosphatase are antigenically active, though they show little cross-reaction with the dimeric enzyme molecule.

The alkaline phosphatases of *E. coli* have also been separated on DEAE-cellulose columns and by acrylamide disc and starch-gel electrophoresis. Three forms, considered to be AA, BB and AB dimers, were found to have similar amino-acid compositions and similar kinetic properties (Lazdunski and Lazdunski, 1967).

E. coli alkaline phosphatase also possesses pyrophosphatase and glucose phosphotransferase activity (Anderson and Nordlie, 1967), and in this respect it resembles the mammalian enzyme (Moss *et al.*, 1966) (p. 254).

Acid phosphatase
Although differences in sensitivity to ethanol of the acid phosphatases of the prostate and the erythrocytes were reported by Herbert in 1946, other investigators found the enzyme from both sources to be equally inactivated. Other inhibitors have proved to be much more selective, and Abul-Fadl and King (1949) found the greatest differential effect to be exerted by formaldehyde (0·5%), which has little action on the prostatic enzyme while almost completely inhibiting the red-cell enzyme, and by 0·01M-tartrate, which by contrast has no effect on erythrocyte acid phosphatase yet almost completely inactivates the pro-

static enzyme. Determination of formaldehyde-stable or tartrate-labile acid phosphatase of the serum has since become an important diagnostic test for carcinoma of the prostate.

The importance of this test in cases of suspected prostatic carcinoma has prompted the application of electrophoretic and chromatographic techniques to the search for multimolecular forms of the enzyme.

Electrophoretic separation
The first attempt at the separation of human serum acid phosphatase was that of Estborn and Swedin (1959), who found that it migrated as a single band on starch-gel electrophoresis in borate buffer at pH 8·9. This result, however, has been attributed by Sur, Moss and King (1962*a*) to the instability of acid phosphatase at the high pH value employed: minor bands, if present, might be destroyed at pH 8·9. At pH 8·6, however, Lawrence, Melnick and Weimer (1960) separated two components and Dubbs, Vivonia and Hilburn (1960) three zones of acid phosphatase activity in human serum by starch-gel electrophoresis. At the same pH, Gerhardt, Clausen, Christensen and Riishede (1963) succeeded in separating two main bands of acid phosphatase activity from normal brain tissue by agar-gel electrophoresis. One of these was a diffuse band in the α area, while the other was a sharper more intensely stained band in the β zone. A third, much faster band was found in the pre-albumin zone in certain tumour tissues (neurinomas).

The risk of decomposition may be largely eliminated by carrying out starch-gel electrophoresis in 0·2M-citrate buffer at pH 6·2 (Sur *et al.*, 1962*a*). The fluorescent technique of Moss *et al.* (1961*a*) may be used at pH 4·9 for the detection of acid phosphatase activity on starch gel. Under these conditions prostatic acid phosphatase gives three main zones of activity: a major zone migrating close behind serum albumin, followed by a narrow, intense band and a slow-moving, weak band. Since the bands cannot be detected at pH 10, they cannot be due to alkaline phosphatase, and since they are inhibited by tartrate but not by formaldehyde, they appear to consist solely of the prostatic enzyme (Sur *et al.*, 1962*a*). Somewhat different patterns containing four zones of activity have been obtained by acrylamide-gel electrophoresis (Kaschnitz, 1967).

Unlike the alkaline phosphatase fractions discussed in the previous section, the various bands of acid phosphatase exhibit different K_m values for β-naphthyl phosphate (Table 38). The slowest band has a pH optimum at 5·5, but the middle and fast fractions have broad pH optima curves (Sur *et al.*, 1962*a*).

Increased resolution of prostatic acid phosphatase is attainable by prolonging the period of electrophoresis and maintaining the temperature at 4°: by such means at least 13 fractions have been separated (Sur *et al.*, 1962*a*). Since older extracts exhibit fewer bands than fresh extracts, the authors consider that the multiple bands may not be artifacts. It is possible that these might differ in their degrees of association

TABLE 38 *Michaelis constants (K_m) for β-naphthyl phosphate of the principal prostatic acid phosphatase bands separated by starch-gel electrophoresis*

(Sur, Moss and King, 1962*a*)

Band	K_m (mM)	
	pH 5·5	pH 6·2
Fastest	0·119	0·155
Middle	0·154	0·140
Slow	0·194	0·226

with cofactors, and a further report from the same group indicates that the major fractions eluted from starch gel have similar pH optima (about 5·5) and K_m values for β-naphthyl phosphate. They had almost identical thermal stabilities and behaved similarly towards butanol, proteolytic agents and EDTA. It therefore seems unlikely that they are bound to other charged molecules, including proteins, or metallic ions (Sur, Moss and King, 1962*b*). More recently as many as seventeen distinct acid phosphatase components have been detected in human liver, kidney and prostate, and in the corresponding tissues of other species (Lundin and Allison, 1966).

Beckman and Beckman (1967) observed four electrophoretically different zones of activity (A, B, C and D) in various human tissues, but the patterns differed greatly according to their tissues of origin. The C band occurs only in the placenta, A, B and D in the kidney, B and D in liver, intestine, heart and skeletal muscle, B in skin and D in the pancreas. The various bands differ in their heat stabilities, pH optima and sialic acid content, but all are inhibited by L-tartrate.

The four bands have since been observed in cell cultures established from skin biopsies, but the BD combination was by far the most frequently observed pattern (Beckman, Beckman, Bergman and Lundgren, 1968).

Three zones of acid phosphatase have been separated from rat-kidney cortex by Pla, Papadopoulos and Rosen (1967) using agar-gel electrophoresis, a fast-moving major band, a lesser band remaining near the origin and a weak band with intermediate mobility. The intensity of the major components is greatly increased by treatment with Triton X-100, but they are markedly inhibited by fluoride. The intermediate band is less sensitive to Triton and fluoride, and appears to be absent in the renal medulla.

Acid phosphatases from different sources differ in their substrate specificities. Acid phenylphosphatase and acid β-glycerophosphatase activity has been detected in the lysosomes of rat bone and liver, but the supernatant fraction contains significant amounts of glycerophosphatase activity only (Vaes and Jacques, 1965). Two electrophoretically distinct fractions of acid phosphatase have also been obtained by Allen and Gockerman (1964) from the mitochondrial–lysosomal fraction of rat liver: one is released during homogenization, but the second is separated from the particulate matter only after treatment with a detergent such as Triton X-100.

Variations in their activities against phenyl phosphate and α- and β-glycerophosphates have been reported for the human plasma acid phosphatases (Gründig, Czitober and Schobel, 1965). In normal controls, osteopetrosis, Gaucher's disease and carcinoma of the prostate, phenyl phosphate is the most rapidly hydrolysed, followed by β-glycerophosphate, while α-glycerophosphate is the least rapidly hydrolysed substrate. In primary hyperparathyroidism, however, α-glycerophosphate is the most effective substrate. The enzymes also differ in their behaviour with 10^{-4}M-cysteine, for while in osteopetrosis the serum enzyme is inhibited to about 70% of the control activity, it is activated by about 70% in Gaucher's disease and by 30% in carcinoma of the prostate. In primary hyperparathyroidism, 2·5mM-EDTA increases the acid phosphatase activity by 40%, but not in other conditions (Gründig *et al.*, 1965).

By means of starch-gel electrophoresis, Gründig *et al.* (1965) separated the serum enzyme into three zones of activity in primary hyperparathyroidism, osteopetrosis and normal controls, four in carcinoma of the prostate and five in Gaucher's disease. Goldberg, Takakura and Rosenthal (1966), using acrylamide-gel electrophoresis, found three slow-moving minor bands in the serum of patients with Gaucher's disease in addition to the normal fast component. Patients who had undergone splenectomy, however, did not exhibit the minor bands. which suggests that they may have originated in the spleen.

Chromatographic investigations

The heterogeneity of the acid phosphatase of animal tissues has been demonstrated chromatographically on DEAE-cellulose columns by Moore and Angeletti (1961). Rat liver separates into three major and one minor fractions which differ in their substrate affinities and sensitivity to fluoride inhibition (Table 39). The three fractions of rat brain closely resemble those of rat liver. Rabbit liver also contains four separate forms of acid phosphatase.

TABLE 39 *Properties of acid phosphatase fractions separated by chromatography on DEAE-cellulose columns*

(From Moore and Angeletti, 1961)

Tissue	Fraction	K_m (p-nitrophenyl phosphate) mM	pH optimum	Inhibition by 0·02M-F⁻
				%
Rat liver	1	4·1	4·0	100
	2 (minor)	3·7	4·9	80
	3	0·4	5·2	2
	4	0·1	6·0	6
Rat brain	1	4·6	4·0	87
	2	0·4	4·8	8
	3	0·1	5·8	0
Rabbit liver	1	2·8	4·0	64
	2	0·5	5·2	0
	3	0·5	5·6	5
	4	—	4·7	55

DEAE-cellulose columns have also been employed in the study of rat-brain acid phosphatase. Sucrose homogenates show four components, the activities of which are greatly increased by treatment during the extraction process with Triton X-100 or by autolysis (Anderson, 1965).

Human placental acid phosphatases have been shown to consist of three chromatographically distinct components, I, II and III, separable on Sephadex G-200 columns (DiPietro and Zengerle, 1967). Their properties are summarized in Table 40. Isoenzyme III differs markedly from the other two components in its behaviour and in being activated by a variety of purines. It is also peculiarly unreactive with a series of physiologically important organic phosphates.

Genetic variants of human acid phosphatases

The first report of genetic variation of human acid phosphatase was that of Hopkinson, Spencer and Harris (1963, 1964), who subjected haemolysates to starch-gel electrophoresis and detected the zones of enzyme activity with phenolphthalein diphosphate. The diazo technique of Barka (1961) is unsuitable for use with red-cell haemolysates. Six different phenotypes (A, AB, B, AC, BC and C) appear to be controlled by three autosomal allelic genes (P^a, P^b and P^c). A, B and C are homozygous phenotypes, while AB, AC and BC represent the corresponding heterozygotes. The C phenotype is relatively rare and was not observed in the original study, but its existence has been confirmed by Lai, Nevo and Steinberg (1964) in a Brazilian population.

Several investigators have studied the genotype distribution in many different racial groups, including Tristan da Cunhans (Hopkinson *et al.*, 1964), Italians (Modiano, Brunelli, Ferrari and Frattaroli, 1965), Japanese and Oriental, Negro and white Americans (Giblett and Scott, 1965), white Australians and natives of New Guinea (Lai, 1966) and Chinese and Malay residents of Singapore (Lai and Kwa, 1968). In all these populations P^b is the commonest genotype, while the P^c genotype is by far the rarest, especially in Asian groups (Lai and Kwa, 1968).

Giblett and Scott (1965) used a formate-buffer system which enabled them to differentiate a rare RA phenotype in two members of an American Negro family.

TABLE 40 *Properties of human placental acid phosphatases*
(From DiPietro and Zengerle, 1967)

	Acid phosphatases		
	I	II	III
Molecular weights	> 200,000	105,000	35,000
pH optima	5·5	4·0	5·5
(*p*-nitrophenyl phosphate)			
Inhibition			
20mM-L-(+)tartrate	41%	90%	0%
1 *μ*M-chloromercuribenzoate	14%	7%	100%
1mM-molybdate	45%	47%	2%
50mM-fluoride	51%	23%	5%
Purines	—	—	Enhanced
Pyridoxine-5-phosphate	Substrate	Substrate	Inhibited

274 · *Isoenzymes*

REFERENCES

ABUL-FADL, M. A. M. and KING, E. J. (1949), *Biochem. J.* **45**, 51.
AHMED, Z., ABUL-FADL, M. A. M. and KING, E. J. (1959), *Biochim. biophys. Acta* **36**, 228.
AHMED, Z. and KING, E. J. (1959), *Biochim. biophys. Acta* **34**, 313.
AKHTAR, A., HANSEN, A. and KAERCHER, K. H. (1968), *Z. klin. Chem.* **6**, 334.
ALLEN, J. M. (1963), *J. Histochem. Cytochem.* **11**, 542.
ALLEN, J. M. and GOCKERMAN, J. (1964), *Ann. N.Y. Acad. Sci.* **121**, 616.
ALLEN, J. M. and HYNCIK, G. (1963), *J. Histochem. Cytochem.* **11**, 169.
ANDERSON, P. J. (1965), *J. Neurochem.* **12**, 919.
ANDERSON, W. B. and NORDLIE, R. C. (1967), *J. biol. Chem.* **242**, 114.
ARFORS, K. E., BECKMAN, L. and LUNDIN, L. G. (1963a), *Acta Genet.* **13**, 89.
ARFORS, K. E., BECKMAN, L. and LUNDIN, L. G. (1963b), *Acta Genet.* **13**, 366.
ARMSTRONG, A. R. and BANTING, F. G. (1935), *Canad. med. Assoc. J.* **33**, 243.
ARMSTRONG, A. R., KING, E. J. and HARRIS, R. I. (1934), *Canad. med. Assoc. J.* **31**, 14.
BABSON, A. L. (1965), *Clin. Chem.* **11**, 789.
BABSON, A. L., GREELEY, S. J., COLEMAN, C. M. and PHILLIPS, G. E. (1966), *Clin. Chem.* **12**, 482.
BACH, M. L., SIGNER, E. R., LEVINTHAL, C. and SIZER, I. W. (1961), *Federation Proc.* **20**, 255.
BAHR, M. and WILKINSON, J. H. (1967), *Clin. chim. Acta* **17**, 367.
BAKER, R. W. R. and PELLEGRINO, C. (1954), *Scand. J. clin. Lab. Invest.* **6**, 94.
BAMFORD, K. F., HARRIS, H., LUFFMAN, J. E., ROBSON, E. B. and CLEGHORN, T. E. (1965), *Lancet* **i**, 530.
BARKA, T. (1961), *J. Histochem. Cytochem.* **9**, 564.
BECKMAN, L. (1964), *Acta Genet.* **14**, 286.
BECKMAN, L. (1966), *Isozyme Variations in Man*, Basel and New York: Karger.
BECKMAN, L. and BECKMAN, G. (1967), *Biochem. Genet.* **1**, 145.
BECKMAN, L., BECKMAN, G., BERGMAN, S. and LUNDGREN, E. (1968), *Acta Genet.* **18**, 409.
BECKMAN, L., BECKMAN, G., CHRISTODOULOU, C. and IFEKWUNIGWE, A. (1967), *Acta Genet.* **17**, 406.
BECKMAN, L., BJÖRLING, G. and CHRISTODOULOU, C. (1966), *Acta Genet.* **16**, 59.
BECKMAN, L. and GRIVEA, M. (1965), *Acta Genet.* **15**, 218.
BECKMAN, L. and JOHANNSSON, E. O. (1967), *Acta Genet.* **17**, 413.
BECKMAN, L. and JOHNSON, F. M. (1964), *Hereditas* **51**, 221.
BESSEY, O. A., LOWRY, O. H. and BROCK, M. J. (1946), *J. biol. Chem.* **164**, 321.
BIRKETT, D. J., CONYERS, R. A. J., NEALE, F. C., POSEN, S. and BRUDENELL-WOODS, J., (1967), *Arch. Biochem. Biophys.* **121**, 470.
BIRKETT, D. J., DONE, J., NEALE, F. C. and POSEN, S. (1966), *Br. med. J.* **i**, 1210.
BODANSKY, O. (1933), *J. biol. Chem.* **101**, 93.
BODANSKY, O. (1936), *J. biol. Chem.* **115**, 101.
BODANSKY, O. (1937), *J. biol. Chem.* **118**, 341.
BODANSKY, O. (1948), *J. biol. Chem.* **174**, 465.
BÖRNIG, H., ŠTĚPÁN, J., HORN, A., GIERTLER, R., THIELE, G. and VEČEREK, B. (1967), *Hoppe-Seyler's Z. physiol. Chem.* **348**, 1311.
BOYER, S. H. (1961), *Science*, **134**, 1002.

BURKE, J. O. (1950), *Gastroenterology* **16**, 660.

BUTTERWORTH, P. J. (1968a), *Biochem. J.* **107**, 467.

BUTTERWORTH, P. J. (1968b), *Biochem. J.* **108**, 243.

BUTTERWORTH, P. J. and MOSS, D. W. (1966), *Nature, Lond.* **209**, 805.

BUTTERWORTH, P. J. and MOSS, D. W. (1967), *Enzymologia* **32**, 269.

BUTTERWORTH, P. J., MOSS, D. W., PITKÄNEN, E. and PRINGLE, A. (1965), *Clin. chim. Acta* **11**, 212.

BUTTERWORTH, P. J., PITKÄNEN, E. and MOSS, D. W. (1963), *Biochem. J.* **88**, 19P.

CHIANDUSSI, L., GREENE, S. F. and SHERLOCK, S. (1962), *Clin. Sci.* **22**, 425.

CLOETENS, R. (1939), *Enzymologia* **6**, 46.

CLUBB, J. S., NEALE, F. C. and POSEN, S. (1965), *J. Lab. clin. Med.* **66**, 493.

CONYERS, R. A. J., BIRKETT, D. J., NEALE, F. C. and POSEN, S. (1966). *Biochim. biophys. Acta* **139**, 363.

COOKE, K. B. and ZILVA, J. F. (1961), *J. clin. Path.* **14**, 500.

COX, R. P. and GRIFFIN, M. J. (1965), *Lancet* **ii**, 1018.

CUNNINGHAM, V. R. and RIMER, J. G. (1963), *Biochem. J.* **89**, 50P.

CURZEN, P. and MORRIS, I. (1966), *J. Obst. Gyn. Brit. Cwlth.* **73**, 640.

DI BELLA, S., RICHETTA, G. and PICHIERRI, U. (1963), *Clin. chim. Acta* **8**, 788.

DiPIETRO, D. L. and ZENGERLE, F. S. (1967), *J. biol. Chem.* **242**, 3391.

DIXON, T. F. and PURDOM, M. (1954), *J. clin. Path.* **7**, 341.

DUBBS, C. A., VIVONIA, C. and HILBURN, J. M. (1960), *Science,* **131**, 1529.

DUNNE, J., FENNELLY, J. J. and MCGEENEY, K. F. (1968), *Biochem. J.* **110**, 12P.

EATON, R. H. and MOSS, D. W. (1967a), *Biochem. J.* **102**, 917.

EATON, R. H. and MOSS, D. W. (1967b), *Nature, Lond.* **214**, 842.

EATON, R. H. and MOSS, D. W. (1967c), *Biochem. J.* **104**, 65P.

EATON, R. H. and MOSS, D. W. (1968a), *Enzymologia* **35**, 31.

EATON, R. H. and MOSS, D. W. (1968b), *Enzymologia* **35**, 168.

EISFELD, G. and KOCH, E. (1954), *Z. ges. inn. Med.* **9**, 514.

EPSTEIN, E., WOLF, P. I., HORWITZ, J. P. and ZAK, B. (1967), *Am. J. clin. Path.* **48**, 530.

ESTBORN, B. (1959), *Nature, Lond.* **184**, 1636.

ESTBORN, B. and SWEDIN, B. (1959), *Scand. J. clin. Lab. Invest.* **11**, 235.

EVANS, D. A. P. (1965), *J. med. Genet.* **2**, 126.

FAHEY, J. L., MCCOY, P. F. and GOULIAN, M. (1958), *J. clin. Invest.* **37**, 272.

FAN, D. F., SCHLESINGER, M. J., TORRIANI, A., BARRETT, K. and LEVINTHAL, C. (1966), *J. molec. Biol.* **15**, 32.

FERNLEY, H. N. and WALKER, P. G. (1966), *Biochem. J.* **99**, 39P.

FISHMAN, W. H. and GHOSH, N. K. (1967), in *Advances in Clinical Chemistry*, ed. BODANSKY, O. and STEWART, C. P., New York and London: Academic Press, Vol. 10, p. 256.

FISHMAN, W. H., GREEN, S. and INGLIS, N. I. (1962), *Biochim. biophys. Acta* **62**, 363.

FISHMAN, W. H., INGLIS, N. I. and GHOSH, N. K. (1968), *Clin. chim. Acta* **19**, 71.

FISHMAN, W. H., INGLIS, N. I. and KRANT, M. J. (1965), *Clin. chim. Acta* **12**, 298.

FISHMAN, W. H., INGLIS, N. I., SARKE, F. and GHOSH, N. K. (1966), *Federation Proc.* **25**, 748.

FISHMAN, W. H., INGLIS, N. I., STOLBACH, L. L. and KRANT, M. J. (1968), *Cancer Res.* **28**, 150.

FOLLEY, S. J. and KAY, H. D. (1936), *Ergeb. Enzymforsch.* **5**, 159.

GAHNE, B. (1963), *Nature, Lond.* **199**, 305.

GERHARDT, W., CLAUSEN, J., CHRISTENSEN, E. and RIISHEDE, J. (1963), *Acta neurol. Scand.* **39**, 2.

GHOSH, N. K. and FISHMAN, W. H. (1966), *J. biol. Chem.* **241**, 2516.

GHOSH, N. K. and FISHMAN, W. H. (1968), *Biochem. J.* **108**, 779.

GHOSH, N. K., GOLDMAN, S. S. and FISHMAN, W. H. (1967), *Enzymologia* **33**, 113.

GIBLETT, E. R. and SCOTT, N. M. (1965), *Am. J. Hum. Genet.* **17**, 425.

GOLDBERG, A. F., TAKAKURA, K. and ROSENTHAL, R. L. (1966), *Nature, Lond.* **211**, 41.

GROSSBERG, A. L., HARRIS, E. H. and SCHLAMOWITZ, M. (1961), *Arch. Biochem. Biophys.* **93**, 267.

GRÜNDIG, E., CZITOBER, H. and SCHOBEL, B. (1965), *Clin. chim. Acta* **12**, 157.

GRYDEN, R. M., FRIEDENWALD, J. S. and CARLSON, C. (1955), *Arch. Biochem. Biophys.* **54**, 281.

GUTMAN, A. B. (1959), *Am. J. Med.* **27**, 875.

HAIJE, W. G. and DE JONG, M. (1963), *Clin. chim. Acta* **8**, 620.

HENION, W. F. and SUTHERLAND, E. W. (1957), *J. biol. Chem.* **224**, 477.

HERBERT, F. K. (1946), *Quart. J. Med.* **15**, 221.

HODSON, A. W., LATNER, A. L. and RAINE, L. (1962), *Clin. chim. Acta* **7**, 255.

HOPKINSON, D. A., SPENCER, N. and HARRIS, H. (1963), *Nature, Lond.* **199**, 969.

HOPKINSON, D. A., SPENCER, N. and HARRIS, H. (1964), *Am. J. Hum. Genet.* **16**, 141.

HORNE, M., CORNISH, C. J. and POSEN, S. (1968), *J. Lab. clin. Med.* **72**, 905.

HOSOYAMA, Y., FUKUDA, Y. and SHIMADATE, T. (1967), *Clin. chim. Acta* **18**, 141.

INGLIS, N. I., GHOSH, N. K. and FISHMAN, W. H. (1968), *Anal. Biochem.* **22**, 382.

JOHNSON, F. M. (1966), *Nature, Lond.* **212**, 843.

KANEKO, A. (1968), *J. Biochem., Japan* **64**, 785.

KASCHNITZ, R. (1967), *Z. Klin. Chem.* **5**, 126.

KASCHNITZ, R., PATSCH, J. and PETERLIK, M. (1968), *Europ. J. Biochem.* **5**, 51.

KEIDING, N. R. (1959), *Scand. J. clin. Lab. Invest.* **11**, 106.

KEIDING, N. R. (1964), *Clin. Sci.* **26**, 291.

KEIDING, N. R. (1966), *Scand. J. clin. Lab. Invest.* **18**, 134.

KERKHOFF, J. F. (1968), *Clin. chim. Acta* **22**, 231.

KING, E. J. (1938), *J. Path. Bact.* **57**, 85.

KING, E. J. and ARMSTRONG, A. R. (1934), *Canad. med. Assoc. J.* **31**, 376.

KING, E. J. and DELORY, G. (1939), *Biochem. J.* **33**, 1185.

KING, E. J. and WOOTTON, I. D. P. (1956), *Micro-Analysis in Medical Biochemistry*, 3rd ed., London: Churchill.

KITCHENER, P. N., NEALE, F. C., POSEN, S. and BRUDENELL-WOODS, J. (1965), *Am. J. clin. Path.* **44**, 654.

KLEIN, B., READ, P. A. and BABSON, A. L. (1960), *Clin. Chem.* **6**, 269.

KORNER, N. H. (1962), *J. clin. Path.* **15**, 195.

KOWLESSAR, O. D., HAEFFNER, L. J. and RILEY, E. M. (1961), *Ann. N.Y. Acad. Sci.* **94**, 836.

KOWLESSAR, O. D., PERT, J. H., HAEFFNER, L. J. and SLEISENGER, M. H. (1959), *Proc. Soc. exp. Biol. N.Y.* **100**, 191.

LAI, L. Y. C. (1966), *Acta Genet.* **16**, 313.

LAI, L. Y. C. and KWA, S. B. (1968), *Acta Genet.* **18**, 45.

LAI, L. Y. C., NEVO, S. and STEINBERG, A. G. (1964), *Science,* **145**, 1187.

LATNER, A. L. (1961), *Proc. 9th Colloq. Protides of Biological Fluids*, ed. PEETERS, H., Amsterdam: Elsevier.

LATNER, A. L. (1965), in *Enzymes in Clinical Chemistry*, ed. RUYSSEN, R. and VANDENDRIESSCHE, L., Amsterdam: Elsevier, p. 110.

LATNER, A. L. and SKILLEN, A. W. (1962), *Proc. Assoc. clin. Biochem.* **2**, 3.

LAW, G. R. J. (1967), *Science*, **156**, 1106.

LAW, G. R. J. and MUNRO, S. S. (1965), *Science*, **149**, 1518.

LAWRENCE, S. H., MELNICK, P. J. and WEIMER, H. E. (1960), *Proc. Soc. exp. Biol.*, *N.Y.* **105**, 572.

LAZDUNSKI, C. and LAZDUNSKI, M. (1967), *Biochim. biophys. Acta* **147**, 280.

LEVINTHAL, C., SIGNER, E. R. and FETHEROLF, K. (1962), *Proc. Natl. Acad. Sci.*, *Wash.* **48**, 1230.

LUNDIN, L. G. and ALLISON, A. C. (1966), *Biochim. biophys. Acta* **127**, 527.

MCMASTER, Y., TENNANT, R., CLUBB, J. S., NEALE, F. C. and POSEN, S. (1964), *J. Obst. Gyn. Brit. Cwlth.* **71**, 735.

MARKERT, C. L. and MØLLER, F. (1959), *Proc. Natl. Acad. Sci.*, *Wash.* **45**, 753.

MESSER, R. H. (1967), *Am. J. Obstet. Gynec.* **98**, 459.

MODIANO, G., BRUNELLI, F., FERRARI, A. and FRATTAROLI, W. (1965), *Atti Assoc. Genet. Ital.* **10**, 296.

MOOG, F., VIRE, H. E. and GREY, R. D. (1966), *Biochim. biophys. Acta* **113**, 366.

MOORE, B. W. and ANGELETTI, P. U. (1961), *Ann. N.Y. Acad. Sci.* **94**, 659.

MORTON, R. K. (1955), *Biochem. J.* **61**, 232.

MOSS, D. W. (1960a), *Biochem. J.* **76**, 32P.

MOSS, D. W. (1960b), *Clin. chim. Acta* **5**, 283.

MOSS, D. W. (1962a), *Proc. Assoc. clin. Biochem.* **2**, 5.

MOSS, D. W. (1962b), *Nature, Lond.* **193**, 981.

MOSS, D. W. (1963), *Nature, Lond.* **200**, 1206.

MOSS, D. W. (1964a), *Proc. Assoc. clin. Biochem.* **3**, 132.

MOSS, D. W. (1964b), *Biochem. J.* **92**, 16P.

MOSS, D. W. (1965), *Biochem. J.* **94**, 458.

MOSS, D. W., CAMPBELL, D. M., ANAGNOSTOU-KAKARAS, E. and KING, E. J. (1961a), *Biochem. J.* **81**, 441.

MOSS, D. W., CAMPBELL, D. M., ANAGNOSTOU-KAKARAS, E. and KING, E. J. (1961b), *Pure and appl. Chem.* **3**, 397.

MOSS, D. W., EATON, R. H., SMITH, J. K. and WHITBY, L. G. (1967), *Biochem. J.* **102**, 53.

MOSS, D. W. and KING, E. J. (1962), *Biochem. J.* **84**, 192.

MOTZOK, I. (1959), *Biochem. J.* **72**, 169.

MÜLLER-EBERHARD, H. J. (1960), *Scand. J. clin. Lab. Invest.* **12**, 33.

NAYUDU, P. R. V. and MOOG, F. (1967), *Biochem. Genet.* **1**, 155.

NEALE, F. C., CLUBB, J. S., HOTCHKIS, D. and POSEN, S. (1965), *J. clin. Path.* **18**, 359.

NEWTON, M. A. (1966), *J. clin. Path.* **19**, 491.

NISSELBAUM, J. S., SCHLAMOWITZ, M. and BODANSKY, O. (1961), *Ann. N.Y. Acad. Sci.* **94**, 970.

NORDENTOFT-JENSEN, B. (1964), *Clin. Sci.* **26**, 299.

OHMORI, Y. (1937), *Enzymologia* **4**, 217.

PAUL, J. and FOTTRELL, P. F. (1961), *Ann. N.Y. Acad. Sci.* **94**, 668.

PEACOCK, A. C., REED, R. A. and HIGHSMITH, E. M. (1963), *Clin. chim. Acta* **8**, 914.

PLA, G. W., PAPADOPOULOS, N. M. and ROSEN, S. (1967), *Enzymologia* **34**, 40.

POPE, C. E. and COOPERBAND, S. R. (1966), *Gastroenterology* **50**, 631.

POSEN, S. (1967), *Ann. int. Med.* **67**, 183.

POSEN, S., NEALE, F. C., BIRKETT, D. J. and BRUDENELL-WOODS, J. (1967), *Am. J. clin. Path.* **48**, 81.

POSEN, S., NEALE, F. C. and CLUBB, J. S. (1965), *Ann. int. Med.* **62**, 1234.

RASMUSEN, B. A. (1965), *Genetics* **51**, 767.

RENDEL, J., AALUND, O., FREEDLAND, R. A. and MØLLER, F. (1964), *Genetics* **50**, 973.

RENDEL, J. and GAHNE, B. (1963), *Immunogenet. Lett.* **3**, 38.

RENDEL, J. and STORMONT, C. (1964), *Proc. Soc. exp. Biol., N.Y.* **115**, 853.

ROBINSON, J. C., LEVENE, C., BLUMBERG, B. S. and PIERCE, J. E. (1967), *J. med. Genet.* **4**, 96.

ROBINSON, J. C. and PIERCE, J. E. (1964), *Nature, Lond.* **204**, 472.

ROBSON, E. B. and HARRIS, H. (1965), *Nature, Lond.* **207**, 1257.

ROMEL, W. C., LaMANCUSA, S. J. and DUFRENE, J. K. (1968), *Clin. Chem.* **14**, 47.

ROSENBERG, I. N. (1959), *J. clin. Invest.* **38**, 630.

ROSS, R. S., IBER, F. L. and HARVEY, A. M. (1956), *Amer. J. Med.* **21**, 850.

ROTHMAN, F. and BYRNE, R. (1963), *J. molec. Biol.* **6**, 330.

RUSSELL, R. G. G. (1965), *Lancet* ii, 461.

SANDLER, M. and BOURNE, G. H. (1961), *Exptl. Cell Res.* **24**, 174.

SANDLER, M. and BOURNE, G. H. (1963), *Nature, Lond.* **194**, 389.

SARASWATHI, S. and BACHHAWAT, B. K. (1966), *J. Neurochem.* **13**, 237.

SARASWATHI, S. and BACHHAWAT, B. K. (1968), *Biochem. J.* **107**, 185.

SCHLAMOWITZ, M. (1954a), *J. biol. Chem.* **206**, 361.

SCHLAMOWITZ, M. (1954b), *J. biol. Chem.* **206**, 369.

SCHLAMOWITZ, M. and BODANSKY, O. (1959), *J. biol. Chem.* **234**, 1433.

SCHLESINGER, M. J. (1967), *J. biol. Chem.* **242**, 1599, 1604.

SCHLESINGER, M. J. and LEVINTHAL, C. (1963), *J. molec. Biol.* **7**, 1.

SCHNEIDERMAN, H. (1967), *Nature, Lond.* **216**, 604.

SCHNEIDERMAN, H., YOUNG, W. J. and CHILDS, B. (1966), *Science*, **151**, 461.

SCUTT, P. B. and MOSS, D. W. (1968), *Enzymologia* **35**, 157.

SHAPER, A. G. and PATEL, I. (1969), *Am. J. clin. Path.* **51**, 393.

SHERLOCK, S. and WALSHE, V. (1947), *J. Path. Bact.* **59**, 615.

SHINOWARA, G. Y., JONES, L. M. and REINHART, H. L. (1942), *J. biol. Chem.* **142**, 921.

SHREFFLER, D. C. (1965), *Am. J. Hum. Genet.* **17**, 71.

SMITH, J. K. and MOSS, D. W. (1968), *Biochem. J.* **109**, 44P.

STOLBACH, L. L., KRANT, M. J., INGLIS, N. I. and FISHMAN, W. H. (1967), *Gastroenterology* **52**, 819.

SUR, B. K., MOSS, D. W. and KING, E. J. (1962a), *Proc. Assoc. Clin. Biochem.* **2**, 11.

SUR, B. K., MOSS, D. W. and KING, E. J. (1962b), *Biochem. J.* **84**, 55P.

SUSSMAN, H. H. (1968), *Clin. Res.* **16**, 130.

SUSSMAN, H. H. (1970), *Clin. Chim. Acta* **27**, 121.

SUSSMAN, H. H., BOWMAN, M. and LEWIS, J. L. (1968), *Nature, Lond.* **218**, 359.

SUSSMAN, H. H., SMALL, P. A., JR and COTLOVE, E. (1968), *J. biol. Chem.* **243**, 160.

TALEISNIK, A., PAGLINI, S. and ZEITUNE, V. (1955), *Compt. rend. Soc. Biol., Paris* **149**, 1790.

TASWELL, H. F. and JEFFERS, D. M. (1963), *Am. J. clin. Path.* **40**, 349.

THANNHAUSER, S. J., REICHEL, M. and GRATTAN, J. F. (1937), *J. biol. Chem.* **121,** 697.

VAES, G. and JACQUES, P. (1965), *Biochem. J.* **97,** 389.

WALLACH, D. P. and KO, H. (1964), *Canad. J. Biochem.* **42,** 1445.

WALLIS, B. B. and FOX, A. S. (1968), *Biochem. Genet.* **2,** 141.

WARNOCK, M. L. (1966), *Clin. chim. Acta* **14,** 156.

WARNOCK, M. L. and REISMAN, R. (1969), *Clin. chim. Acta* **24,** 5.

WATANABE, K. and FISHMAN, W. H. (1964), *J. Histochem. Cytochem.* **12,** 252.

WEBB, E. C. (1964), *Nature, Lond.* **203,** 821.

WILCOX, F. H. (1966), *Genetics* **53,** 799.

WILKINSON, J. H. and VODDEN, A. V. (1966), *Clin. Chem.* **12,** 701.

WOLFSON, W. Q. (1957), *Nature, Lond.* **180,** 550.

YONG, J. M. (1966), *Lancet* **i,** 1132.

YOUNG, I. I. (1958), *Ann. N.Y. Acad. Sci.* **75,** 357.

CHAPTER 10

Multiple Forms of Esterases

In this chapter the heterogeneity of esterases other than phosphatases is discussed, but it must be emphasized at the outset that the problems of definition of the term *isoenzyme* apply with particular force to this group of enzymes. With few exceptions, esterases have relatively wide substrate specificities, and most have not been fully characterized. They range from lipases and other aliesterases through arylesterases, to the various 'azolesterases', which include the specific and non-specific cholinesterases. This classification can only be regarded as provisional, since esterases of all groups show activity with many substrates, and some aliesterases actually hydrolyse certain aromatic esters more readily than the corresponding aliphatic esters (Aldridge, 1953; Augustinsson, 1958a).

At the present time it is difficult to assess the isoenzyme status of esterases, since many studies of their electrophoretic separation from blood serum and tissue extracts reported in the literature appear to have

TABLE 41 *Principal features of the various classes of esterases*
(Augustinsson, 1961)

	Aliesterases	Cholinesterases	Arylesterases
Specificity:			
Esters hydrolysed	Acetates or butyrates	Acetates or butyrates	Acetates or butyrates
Phenyl esters	+	+	+++
Aliphatic esters	+++	++	—
Choline esters	—	+++	—
Inhibitors:			
Organophosphorus	+++	+++	—
Eserine	—	+++	—
p-Hydroxymercuribenzoate	—	—	++
o-Iodosobenzoate	—	+	+
EDTA	—	—	+++
Activator:			
Ca^{2+}	—	—	+
Electrophoretic mobility	α-Globulin	α-β-Globulin	Albumin

been carried out with a single substrate, and hence the nature of the fractions obtained cannot yet be fully defined. The principal features of the three main classes of esterases have been summarized by Augustinsson (1961) (Table 41), and for convenience they will be discussed under these headings.

Attention might be drawn to the differences in electrophoretic mobility exhibited by the various groups of esterases found in the plasmata of various species. The fastest moving are the arylesterases which accompany albumin; these are closely followed by the aliesterases in the α_1-globulin region; while the slowest are the cholinesterases, which are usually to be found between the α_2- and β-globulins (Kekwick, Mackay and Martin, 1953; Surgenor and Ellis, 1954; Augustinsson, 1958a). Five distinct proteins with esterase activity have

TABLE 42 *Classification of esterases in the guinea-pig*
(Holmes and Masters, 1967a)

	Slow carboxyl-esterases	Fast carboxyl-esterases	Cholin-esterase	Acetyl-cholin-esterase	Acetyl-esterase	Aryl-esterase
Substrates:						
α-Naphthyl acetate	+ + +	+ + +	+ + +	+ + +	+ +	+ + +
α-Naphthyl butyrate	+ +	+ +	+ + +	+ +	(+)	+
Indoxyl acetate	—	+ +	+ +	+ +	+ +	+ +
Inhibitors:						
10⁻⁴M-DFP*	+ + +	+ + +	+ + +	+ + +	—	—
10⁻⁵M-Eserine	—	—	+ + +	+ + +	—	—
10⁻³M-PCMB⁺	—	—	—	—	—	+ +
10⁻³M-Acetylcholine iodide	—	—	—	+ +	—	—
10M-urea	+	+ +	+	+	—	+ + +
Heating at 55	—	+	+ + +	+ + +	—	+ + +

* Diisopropylfluorophosphate.
⁺ p-Chloromercuribenzoate.

been isolated from human serum by starch-block electrophoresis by Pilz and Boo (1967), who classify them according to their substrate affinities and pH optima into the following categories: endogenous lipoprotein lipase, acylesterase I, acetylcholinesterase I, acetylcholinesterase II and a group of eleven aliesterases (Pilz, 1966; Pilz and Horlein, 1964).

Previous classifications have recently been extended by Holmes and Masters (1967), who distinguished six different groups in the guinea-pig (Table 42).

Aliesterases

This group of enzymes is closely related to the cholinesterases, since they hydrolyse a wide range of aliphatic esters, including choline esters. They may be distinguished, however, by their resistance to 10^{-5}M-eserine, which completely inhibits cholinesterases (Richter and Croft, 1942). They were originally described as simple esterases to differentiate them from lipases, which are especially active with long-chain triglycerides, but such a distinction cannot readily be sustained owing to their lack of substrate specificity.

The properties of aliesterases have been reviewed by Augustinsson (1961), who found aliesterase to be the principal esterase of the blood plasma of lower vertebrates. The aliesterases of cat and horse plasma are more effective against the lower alkyl propionates and butyrates than against the corresponding acetates, whereas those of certain fish are especially active against acetates.

In some species there is evidence that aliesterase occurs in multiple forms. Electrophoresis of guinea-pig plasma leads to the separation of one aliesterase which migrates as an α-globulin and a second which travels with the β-globulins. Both are resistant to eserine, both are inhibited by organophosphorus compounds and both hydrolyse propionates and butyrates, but the fast fraction differs from the slow component in being particularly active with triglycerides and diacetylmorphine (Augustinsson, 1961).

Numerous esterase bands have been observed after the electrophoresis of human serum and tissues, but apart from the cholinesterases their classification is uncertain at the present time. Among the tissues which appear to contain groups of aliesterases are the gastro-intestinal tract (Markert and Hunter, 1959; Weiser, Bolt and Pollard, 1964), liver (Ecobichon and Kalow, 1961, 1962), brain (Barron, Bernsohn and Hess, 1963a, 1963b; Barron and Bernsohn, 1965; Ecobichon, 1966), kidney (Ecobichon and Kalow, 1964) and skeletal muscle (Ecobichon and Kalow, 1965). Ecobichon (1965) has studied the various groups of human tissue esterases by electrophoresis in a series of starch gels of varying starch concentration, from which he concluded that the various components within each group had similar molecular weights but differed in their electric charges.

At least eleven anionic bands and five cationic bands may be detected in extracts of human brain by starch-gel electrophoresis at pH 8·5 with α-naphthyl acetate as substrate. Anionic bands 6 and 7 are among the most prominent with this substrate, but they do not react with α-naphthyl propionate or butyrate. They are of particular interest, since they are not found in the demyelinated white matter in multiple sclerosis (Barron and Bernsohn, 1965).

Tissue esterase patterns have also been investigated in a number of other pathological conditions, e.g. non-tropical sprue (Weiser *et al.*, 1964) and carcinoma of the lung (Oort and Willighagen, 1961), but the results observed have not shown any clearly defined distinction from those obtained with normal tissues.

By means of acrylamide-gel (disc) electrophoresis Holmes and Masters (1967*a*) have separated cavian aliesterases into two groups of carboxyl-esterases (Table 42). A total of ten such components occurs in various tissues. Eight carboxylesterases account for more than 50% of the total esterase activity of the liver: most of this is found in the slow group. More than 90% of the total serum esterase activity is distributed among three bands of slow carboxylesterase and a single band of fast carboxylesterase. About 36% of the heart esterase occurs as a single band of fast carboxylesterase. Similar studies have been made in the rat (Holmes and Masters, 1967*b*) and in the pig and duck (Holmes and Masters, 1968).

Of the ten zones of esterase activity against naphthyl esters detected in mouse plasma after starch-gel electrophoresis, Hunter and Strachan (1961) consider that four may provisionally be regarded as aliesterases. The fastest of these occurs in the pre-albumin region, the second between the fast α_2 and β fractions, the third in association with the β-globulins, while the slowest band migrates with the slow α_2-globulins. The authors point out, however, that inhibition studies with organophosphorus compounds did not give clear-cut differentiation into classes of esterases.

With the aid of both starch-gel and acrylamide-gel electrophoresis, Hunter, Rocha, Pfrender and DeJong (1964) have studied the effects of various stimuli on the esterases of mouse and rabbit tissues and serum. The nine principal components of mouse serum were little affected by partial hepatectomy, but the fastest component of intestinal mucosal extracts was markedly reduced by a previous injection of pilocarpine. During pregnancy in the rabbit there is progressive diminution in all the esterase bands, especially the slowest.

There have been several reports of multiple esterase components in non-mammalian vertebrates. Eel plasma also contains two aliesterases, a fast-migrating butyryl- or propionyl-esterase, which is strongly inhibited by organophosphorus compounds, and a slow-moving acetyl-esterase, which is less sensitive to these inhibitors (Augustinsson, 1961). Holmes and Masters (1968) found five carboxylesterases and four acetylesterases in the White Muscovy duck. Tissues of the lizard and frog contain single carboxylesterases of slow or intermediate electrophoretic mobility, while the catfish exhibits a group of three such enzymes, but acetylesterases are absent from all three species (Holmes, Masters and Webb, 1968).

Esterases also occur in multiple forms in insects and plants. Particularly interesting is the observation of Johnson and Bealle (1968) that in one subgenus (*Sophophora*) and certain other primitive species of *Drosophila* the ejaculatory bulbs of male insects contain esterases which do not react with β-naphthyl acetate, though capable of hydrolysing the α-naphthyl ester. Other members of the genus contain esterases which react with both substrates.

During the course of a study of the biochemistry of plant growth Mäkinen and his co-workers have described multiple forms of esterases, leucine aminopeptidase and acid phosphatases in several plant species. Pollen grains contain a wide variety of esterase patterns ranging from none at all in *Petunia inflata* to ten components or more in *Crotolaria juncea* (Mäkinen and Macdonald, 1968). Esterases readily diffuse from the pollen grains of *Oenothera organensis* into the medium whether or not they are germinating. The number of esterases detected by starchgel electrophoresis and the intensity of staining increases with the diffusion time (Mäkinen and Brewbaker, 1967). The number and activity of the esterase isoenzymes appear to correlate closely with mitotic activity at various stages of development in onion seedlings. Mäkinen (1968) reported a total of twenty-three esterases, practically all of them in zones of most intense growth, such as the coleoptile, but relatively few in differentiated zones. Numerous different esterase patterns have been observed in oat seeds (*Avena sativa*), and it is suggested that these are probably under genetic control (Williamson, Kleese and Snyder, 1968).

Genetic variations of non-specific esterases
There is evidence that esterase isoenzymes (other than those of cholinesterase, discussed in the next section) are under genetic control in animals also. Genetic variation is particularly well shown in the protozoan, *Tetrahymena pyriformis*. Each homozygote for the urea-stable esterase I pattern has a set of six or more isoenzymes, while the heterozygote shows both groups. A second type of esterase found in this species is the urea-sensitive esterase II, which occurs as either a single anionic or a single cationic band in homozygotes and as a double band in the heterozygote (S. L. Allen, 1961, 1965, 1968).

The zones of esterase activity have been detected in *Drosophila melanogaster* by Wright (1963). One of these, described as 'Est 6', occurs in fast or slow forms in homozygous insects and as both in the heterozygote. These observations have been confirmed by Beckman and Johnson (1964), who also observed genetic variation in a second isoenzyme.

Ih the mouse some esterase isoenzymes appear to be under autosomal allelic control (Popp and Popp, 1962; Petras, 1963), and Center, Hunter and Dodge (1967) have recently observed a relation between the presence of the luxoid gene (lu) and the liver esterase patterns in this species.

The electrophoretically fast esterases of the serum of naturally occurring catostomid fish have been shown to consist of either one or three bands of the A, AB and B type, apparently under genetic control. This view is supported by their geographical distribution, since areas in which the homozygotes occur are separated by an area in which the heterozygote is found (Koehn and Rasmussen, 1967).

Ontogenetic observations

In an extensive ontogenetic study of the esterase patterns of various cavian tissues, Holmes and Masters (1967*a*) have reported the most marked changes to occur in the liver, kidney and intestine. The foetal liver contains small amounts of multiple fast and slow carboxylesterases, along with single bands of cholinesterase and arylesterase. This pattern persists until birth, when there is a marked increase in the number and activity of the slow carboxylesterases and the cholinesterases, while the fast carboxylesterases and the arylesterase remain substantially unchanged. In the kidney the principal change during foetal and post-natal development is a substantial increase in the three slow carboxylesterases which gradually fall to adult levels after the sixth week of life. The fast arylesterases are most prominent in the foetal kidney, while the cholinesterases change but little during ontogeny. In the intestine, however, the cholinesterases show the most marked change, gradually increasing during foetal development and in the first few weeks of life. Arylesterase is the most prominent esterase in the foetus and new-born, while the single slow and three fast carboxylesterases of this tissue show only minor changes.

Hunter *et al.* (1964) have described the changes which occur during the development of the liver and kidney in the foetal and weanling mouse. In the liver, fractions 2, 7 and 9 are the principal forms detectable in the foetus, but after birth isoenzymes 1, 3 and 4 soon become apparent. Isoenzymes 1, 2 and 3 are the most prominent during the first three weeks, after which the complete adult pattern appears. The foetal kidney exhibits bands 2, 4 and 6, which fade after birth so that during the first four days of life only traces can be observed. Band 2 again becomes prominent at the end of the first week, followed by bands 1 and 6 and others until after five weeks all the adult components can be seen.

Relatively few studies have been made of the ontogeny of human-tissue esterases. Paul and Fottrell (1961) found foetal human-tissue esterase isoenzyme patterns to resemble those of adults, but Blanco and Zinkham (1966) reported an increase both in the number of isoenzymes and in their activities during development.

Cholinesterases

The cholinesterases are distinguished from other esterases by their relatively high activities with choline esters and by their sensitivity to inhibition by eserine. They form a heterogeneous group, the individual enzymes of which differ from each other in a variety of ways. Two distinct types occur in human and animal tissues: acetylcholinesterase ('true' cholinesterase), found in erythrocytes and nervous tissue, and 'non-specific' or pseudocholinesterase, which occurs in blood plasma and many tissues.

Neither of these is highly specific in its substrate requirements; both hydrolyse various non-choline esters, including the carbon analogue of acetylcholine, 3,3-dimethylbutyl acetate. Acetylcholinesterase hydrolyses acetyl-β-methylcholine but not benzoylcholine, while pseudocholinesterase has little action against acetyl-β-methylcholine, though it readily hydrolyses benzoylcholine (Mendel and Rudney, 1943). Acetylcholinesterase shows greatest activity with acetates and little with butyrates, whereas pseudocholinesterase exhibits optimal activity with butyrates (Adams and Whittaker, 1949). Acetylcholinesterase is strongly inhibited by excess of the 'natural' substrate, acetylcholine, which exerts no such action with pseudocholinesterase.

Cholinesterases are widely distributed throughout practically all animal tissues, and in the rat Ord and Thompson (1950) found that acetylcholinesterase predominates in the brain, skeletal muscle and the adrenal, while pseudocholinesterase is the principal form in the intestine and the heart. Both enzymes occur in the liver, stomach and lung. A purified rat-heart cholinesterase preparation has been shown by Ord and Thompson (1951) to contain at least two forms of the enzyme, the principal one of which differs in several respects from both acetylcholinesterase and pseudocholinesterase.

Plasma cholinesterases

Most recent work on the multiple forms of cholinesterase has been carried out on blood plasma with the aid of a number of electrophoretic techniques. The observation by Ord and Thompson (1951) that rat-heart cholinesterase consists of two or more components has been followed by studies of the human plasma enzyme, which may be resolved into several fractions by electrophoresis (Pinter. 1957; Heilbronn, 1958; Dubbs, Vivonia and Hilburn, 1960). In the course of a study of the reaction kinetics with a number of different substrates and inhibitors, Berry (1960) deduced that three separate cholinesterases occur in human serum.

The resolution obtained with paper electrophoresis is similar to that found with cellulose-column electrophoresis, but on starch gel a much more complex pattern emerges, presumably because the high molecular weights of cholinesterases $(2-12 \times 10^6)$ render them particularly susceptible to the molecular sieving action of this medium (Chapter 2). Although De Grouchy (1958) found only a single zone of activity in the β-globulin region. Dubbs *et al.* (1960) succeeded in resolving this fraction into a doublet. However, with acetyl- or butyryl-thiocholine as substrate, Bernsohn, Barron and Hess (1961, 1962) found seven bands of activity in starch-gel electropherograms of human serum and six in rat serum. With the exception of the fastest band in human serum, which was only partially inhibited, all were completely inhibited by eserine. These investigations have been extended by Hess. Angel, Barron and Bernsohn (1963) to the sera of other species. The monkey and cat show two and three bands of eserine-sensitive activity respectively with both substrates, but the rabbit, which exhibits three bands of eserine-sensitive activity with acetylthiocholine as substrate, displays no activity with butyrylthiocholine. Several of the bands also show esterase activity with α-naphthyl acetate as substrate: some, especially those near the origin, are only partially inhibited by eserine, a finding which suggests that they may consist of more than one enzyme type (cf. Marton and Kalow, 1959).

The application of two-dimensional electrophoresis on paper and starch gel by Harris, Hopkinson and Robson (1962) has produced results of considerable interest. The single band of human serum cholinesterase obtained after paper electrophoresis separates into at least four fractions when subjected to starch-gel electrophoresis (Fig. 61*a*). All of these zones of esterase activity, detected with α-naphthyl acetate as substrate, are completely inhibited by eserine and by tetraethyl pyrophosphate (TEPP), and all show activity with choline esters. A fifth zone of esterase activity which migrates with albumin in both buffers appears to be the albumin esterase of Wilde and Kekwick (1962), since it is inactive with β-carbonaphthoxycholine and is not sensitive to eserine or TEPP.

The four zones of cholinesterase activity (C_1, C_2, C_3, C_4 in order of decreasing mobility in starch gel) all have electrophoretic mobilities on paper similar to those of α_2-β-globulins, but in this medium C_2 moves at a slightly greater rate than the other fractions. C_4 is always present in

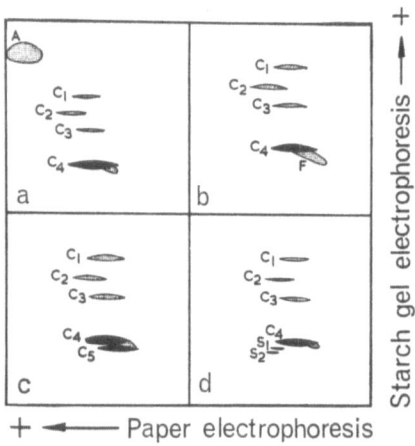

Fig. 61. Two-dimensional electrophoresis of human serum cholinesterases. (*a*) Normal human serum. A, eserine-resistant albumin component; C_1, C_2, C_3, C_4, cholinesterase fractions. (*b*) α-β-globulins of cord blood showing the presence of component F. (*c*) α-β-globulins of normal human serum showing the presence of the genetically determined variant C_5. (*d*) α-β-globulins of stored blood showing the presence of storage components S_1 and S_2. (From Harris, Hopkinson and Robson, 1962.)

greatest amount, and several variants of this component occur. In cord-blood serum, closely associated with C_4, is an additional 'foetal' zone which, it is considered, might arise as the result of the synthesis of cholinesterase molecules lacking the normal complement of sialic acid residues (Fig. 61*b*). This is suggested by the finding that a zone very similar in appearance to the 'foetal' zone appears after the C_4 fraction of adult serum has been subjected to limited action by neuraminidase.

Using the same technique in population studies, Harris *et al.* (1962) found evidence of polymorphism of the serum cholinesterases. In the sera of 14 of 300 randomly selected adults an additional zone (C_5) having slightly lower mobilities than C_4 both in starch gel and on paper was repeatedly found (Fig. 61c). It appears to be genetically determined, since the peculiarity occurs more frequently among relatives of subjects with it than in the general population. The authors suggest that the individuals displaying C_5 are probably heterozygous for an uncommon gene which determines its formation. C_5 seems to be quite distinct from the dibucaine- and fluoride-sensitive variants discussed below (p. 290).

Two further zones running slightly faster on paper but slightly slower in starch gel than C_4 appear in sera which have been stored for some days (Fig. 61d).

Five interconvertible forms of cholinesterase activity have been separated by starch-gel electrophoresis from a concentrate of human serum by LaMotta, McComb and Wetstone (1965), who, as discussed in Chapter 4 (p. 79), attributed their multiplicity to reversible polymerization. Subsequently the same group observed two additional eserine-sensitive forms in human serum, which like the original five could be converted into the same single band on concentration (LaMotta, McComb, Noll, Wetstone and Reinfrank, 1968).

GENETIC POLYMORPHISM OF SERUM CHOLINESTERASES. The introduction of the muscle relaxant, suxamethonium, led to the discovery of inherited variations in human serum cholinesterase activity. Suxamethonium (succinyldicholine) is closely related chemically to acetylcholine, and in normal subjects it is rapidly hydrolysed by the serum cholinesterase, though Pilz (1967) has reported evidence suggesting that it is hydrolysed in the lungs rather than the serum.

$$(CH_3)_3 \overset{+}{N} \cdot CH_2 \cdot CH_2 O \cdot CO \cdot CH_2 \cdot CH_2 \cdot CO \cdot OCH_2 \cdot CH_2 \cdot \overset{+}{N}(CH_3)_3$$
Succinyldicholine (Suxamethonium)

A small number of individuals, however, are abnormally sensitive and experience prolonged apnoea when the drug is administered during surgery. Such persons have abnormally low serum cholinesterase activity, and their relatives are found to fall into three groups: those with low values, who appear to be homozygous for an 'atypical' gene; those with normal values, who are homozygous for the 'usual' gene; and heterozygotes, who have serum cholinesterase levels somewhat lower than the normal range (Bourne, Collier and Somers, 1952; Evans, Gray, Lehmann and Silk, 1952, 1953; Lehmann and Ryan, 1956; Lehmann and Simmons, 1958; Kaufman, Lehmann and Silk, 1960). There is appreciable overlap between the enzyme levels of the three groups, and it seems that there is considerable variation within each genotype.

Consistent with these views is the observation that the serum cholinesterase of suxamethonium-sensitive individuals is inhibited by 10^{-5}M-dibucaine to a markedly lesser extent than that of non-sensitive subjects (Kalow and Genest, 1957). The 'dibucaine number' (i.e. the percentage inhibition of the serum cholinesterase by dibucaine measured under standard conditions) for an individual appears to be constant and independent of variations in the enzyme level. Its distribution among the general population is trimodal: most persons have a dibucaine number of about 79, while a second group, amounting to about 3% of the population, have values about 62, and comparatively rare individuals have dibucaine numbers about 16. Suxamethonium-sensitive patients belong to this third group, so it seems that there is a qualitative difference between the serum cholinesterase of these persons and that of normal subjects (Kalow and Staron, 1957; Kalow, 1959).

This has been confirmed by the finding of lower substrate affinities (higher Michaelis constants) for a number of choline esters with the atypical as compared with the normal enzyme (Kalow, 1959), an observation which Harris, Whittaker, Lehmann and Silk (1960) interpret as explaining the familial serum cholinesterase 'deficiency' of suxamethonium-sensitive subjects.

A rapid screening technique for the atypical serum cholinesterase based upon its failure to exhibit enhanced activity in the presence of $0 \cdot 1$M-sodium chloride has been devised by Swift and La Du (1966), but the authors point out that this procedure cannot detect the heterozygote.

Harris *et al.* (1960) determined the serum cholinesterase levels and dibucaine numbers of 69 members of 11 unrelated families in which at least one individual had been shown to be suxamethonium-sensitive. They found that, though the dibucaine numbers formed three clearly defined groups, there was no clear-cut parallel separation of the enzyme levels. However, when the enzyme activities were separated according to the respective dibucaine numbers they fell into three distinct but overlapping distributions.

In some families the segregation pattern of the serum cholinesterase appears to be more complicated, and both Kalow and Staron (1957) and Lehmann, Silk, Harris and Whittaker (1960) have reported families in which relatives with atypical cholinesterases have anomalous dibucaine numbers. Harris and Whittaker (1961) have since shown that fluoride differentially inhibits normal and atypical serum cholinesterase: the mean percentage inhibitions with 5×10^{-5}M-sodium fluoride were $61 \cdot 4$, $47 \cdot 8$ and $23 \cdot 1$ for the usual, intermediate and atypical phenotypes respectively. Although there is, in general, a close correlation between dibucaine numbers and fluoride numbers, there occur certain exceptions from which it is deduced that both usual and intermediate phenotypes include small sub-groups.

Two families, one Cypriot and one English, have been described in which the homozygotes have had unusually low fluoride numbers in the presence of normal dibucaine numbers (Lehmann, Liddell, Blackwell, O'Connor and Daws, 1963; Griffiths, Davies and Lehmann, 1966).

Yet another variant is the 'silent' gene which leads to the almost complete absence of serum cholinesterase in the homozygote. This was first detected by Liddell, Lehmann and Silk (1962), but according to Kattamis, Davies and Lehmann (1967) a total of eleven such cases have now been recorded. Two of these were further investigated by Goedde, Gehring and Hofmann (1965), who found traces of activity with benzoylcholine as substrate. Goedde and Altland (1968) have recently confirmed these findings, and in addition have demonstrated either no antigen–antibody reaction or only a trace on immunoelectrophoresis with antisera prepared in rabbits against the usual human serum cholinesterase.

Kattamis *et al.* (1967) suggest that four separate allelic genes control the four enzyme variants: the usual gene, the dibucaine-resistant gene, the fluoride-resistant gene and the silent gene. To these may be added the gene producing an extra C_5 component (Wetstone, Honeyman and McComb, 1965; Whittaker, 1968*b*). Additional bands described as C_6, C_{7a} and C_{7b} have also been detected in African Negroes by unidimensional agar-gel and paper–starch-gel two-way electrophoresis. These components have slower mobilities than the normal and slow C_5 isoenzymes (Van Ros and Druet, 1966).

An investigation of the eserine-sensitive serum cholinesterases of sixty-four pairs of twins, their parents and siblings by Wetstone *et al.* (1965) has provided evidence for the genetic control of the quantity of enzyme activity in the serum. *o*-Nitrobenzoylcholine was used as substrate (Main, Miles and Braid, 1961; McComb, LaMotta and Wetstone, 1964), and all qualitative variants with low dibucaine and succinylcholine numbers and the extra C_5 component were excluded.

Certain other inhibitors may aid the solution of such genetic problems, and in this connection it is interesting to note the demonstration by Harris and Whittaker (1959) that the cholinesterase inhibitor found in the potato by Orgell, Vaidya and Dahm (1958) exerts a differential action in the usual, intermediate and atypical phenotypes similar to that produced by dibucaine.

Alkyl alcohols, especially *n*-butanol, activate serum cholinesterase

(Main, 1961; Main *et al.*, 1961), and Whittaker (1968*a*) has recently applied alcohol activation in the study of normal and atypical serum enzymes. Although methanol and ethanol activate both groups of enzymes, *n*-butanol and *tert*-amyl alcohol are effective at lower concentrations. Whittaker (1968*b*) has suggested that alcohol numbers determined with *n*-butanol may prove useful in identifying new phenotypes.

Cholinesterase isoenzymes in other species

Reference has already been made (p. 285) to the five cholinesterase isoenzymes detected in tissues of the guinea-pig (Holmes and Masters, 1967), but other species have also been studied. Holmes *et al.* (1968) found up to six electrophoretically slow cholinesterases in various tissues of the frog. The intestine of the lizard contains a single slow band, two fast bands and three bands of intermediate mobility. The liver, however, is by far the richest source of cholinesterase in this species, but consists of the three intermediate bands only. One, two or three slow bands occur in the pig, horse, sheep and ox, but the rat possesses three slow bands and two fast components. No cholinesterase is seen in various tissues of the catfish.

Oki, Oliver and Funnell (1964) found four usual cholinesterase isoenzymes in horse sera, but in some specimens an extra slow zone was also detected. This is believed to be genetically determined.

An interesting difference between the electrophoretic mobilities of the butyrylcholinesterases of sow's colostrum and milk has been reported (Augustinsson, 1958*b*; Augustinsson and Olsson, 1959*b*). The enzyme of mature milk migrates almost entirely as a single peak at a slower rate than the bulk of the milk proteins, whereas at parturition the colostrum contains in addition a much faster zone of activity, which gradually disappears over a period of 2–3 weeks.

The two fractions of colostrum display identical substrate specificities and sensitivities to inhibitors, and the mobility of the faster component is the same as that of pig-plasma cholinesterase. Augustinsson (1961) has suggested that the slow component, which is identical with that of mature milk, is a microsomal enzyme. He considers, moreover, that the two forms consist of a single enzyme protein and that the slower mobility of one is due to combination with other proteins or cellular constituents.

Arylesterases

The term 'arylesterase' was introduced by Augustinsson (1958*b*) to define the group of eserine-resistant human plasma esterases of high electrophoretic mobility, previously described by Aldridge (1953) as A-esterases. They readily hydrolyse a variety of aromatic esters, but have little action with aliphatic esters. Acetates are generally hydrolysed more rapidly than propionates or butyrates, but there are several exceptions. Unlike the cholinesterases, they are relatively resistant to DFP (Mounter and Whittaker, 1953), and they can catalyse the hydrolysis of certain phosphate esters such as diethyl *p*-nitrophenyl phosphate (Aldridge, 1954).

They are inhibited by heavy metallic cations, such as Mn^{2+}, Cu^{2+} and Hg^{2+}, and some are particularly sensitive to La^{3+}, while other aryl-esterases are activated by the alkaline-earth cations, Ca^{2+}, Sr^{2+} and Ba^{2+}. Ca^{2+}, indeed, appears to be an essential co-factor, since the addition of Ca^{2+} restores the activity of several arylesterases deactivated by dialysis against distilled water (Augustinsson, 1961). Ca^{2+} has also been shown to be an essential co-factor for human serum arylesterase (Erdös, Debay and Westerman, 1959, 1960). Arylesterases are sensitive to many SH reagents, such as *p*-chloromercuribenzoate, and the reversal of such inhibition by cysteine led Augustinsson (1961) to postulate that during enzymic hydrolysis there is formed a thioester involving an enzyme thiol group and the acyl group of the substrate.

Like the cholinesterases, arylesterases are widely distributed in mammalian sera and tissues, but they appear to be absent from the sera of other vertebrates. There is evidence suggesting an endocrinological influence on their synthesis, for after becoming sexually mature boars show a fall in plasma arylesterase activity, which rises again if the animals are castrated. Furthermore, administration of testosterone to a castrated boar brings about a fall to the low levels of the normal adult (Augustinsson and Olsson, 1960).

As is the case with the cholinesterases (Lehmann, Cook and Ryan, 1957; Brody, 1960), new-born infants have low plasma arylesterases. These increase steadily to reach normal adult levels when the child is about two years old (Augustinsson and Barr, 1963).

Serum arylesterases
Augustinsson (1959) demonstrated that, after electrophoresis on a cellulose column, the main part of the eserine- and organophosphorus-resistant human serum arylesterases (hydrolysing phenyl acetate) travels with the albumin. Since the esterase peak is not uniform, he suggested that a second faster component might be present, and the heterogeneity of human serum arylesterase has since been confirmed by starch-gel electrophoresis. The principal zone of activity with β-naphthyl acetate

as substrate occurs in the fast α_2 region, while a smaller band of activity is found in the albumin fraction (Dubbs *et al.*, 1960). The presence of two peaks of arylesterase has also been demonstrated in the plasmata of mothers and their new-born infants by cellulose-column electrophoresis (Augustinsson and Brody, 1962). One of these occurs in the albumin fraction, while the other migrates at a somewhat faster rate.

The association of arylesterases with albumin raises the possibility that such enzyme activity may be an integral feature of the albumin molecule *per se*. Much evidence in support of this view has accumulated, but none of it is conclusive. Human serum albumin, purified by the cold ethanol technique (Cohn fraction V), exhibits esterolytic activity with α-naphthyl acetate and p-nitrophenyl acetate (Casida and Augustinsson, 1959), and commercial crystalline preparations have been shown to hydrolyse a variety of β-naphthyl esters (Tove, 1962). Moreover, arylesterase activity is retained in the albumin fraction after two-dimensional electrophoresis of human serum on paper and starch gel (Harris *et al.*, 1962). The question has recently been re-examined by Wilde and Kekwick (1964), who found arylesterase activity in each of three different albumin preparations, two of which were highly purified. Since there was a close concordance between the albumin monomer and arylesterase on chromatography on a hydroxylapatite column, these authors consider that arylesterase activity may be an intrinsic property of albumin monomer.

In addition to the albumin arylesterase, two other active fractions may be separated by cellulose-column electrophoresis (Augustinsson and Brody, 1962; Wilde and Kekwick, 1964). The major component of human serum migrates at a faster rate than albumin on a cellulose column, but is slower on starch-gel electrophoresis. The third component is eserine-sensitive, and since it is also active with choline esters, it is a cholinesterase. The albumin fraction is shown to be enzymically homogeneous by DEAE-cellulose column chromatography, while the pre-albumin and cholinesterase fractions appear to contain several distinct components.

The properties of the three groups of arylesterases are summarized in Table 43. The pre-albumin arylesterase differs from the others in requiring Ca^{2+} ions as a co-factor, and the use of a phosphate buffer causes the virtual disappearance of this component from the pattern obtained. Dialysis against phosphate buffer leads to its inactivation, an effect which can be reversed by the addition of Ca^{2+} ions. EDTA also completely inhibits this fraction, but has no effect upon the activities of the albumin component and cholinesterase. Other differences include the much greater sensitivity of the pre-albumin fraction to reagents for thiol groups, and the somewhat higher figure for the pH optimum for the hydrolysis of phenyl acetate by the cholinesterase. The albumin arylesterase gives a linear Lineweaver–Burk (1934) reciprocal plot for the hydrolysis of phenyl acetate ($K_m = 0.92$ mM), whereas the corresponding plot for the pre-albumin fraction is not linear and provides confirmation that two or more enzymes are present in the fraction (Wilde and Kekwick, 1964).

TABLE 43 *Summary of properties of human serum arylesterase*
(Based upon Wilde and Kekwick, 1964)

	Pre-albumin fraction	Albumin fraction	Cholinesterase fraction
Effects of inhibitors:			
10 μM-eserine	—	—	Sensitive
10 μM-DFP	—	—	Sensitive
0·1 mM-EDTA	Sensitive	—	—
0·1 mM-CuCl$_2$	Complete	Partial	Almost complete
1 mM-HgCl$_2$	Complete	Slight	Partial
1 mM-p-chloromercuribenzene-sulphonic acid	Complete	—	—
1 mM-Na$_3$AsO$_3$	Complete	—	—
Dependence on Ca^{2+} ions	Essential	—	—
pH optimum (phenyl acetate)	7·9	7·9	8·5
Optimal temperature	32°	55°	47°

Arylesterases are found in the sera of other species, including that of the pig, which contains an arylesterase which migrates with the albumin on cellulose-column electrophoresis (Augustinsson and Olsson, 1959a). Of the nine (or occasionally ten) bands of esterase activity found in mouse plasma by starch-gel electrophoresis, with α-naphthyl acetate, propionate or butyrate as substrate, only three fractions are not inhibited by eserine, and of these only one, which migrates with the albumin, retains activity in the presence of DFP (Hunter and Strachan, 1961). This albumin bands is therefore regarded as an arylesterase.

Tissue esterases
Extensive starch-gel electrophoretic studies of the esterases of the tissues from several species have shown highly reproducible species-specific patterns (Hunter and Markert, 1957; Paul and Fottrell, 1961a, 1961b; Oort and Willighagen, 1961). Distinctive patterns of the

esterases of human sera and those of the mouse, rat, guinea-pig, rabbit, cow, lamb, chicken and horse have been observed as well as of liver extracts of the mouse, rat, guinea-pig and man. It is difficult, however, to relate the results to the main serum proteins, and the limited studies with inhibitors do not permit the various bands to be fully classified as aliesterases, arylesterases or cholinesterases. Nevertheless, Paul and Fottrell (1961*b*) have obtained convincing evidence of a species specificity which persists even after prolonged tissue culture *in vitro*.

The occurrence of arylesterases in the guinea-pig and the changes which occur in various tissues during development have been described by Holmes and Masters (1967) (see p. 285). In the lizard four bands of arylesterase occur in the testis, oviduct and brain, but only traces in other tissues. Three or four weak bands are detectable in various tissues of the catfish, but arylesterase is not found in frog tissues (Holmes *et al.*, 1968).

About ten zones of esterase activity may be detected in human brain by starch-gel or agar-gel electrophoresis (Barron, Bernsohn and Hess, 1961, 1962; Eränkö, Kokko and Söderholm, 1962; Gerhardt, Clausen and Andersen, 1963). Gerhardt, Clausen, Christensen and Riishede (1963) have recently shown that, on agar-gel electrophoresis, the frontal cortex shows two pre-albumin zones, two in the α region, two in the β region and five closely grouped bands in the γ region. The intense slow α band is definitely identified as an arylesterase, while the two β bands and the slow pre-albumin band appear to be cholinesterases. The pattern varies somewhat in other parts of the brain, and in brain tumours the distribution of the esterases is abnormal.

Barron and Bernsohn (1965) found that their fast bands 9 and 10 are strongly inhibited by *p*-chloromercuribenzoate, and hence appear to be arylesterases, but band 11 is activated by this reagent. Band 9 is markedly reduced in demyelinated brain tissue from patients with multiple sclerosis.

The esterase patterns of rat brain and blood serum, obtained by electrophoresis in acrylamide gel, are markedly affected by cranial irradiation with X-rays (Masurovsky and Noback, 1963). The differences between the patterns observed in irradiated and control animals are greatest in the albumin and pre-albumin regions, but are also quite distinct in the slow α_2 region.

Esterases and isoenzymes

At the beginning of this chapter the difficulty of applying the term 'isoenzyme' to the electrophoretically distinct esterases was discussed, but mention must be made of two important suggestions concerning the nature of isoenzymes which have arisen during the investigation of the multiple forms of these enzymes.

The esterases of the protozoan, *Tetrahymena pyriformis*, have been resolved by starch-gel electrophoresis into a complex pattern of at least eighteen forms, nine of which are eserine-sensitive (Allen, 1960, 1961). The eserine-sensitive esterases occur in two groups which differ considerably in their electrophoretic mobilities, and are characteristic of certain strains. Pure *B* strains contain only the slow (cathodic) esterases, while pure *C* strains display only the fast (anodic) components. Hybrids of *B* and *C* cells contain both groups of isoenzymes. Further crossing experiments lead to the conclusion that the two groups of isoenzymes are controlled by a pair of alleles at a single locus. Since a single gene appears to be concerned in the synthesis of a set of esterases, it seems that a single protein is involved and that isoenzymes arise as the result of minor modifications.

Allen (1968) has further suggested that esterase heterogeneity may be related to the intracellular location of the various isoenzymes. Individual esterases may be associated with specific intracellular proteins in much the same way as malate dehydrogenase interacts with the mitochondrion (Munkres and Woodward, 1966). At present, however, nothing is known of such associations between esterases and structural proteins.

A somewhat similar conclusion has been reached as a result of studies of the molecular structure of certain esterases. Augustinsson (1961) has proposed that the active centre has a characteristic aminoacid sequence, but that the rest of the molecule may vary. Thus, variations in physicochemical properties may occur which have little or no effect upon enzyme–substrate complex formation. Augustinsson suggests that the term 'isoenzyme' should be restricted to enzymes differing from each other only in this manner.

REFERENCES

ADAMS, D. H. and WHITTAKER, V. P. (1949), *Biochim. biophys. Acta* **3**, 358.
ALDRIDGE, W. N. (1953), *Biochem. J.* **53**, 110.
ALDRIDGE, W. N. (1954), *Biochem. J.* **57**, 693.
ALLEN, S. L. (1960), *Genetics* **45**, 1051.
ALLEN, S. L. (1961), *Ann. N.Y. Acad. Sci.* **94**, 753.
ALLEN, S. L. (1965), *Brookhaven Symps. Biol.* **18**, 27.
ALLEN, S. L. (1968), *Ann. N.Y. Acad. Sci.* **151**, 190.

AUGUSTINSSON, K.-B. (1958a), *Nature, Lond.* **181**, 1786.

AUGUSTINSSON, K.-B. (1958b), *Acta chem. Scand.* **12**, 1150.

AUGUSTINSSON, K.-B. (195ot, *Acta chem. Scand.* **13**, 571.

AUGUSTINSSON, K.-B. (1961), *Ann. N.Y. Acad. Sci.* **94**, 844.

AUGUSTINSSON, K.-B. and BARR, M. (1963), *Clin. chim. Acta* **8**, 568.

AUGUSTINSSON, K.-B. and BRODY, S. (1962), *Clin. chim. Acta* **7**, 560.

AUGUSTINSSON, K.-B. and OLSSON, B. (1959a), *Biochem. J.* **71**, 477.

AUGUSTINSSON, K.-B. and OLSSON, B. (1959b), *Biochem. J.* **71**, 484.

AUGUSTINSSON, K.-B. and OLSSON, B. (1960), *Nature, Lond.* **187**, 924.

BARRON, K. D. and BERNSOHN, J. (1965), *Ann. N.Y. Acad. Sci.* **122**, 369.

BARRON, K. D., BERNSOHN, J. and HESS, A. R. (1961), *J. Histochem. Cytochem.* **9**, 656.

BARRON, K. D., BERNSOHN, J. and HESS, A. R. (1962), *Acta neurol. Scand.* **38**, Suppl. 1.

BARRON, K. D., BERNSOHN, J. and HESS, A. R. (1963a), *Proc. Soc. exp. Biol., N.Y.* **113**, 521.

BARRON, K. D., BERNSOHN, J. and HESS, A. R. (1963b), *J. Histochem. Cytochem.* **11**, 139.

BECKMAN, L., and JOHNSON, F. M. (1964), *Hereditas* **51**, 221.

BERNSOHN, J., BARRON, K. D. and HESS, A. R. (1961), *Proc. Soc. exp. Biol. N.Y.* **108**, 71.

BERNSOHN, J., BARRON, K. D. and HESS, A. R. (1962), *Nature, Lond.* **195**, 285.

BERRY, W. K. (1960), *Biochem. biophys. Acta* **39**, 346.

BLANCO, A. and ZINKHAM, W. H. (1966), *Bull. Johns Hopkins Hosp.* **118**, 27.

BOURNE, J. G., COLLIER, H. O. J. and SOMERS, G. F. (1952), *Lancet* i, 1225.

BRODY, S. (1960), *Acta Obstet. Gynecol. Scand.* **39**, 1.

CASIDA, J. and AUGUSTINSSON, K.-B. (1959), *Biochim. biophys. Acta* **36**, 411.

CENTER, E. M., HUNTER, R. L. and DODGE, A. H. (1967), *Genetics* **55**, 349.

DE GROUCHY, J. (1958), *Rev. franç. Études clin. biol.* **3**, 881.

DUBBS, C. A., VIVONIA, C. and HILBURN, J. M. (1960), *Science,* **131**, 1529.

ECOBICHON, D. J. (1965), *Canad. J. Biochem. Physiol.* **43**, 595.

ECOBICHON, D. J. (1966), *Canad. J. Biochem. Physiol.* **44**, 1277.

ECOBICHON, D. J. and KALOW, W. (1961), *Canad. J. Biochem. Physiol.* **39**, 1329.

ECOBICHON, D. J. and KALOW, W. (1962), *Biochem. Pharmacol.* **11**, 573.

ECOBICHON, D. J. and KALOW, W. (1964), *Canad. J. Biochem. Physiol.* **42**, 277.

ECOBICHON, D. J. and KALOW, W. (1965), *Canad. J. Biochem. Physiol.* **43**, 73.

ERÄNKÖ, O., KOKKO, A. and SÖDERHOLM, U. (1962), *Nature, Lond.* **193**, 778.

ERDÖS, E. G., DEBAY, C. R. and WESTERMAN, M. P. (1959), *Biochem. Pharmacol.* **5**, 173.

ERDÖS, E. G., DEBAY, C. R. and WESTERMAN, M. P. (1960), *Nature, Lond.* **184**, 430.

EVANS, F. T., GRAY, P. W. S., LEHMANN, H. and SILK, E. (1952), *Lancet* i, 1229.

EVANS, F. T., GRAY, P. W. S., LEHMANN, H. and SILK, E. (1953), *Br. med. J.* i, 136.

GERHARDT, W., CLAUSEN, J. and ANDERSEN, H. (1963), *Acta neurol. Scand.* **39**, 31.

GERHARDT, W., CLAUSEN, J., CHRISTENSEN, E. and RIISHEDE, J. (1963), *Acta neurol. Scand.* **39**, 85.

GOEDDE, H. W. and ALTLAND, K. (1968), *Ann. N.Y. Acad. Sci.* **151**, 540.

GOEDDE, H. W., GEHRING, D. and HOFMANN, R. A. (1965), *Biochim. biophys. Acta* **107**, 391.

GRIFFITHS, P. D., DAVIES, D. and LEHMANN, H. (1966), *Br. med. J.* ii, 215.

HARRIS, H., HOPKINSON, D. A. and ROBSON, E. B. (1962), *Nature, Lond.* **196**, 1296.

HARRIS, H. and WHITTAKER, M. (1959), *Nature, Lond.* **183**, 1808.

HARRIS, H. and WHITTAKER, M. (1961), *Nature, Lond.* **191**, 496.

HARRIS, H., WHITTAKER, M., LEHMANN, H. and SILK, E. (1960), *Acta Genet.* **10**, 1.

HEILBRONN, E. (1958), *Acta chem. Scand.* **12**, 1879.

HESS, A. R., ANGEL, R. W., BARRON, K. D. and BERNSOHN, J. (1963), *Clin. chim. Acta* **8**, 656.

HOLMES, R. S. and MASTERS, C. J. (1967a), *Biochim. biophys. Acta* **132**, 379.

HOLMES, R. S. and MASTERS, C. J. (1967b), *Biochim. biophys. Acta* **146**, 138.

HOLMES, R. S. and MASTERS, C. J. (1968), *Biochim. biophys. Acta* **159**, 81.

HOLMES, R. S., MASTERS, C. J. and WEBB, E. C. (1968), *Comp. Biochem. Physiol.* **26**, 837.

HUNTER, R. L. and MARKERT, C. L. (1957), *Science*, **125**, 1294.

HUNTER, R. L., ROCHA, J. T., PFRENDER, A. R. and DeJONG, D. C. (1964), *Ann. N.Y. Acad. Sci.* **121**, 532.

HUNTER, R. L. and STRACHAN, D. S. (1961), *Ann. N.Y. Acad. Sci.* **94**, 861.

JOHNSON, F. M. and BEALLE, S. (1968), *Biochem. Genet.* **2**, 1.

KALOW, W. (1959), in *Biochemistry of Human Genetics*, Ciba Found. Symp., ed. WOLSTENHOLME, G. E. W. and O'CONNOR, C. M., London: Churchill, p. 39.

KALOW, W. and GENEST, K. (1957), *Canad. J. Biochem. Physiol.* **35**, 339.

KALOW, W. and STARON, N. (1957), *Canad. J. Biochem. Physiol.* **35**, 1305.

KATTAMIS, C., DAVIES, D. and LEHMANN, H. (1967), *Acta Genet.* **17**, 299.

KAUFMAN, L., LEHMANN, H. and SILK, E. (1960), *Br. med. J.* i, 166.

KEKWICK, R. G. O., MACKAY, M. E. and MARTIN, N. H. (1953), *Biochem. J.* **53**, xxxvi.

KOEHN, R. K. and RASMUSSEN, D. I. (1967), *Biochem. Genet.* **1**, 131.

LaMOTTA, R. V., MCCOMB, R. B., NOLL, C. R. JR, WETSTONE, H. J. and REINFRANK, R. F. (1968), *Arch. Biochem. Biophys.* **124**, 299.

LaMOTTA, R. V., MCCOMB, R. B. and WETSTONE, H. J. (1965), *Canad. J. Biochem. Physiol.* **43**, 313.

LEHMANN, H., COOK, J. and RYAN, E. (1957), *Proc. Roy. Soc. Med.* **50**, 147.

LEHMANN, H., LIDDELL, J., BLACKWELL, B., O'CONNOR, D. C. and DAWS, A. V. (1963), *Br. med. J.* i, 1116.

LEHMANN, H. and RYAN, E. (1956), *Lancet* ii, 124.

LEHMANN, H., SILK, E., HARRIS, H. and WHITTAKER, M. (1960), *Acta Genet.* **10**, 241.

LEHMANN, H. and SIMMONS, P. H. (1958), *Lancet* ii, 981.

LIDDELL, J., LEHMANN, H. and SILK, E. (1962), *Nature, Lond.* **193**, 561.

LINEWEAVER, H. and BURK, D. (1934), *J. Am. chem. Soc.* **76**, 2842.

MCCOMB, R. B., LaMOTTA, R. V. and WETSTONE, H. J. (1964), *J. Lab. clin. Med.* **63**, 827.

MAIN, A. R. (1961), *Biochem. J.* **79**, 246.

MAIN, A. R., MILES, K. E. and BRAID, P. E. (1961), *Biochem. J.* **78**, 769.

MÄKINEN, Y. (1968), *Physiol. Plant.* **21**, 858.

MÄKINEN, Y. and BREWBAKER, J. L. (1967), *Physiol. Plant.* **20**, 477.

MÄKINEN, Y. and MACDONALD, T. (1968), *Physiol. Plant.* **21**, 477.

MARKERT, C. L. and HUNTER, R. L. (1959), *J. Histochem. Cytochem.* **7**, 42.

MARTON, A. and KALOW, W. (1959), *Canad. J. Biochem. Physiol.* **37**, 1367.

MASUROVSKY, E. B. and NOBACK, C. R. (1963), *Nature, Lond.* **200**, 847.

300 · *Isoenzymes*

MENDEL, B. and RUDNEY, H. (1943), *Biochem. J.* **37**, 53.

MOUNTER, L. A. and WHITTAKER, V. P. (1953), *Biochem. J.* **54**, 551.

MUNKRES, K. D. and WOODWARD, D. O. (1966), *Proc. Natl. Acad. Sci., Wash.* **55**, 1217.

OKI, Y., OLIVER, W. T. and FUNNELL, H. S. (1964), *Nature, Lond.* **203**, 605.

OORT, J. and WILLIGHAGEN, R. G. J. (1961), *Nature, Lond.* **190**, 642.

ORD, M. G. and THOMPSON, R. H. S. (1950), *Biochem. J.* **46**, 346.

ORD, M. G. and THOMPSON, R. H. S. (1951), *Biochem. J.* **49**, 191.

ORGELL, W. H., VAIDYA, K. A. and DAHM, P. A. (1958), *Science*, **128**, 1136.

PAUL, J. and FOTTRELL, P. (1961*a*), *Biochem. J.* **78**, 418.

PAUL, J. and FOTTRELL, P. (1961*b*), *Ann. N.Y. Acad. Sci.* **94**, 668.

PETRAS, M. L. (1963), *Proc. Natl. Acad. Sci., Wash.* **50**, 113.

PILZ, W. (1967), *Z. Klin. Chem.* **5**, 1.

PILZ, W. and BOO, A. T. (1967), *Z. Klin. Chem.* **5**, 173.

PILZ, W. and HORLEIN, H. (1964), *Hoppe-Seyler's Z. physiol. Chem.* **339**, 157.

PINTER, I. (1957), *Acta physiol. Acad. Sci. Hung.* **11**, 39.

POPP, R. A. and POPP, D. M. (1962), *J. Hered.* **53**, 111.

RICHTER, D. and CROFT, P. G. (1942), *Biochem. J.* **36**, 746.

SURGENOR, D. M. and ELLIS, D. (1954), *J. Amer. chem. Soc.* **76**, 6049.

SWIFT, M. R. and LA DU, B. N. (1966), *Lancet* i, 513.

TOVE, S. B. (1962), *Biochim. biophys. Acta* **57**, 230.

VAN ROS, G. and DRUET, R. (1966), *Nature, Lond.* **212**, 543.

WEISER, M. M., BOLT, R. J. and POLLARD, H. N. M. (1964), *J. Lab. clin. Med.* **63**, 656.

WETSTONE, H. J., HONEYMAN, M. S. and MCCOMB, R. B. (1965), *J. Am. med. Assoc.* **192**, 1007.

WHITTAKER, M. (1968*a*), *Acta Genet.* **18**, 325.

WHITTAKER, M. (1968*b*), *Acta Genet.* **18**, 335.

WILDE, C. E. and KEKWICK, R. G. O. (1962), *Biochem. J.* **83**, 39P.

WILDE, C. E. and KEKWICK, R. G. O. (1964), *Biochem. J.* **91**, 297.

WILLIAMSON, J. A., KLEESE, R. A. and SNYDER, J. R. (1968), *Nature, Lond.* **220**, 1134.

WRIGHT, T. R. F. (1963), *Genetics* **48**, 787.

CHAPTER 11
Miscellaneous Enzymes

Several enzymes other than those discussed in the preceding chapters have been shown to occur in multiple forms. Chief among these are arylamidase, creatine kinase, caeruloplasmin, catalase, carbonic anhydrase and amylase, which are briefly considered in this chapter.

Arylamidase
The arylamidase to which most attention has been paid is the human serum enzyme, formerly known as leucine aminopeptidase, which hydrolyses L-leucyl-β-naphthylamide:

$$\text{NH·CO·CH(NH}_2\text{)·CH}_2\text{·CH(CH}_3\text{)}_2 + \text{H}_2\text{O} \rightleftharpoons$$

$$(\text{CH}_3)_2\text{·CH·CH}_2\text{·CH(NH}_2\text{)·COOH} \quad + \quad \text{NH}_2$$

It appears to be distinct from a similar enzyme which hydrolyses L-leucinamide or L-leucylglycine (Smith and Rutenburg, 1963; Hanson, Bohley and Mannsfeldt, 1963). Its activity may be determined by colorimetric measurement of the β-naphthylamine liberated (Goldbarg and Rutenburg, 1958).

Arylamidase isoenzymes in man
Most investigations into the heterogeneity of arylamidase have employed electrophoretic techniques, but its isoenzymes have also been separated by column chromatography. It was first shown to occur in multiple forms by starch-gel electrophoresis (Lawrence, Melnick and Weimer, 1960; Kowlessar, Haeffner and Sleisenger, 1960). In most species the serum enzyme migrates as a fast α_2-globulin, though Lawrence et al. (1960) reported the occurrence in normal human serum of a major band of activity in the slow α_2 region. This contrasts with the

experience of other investigators, who have found only a single zone of activity in normal serum either in the post-albumin region or between the post-albumin and the fast α_2 regions (Kowlessar *et al.*, 1960; Kowlessar, Haeffner and Riley, 1961). Single zones of activity have been found in the α_1-globulins by the electrophoresis of normal sera on other media such as paper (Smith, Pineda and Rutenburg, 1962) or cellulose acetate (Smith and Rutenburg, 1963; Meade and Rosalki, 1964). An additional band of activity in the α_2 region, which appears to originate in the placenta, predominates over the normal component in the sera of pregnant women at term (Kowlessar *et al.*, 1961; Page, Titus, Mohun and Glendening, 1961; Smith and Rutenburg, 1963; Meade and Rosalki, 1964). The presence of two components in pregnancy serum has been confirmed by chromatography on Sephadex G200 columns (Goebelsman and Beller, 1965). These fractions were found to differ in their substrate specificities, for while one hydrolysed L-leucyl-β-naphthylamide, cystine-bis-β-naphthylamide was the preferred substrate of the other.

A third component (C) has been detected in pregnancy serum by Beckman, Bjorling and Christodoulou (1966), and in some individuals an extra fourth band was found to hydrolyse)'-L-glutamyl-β-naphthylamide, with which the other arylamidases do not react.

Arturson, Beckman and Persson (1967) have demonstrated a band (B) in the sera of patients treated with oral contraceptives which is electrophoretically indistinguishable from that found in pregnancy. Such a band also occurs in the sera of patients with burn injuries, obstructive jaundice or taking phenothiazine drugs. Incubation of the serum of a woman taking oral contraceptives with a rabbit antiserum which reacts with all human arylamidases except the pregnancy enzyme (Beckman, 1967) led to the disappearance of the B band. The authors conclude therefore that the B component may originate in the liver, and its appearance may be an indication of interference with the metabolism of this organ.

Multiple forms of arylamidase activity have also been demonstrated in human sera by DEAE-cellulose column chromatography (Dioguardi, Agostoni, Fiorelli, Tittobello, Cirla and Schweizer, 1961). By means of a stepwise gradient-elution technique employing a range of phosphate buffers from 0·008M, pH 7, to 0·2M, pH 6, containing 0–0·2M sodium chloride, normal serum proteins can be separated into nine fractions, three of which (VII, VIII, IX) exhibit arylamidase activity, but the last of these (IX) contains only traces of the enzyme.

Pathological sera, from patients with a variety of liver diseases, show much more complex patterns on elution (Table 44). In liver disease there is a considerable increase in the activities of the normal peaks, but in addition, activity becomes apparent in fractions IV, V and VI. It is interesting to note that when the chromatographic fractions are subjected to electrophoresis arylamidase activity appears invariably to be associated with the α-globulins.

TABLE 44 *DEAE-cellulose column chromatography of arylamidase isoenzymes in the sera of patients with liver diseases*

(From Dioguardi *et al.*, 1961)

Arylamidase activity

($-$ = no activity, tr. = trace, + = slight activity, + + = moderate activity, + + + = high activity)

Diagnosis	Protein fraction								
	I	II	III	IV	V	VI	VII	VIII	IX
	Principal electrophoretic components								
	$\alpha_2\beta_1)'$	—	β	$A\alpha_2\beta$	$A\alpha_1\alpha_2$	$A\alpha_1\alpha_2\beta$	$A\alpha_1\alpha_2\beta_2$	$A\alpha_2\beta_2$	$\alpha_2\beta_2$
Normal serum	—	—	—	—	—	—	+	+	tr.
Cholangitis	tr.	—	—	+	+	+	+ +	+ +	+ +
Carcinoma of head of pancreas without liver metastases	—	—	—	+	+	+	+ +	+ +	+ +
Hepatitis	—	—	—	tr.	+	+	+ + +	+ +	+
Cirrhosis	—	—	—	tr.	tr.	tr.	+ +	+	tr.

The observation that the arylamidase of human-liver extracts has the same mobility as the serum enzyme supports the view of Pineda, Goldbarg, Banks and Rutenburg (1960) that the liver is the principal source of the serum enzyme. The enzymes from other tissues, however, show different behaviour on electrophoresis. Kidney extracts exhibit a broad band of activity extending on cellulose acetate over the α_2- and β-globulins, while the pancreas displays two overlapping bands in the β- and γ-globulin regions. The placental enzyme is associated with α_2-globulin (Smith and Rutenburg, 1963; Meade and Rosalki, 1964). Smith and Rutenburg (1963) fractionated their tissue extracts with ammonium sulphate and found that, unlike other tissues, the kidney contains a minor arylamidase component which is soluble in 65% saturated ammonium sulphate. On electrophoresis this fraction migrates as a β-globulin. Normal urine contains arylamidase, which migrates as an α_2-globulin.

Ono, Eto and Arakawa (1968) have recently demonstrated elevated urinary arylamidase activity in patients with renal tubular diseases. The urinary enzyme appears to be of renal origin, since its electrophoretic mobility is similar to that of the kidney enzyme but differs from that of the serum enzyme.

Beckman *et al.* (1966) have also demonstrated characteristic patterns in various tissues. A fast band (A) occurs in practically all tissues, but the major component of the intestinal enzyme is slow-migrating. Four bands occur in the placenta, three in the kidney and two in the heart, liver, lung and spleen, whereas brain and erythrocytes contain the fast isoenzyme only. On treatment with neuraminidase the mobilities of the slower components towards the anode are retarded, while that of the fast band is unaffected. It seems therefore that the slow fractions contain terminal sialic acid groups which can be removed by neuraminidase.

Additional fast bands have been detected by starch-gel electrophoresis in the jejunum and colon and in lymphoma tissue by Smith and Rutenburg (1966), who also observed that certain tissue components were activated by L-methionine, whereas the serum arylamidase was inhibited by this amino acid.

Unusual bands of activity commonly occur in pathological sera, especially in liver diseases. By the starch-gel technique, Kowlessar *et al.* (1961) found, in addition to the normal zone, bands in the slow α_2-region, and in the fast α_2-β regions in sera from patients with infective hepatitis, Laennec's cirrhosis, metastatic carcinoma of the liver, infectious mononucleosis and post-hepatic biliary obstruction. These abnormal bands were also detected in widely disseminated malignant diseases (lymphoma, Hodgkin's disease and sarcoidosis). As the result of paper-electrophoretic studies, similar findings in liver diseases have been reported by Smith *et al.* (1962).

In view of the difficulty of interpreting for diagnostic purposes the intestinal band of alkaline phosphatase, owing to its relation with blood groups and dietary factors, and especially to its diminution in hospital patients generally, Beckman, Beckman and Dahlgren (1968) have recommended a combination of the serum alkaline phosphatase and the arylamidase B component as an aid in differential diagnosis. A raised serum alkaline phosphatase with an elevation of arylamidase B is indicative of biliary disease, whereas an elevated serum alkaline phosphatase accompanied by a normal arylamidase B suggests bone disease. Beckman's group has previously shown that the B band of arylamidase may be partly genetically controlled and that it is certainly increased in patients treated with phenothiazine drugs (Beckman and Wetterberg, 1967).

Arylamidases in other species

Arylamidase isoenzymes have not been studied so extensively in animal tissues as in humans, but certain observations must be mentioned.

The supernatant fractions of rat liver and bovine lens are rich sources of the enzyme which, however, reacts more readily upon L-leucinamide than upon L-leucyl-β-naphthylamide, but the particulate fractions of rat liver appear to have greater affinity for the naphthylamide (Hanson *et al.*, 1963). The bovine lens enzyme has recently been crystallized, and dissociation studies with urea and sodium dodecyl sulphate have shown it to consist of 10 sub-units (Kretschmer and Hanson, 1968).

Extensive experimental studies have been made by Raekallio and Mäkinen of the arylamidase components of regenerating wound tissue in the rat. The total activities were found to increase in the serum and in the wound tissue within the first post-operative hour. It was established by column chromatography on CM-cellulose, DEAE-cellulose and Sephadex G-200, ammonium sulphate precipitation and determination of substrate specificities with the β-naphthylamides of a series of amino acids, that the arylamidases of wound tissue differ markedly from that of control rat serum (Raekallio, 1960; Raekallio and Mäkinen, 1967*a*, 1967*b*; Mäkinen and Raekallio, 1967).

The presence of arylamidase in insects was first demonstrated by Beckman and Johnson (1964) in the pupae of *Drosophila melanogaster*, when six bands of activity were observed (A, B, C, D, E and F), but the adult fly appears to have the single A band only. Genetic variants of pupal bands A and D each exhibit fast and slow components. Homozygotes for the D band have either the fast or the slow isoenzyme, while heterozygotes show both bands, but no hybrid isoenzyme. The A variants are reciprocally related to the D genotype, since pupae with the fast D band also show the slow A band, while the slow D band is accompanied by the fast A band. D heterozygotes exhibit an A band of intermediate mobility. Some flies homozygous for either the fast or slow D band, however, show complete absence of the A component. Beckman and Johnson (1964) suggest that absence of the A band is a recessive trait, and that the interrelation between the A and D bands is controlled by two independent but closely linked loci.

In *D. busckii*, however, three distinct homozygotes containing single bands only have been detected. Three heterozygous forms each contain two electrophoretically distinguishable bands (Johnson and Sakai, 1964; Johnson, Kanapi, Richardson and Sakai, 1967). The authors suggest that the isoenzyme patterns are controlled by three co-dominant alleles, but single-pair matings produced an excess of heterozygotes which indicates selection against the homozygous genotypes. This conclusion is supported by further pair matings described by Richardson and Johnson (1967).

Alanine aminopeptidase

Five components of alanine aminopeptidase have been detected in human tissues by agar-gel electrophoresis. Each tissue appears to contain a single isoenzyme which, it is suggested, may be of diagnostic value (Peters, Rehfeld and Haschen, 1967; Peters, Rehfeld, Beier and Haschen, 1968). A single band also occurs in normal urine, but in urine from a patient with the nephrotic syndrome a second, faster, component has recently been demonstrated by agar-gel electrophoresis and immunoelectrophoresis (Peters, Rehfeld, Schneider, Battke and Haschen, 1969).

γ-Glutamyl transpeptidase

Human serum γ-glutamyl transpeptidase has recently been separated into several different components by starch-gel electrophoresis and by column chromatography on Sephadex G-200. The major fraction of normal serum has a mobility similar to that of the fast α_2-globulins, while two slow minor components appear in the slow α_2-globulin region. The latter are markedly increased in hepatitis and obstructive jaundice, but opinion is divided concerning the diagnostic value of such isoenzyme separations (Orlowski and Szczejlik, 1967; Kokot and Kuska, 1968).

Creatine kinase

Creatine kinase catalyses the reversible phosphorylation of creatine by adenosine triphosphate (Lohmann, 1934; Lehmann, 1935*a*, 1935*b*,

$$\text{Creatine} + \text{ATP} \rightleftharpoons \text{Phosphocreatine} + \text{ADP}$$

1936). It has been prepared in crystalline form from rabbit muscle (Kuby, Noda and Lardy, 1954*a*) and from ox brain (Wood, 1963*a*), but the pure preparations have been shown to differ in several respects (Kuby, Noda and Lardy, 1954*b*; Wood, 1963*b*).

Creatine kinase may be determined by measuring either the amount of creatine phosphorylated by ATP or the amount of creatine released from phosphocreatine (Ennor and Stocken, 1948; Ennor and Rosenberg, 1954). Activity of the enzyme is dependent upon the presence of free thiol groups in the enzyme molecule, and consequently, thiol reagents, such as p-chloromercuribenzoate and iodoacetate, have an inhibitory effect. The inclusion of thioglycollate or glutathione in the reaction mixture, on the other hand, has an activating effect. It appears that the muscle enzyme molecule contains two reactive thiol groups which play an important part in the catalytic reaction, and that the molecule contains two active centres (Lehmann and Pollak, 1942; Benesch, Lardy and Benesch, 1955; Mahowald and Kuby, 1960; Watts, Rabin and Crook, 1961, 1962).

Creatine kinase is also inhibited by thyroxine (Askonas, 1951), and it is interesting to note that the serum levels of the enzyme encountered in human hypothyroidism are higher than the normal range (Graig and Ross, 1963; Griffiths, 1963).

Creatine kinase isoenzymes
Indications that creatine kinase might occur in multiple molecular forms emerged during studies of the pure muscle enzyme, and Watts *et al.* (1962) reported the presence of a trace of a second component detectable by starch-gel electrophoresis. Purified ox-brain creatine kinase migrates on cellulose acetate electrophoresis as a single band of activity with a mobility similar to that of albumin, but chromatography on a DEAE-Sephadex column separates it into two major and one minor components (Wood, 1963*a*).

These observations have been confirmed and extended by Deul and van Breeman (1964*a*, 1964*b*), who separated components with creatine kinase activity from centrifuged extracts of human brain, heart and skeletal muscle by agar-gel electrophoresis. These investigators detected enzyme activity in three different ways: first, by fractional elution from the gel and determination of activities in the eluates by a conventional procedure. Secondly, the gel was incubated with creatine and ATP, after which the reaction was arrested with perchloric acid, and bands of enzyme activity were located by treatment with ammonium molybdate, which forms a precipitate with phosphocreatine. Adenosine triphosphatase activity may be detected by cutting the gel strip into two and using one half as a blank (creatine being omitted from the substrate mixture). The third procedure depended upon the series of coupled enzyme reactions devised by Tanzer and Gilvarg (1959):

(1) $\text{ATP} + \text{Creatine} \rightleftharpoons \text{Phosphocreatine} + \text{ADP}$

(2) $\text{ADP} + \text{Phosphoenolpyruvate} \xrightleftharpoons[\text{Kinase}]{\text{Pyruvate}} \text{Pyruvate} + \text{ATP}$

(3) $\text{Pyruvate} + \text{NADH} \xrightleftharpoons[\text{Dehydrogenase}]{\text{Lactate}} \text{Lactate} + \text{NAD}^+$

A second gel containing the mixed substrates was superimposed on the electrophoresis gel, and after incubation at $37°$ changes in fluorescence due to oxidation of NADH were recorded photographically.

Another method which has proved useful in the writer's laboratory is that of Rosalki (1965), who adapted the procedure of Oliver (1955) for the determination of creatine kinase activity. This depends upon the following series of reactions:

(1) Phosphocreatine + ADP $\rightleftharpoons$ Creatine + ATP

(2) ATP + Glucose $\xrightarrow{\text{hexokinase}}$ Glucose 6-phosphate + ADP

(3) Glucose 6-phosphate + NADP$^+$ $\overset{\text{glucose 6-P}}{\underset{\text{dehydrogenase}}{\rightleftharpoons}}$ NADPH + 6-Phosphogluconate

The reduction of NADP is coupled to the tetrazolium–formazan system. Blank preparations are used to eliminate bands of ATPase activity. This method is particularly suitable for staining creatine kinase isoenzymes separated by cellulose acetate electrophoresis.

Though human-brain and human-skeletal muscle creatine kinase appear to be homogeneous on electrophoresis, they differ considerably in mobility. The former migrates towards the anode, while the latter has the mobility of a γ-globulin (Fig. 62). Heart muscle contains both of

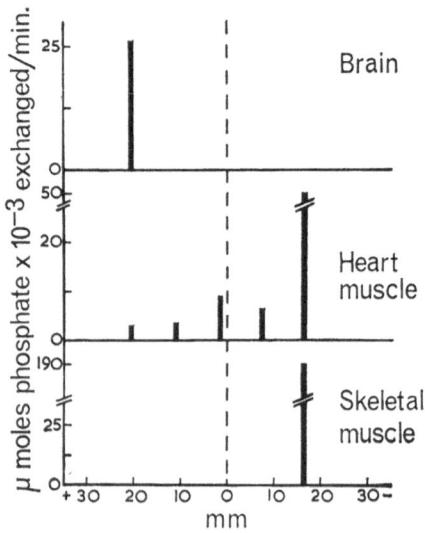

Fig. 62. Separation of the creatine kinase isoenzymes of human tissues by agar-gel electrophoresis. (Duel and van Breeman, 1964*b*.)

these components and, in addition, one or more fractions of intermediate mobility (Deul and van Breeman, 1964*b*; Sjövall and Voigt, 1964; Rosalki, 1965). The presence of more than one component of intermediate mobility found by some investigators has not yet been explained. It has been established (p. 72) that the enzyme molecule is a dimer comprising similar or dissimilar sub-units (Dance and Watts, 1962; Eppenberger, Eppenberger, Richterich and Aebi, 1964; Dawson, Eppenberger and Kaplan, 1967). Such an arrangement could give only a single hybrid isoenzyme.

Though the serum creatine kinase activity is considerably increased after myocardial infarction and in certain forms of progressive muscular dystrophy (Ebashi, Toyokura, Momoi and Sugita, 1959; Fujie, 1960; Dreyfus, Schapira and Demos, 1960; Dreyfus, Schapira, Resnais and Scebat, 1960; Forster and Escher, 1961; Aebi, Richterich, Colombo and Rossi, 1961; Aebi, Richterich, Stillhart, Colombo and Rossi, 1961; Bigazzi, 1962; Hughes, 1962, 1963; Konttinen and Halonen, 1963; Bigazzi and Ciampi, 1963), isoenzyme studies have as yet found few clinical applications.

Nevertheless, several investigators have reported the presence of the intermediate component in the sera of patients with progressive muscular dystrophy (Rosalki, 1965; Van der Veen and Willebrands, 1966; Schapira, Dreyfus and Allard, 1968). Rosalki (1965) found that the 'red' fibres of normal skeletal muscle also exhibited the intermediate band. The fast component is found in the foetal tissues of many species, including man (Schapira, 1966; Schapira *et al.*, 1968), and in experimental neurogenic muscular atrophy in the rabbit there is reversion of the muscle creatine kinase pattern to the foetal type. In human myopathies, both hereditary and acquired, a similar though less pronounced change occurs. This recalls parallel findings with the lactate dehydrogenase isoenzymes of dystrophic muscle (p. 167).

The presence of the brain isoenzyme has rarely been detected in the peripheral circulation even after severe brain damage, presumably because of the impermeability of the blood-brain barrier. The elevated serum creatine kinase has been shown to be of the muscular type after cerebrovascular accidents (Kalbag, Park and Pennington, 1966), and in psychotic disorders (Meltzer, 1968).

The anionic component also occurs in the thyroid (Graig and Smith, 1966; Perkoff, 1968) and in the lung (Perkoff, 1968). The serum creatine kinase activity has sometimes been used to aid the differential diagnosis of myocardial and pulmonary infarction, but Perkoff (1968) suggests that it might be misleading, though in none of his patients with pulmonary disease was the serum activity sufficiently high to permit detection of the anionic isoenzyme.

Arginine kinase

Arginine kinase occurs in the skeletal muscles of invertebrates, where it appears to perform functions similar to those of creatine kinase in vertebrates. Virden and Watts (1964) demonstrated the heterogeneity of arginine kinase in molluscs by starch-gel electrophoresis. Up to six isoenzymes may be detected in various muscles. The most anionic fractions appear to be monomers with molecular weights of about 40,000, while the other components are dimers with molecular weights about 80,000. Moreland and Watts (1967) have suggested that the two forms may have different functions, for in *Cardium edule* and *Tellina tenuis*, both of which show similar distributions, the foot muscle contains only the enzyme of high molecular weight, while muscles of the siphon and adductor contain both types of isoenzyme.

Arginine kinase has also been isolated from the muscles of crustaceans. Blethen and Kaplan (1968) have recently described the purification of two forms of the enzyme from the horseshoe crab, *Limulus polyphemus*, described as the 'neutral' and 'negative' forms according to the charges they have at pH 8·6. The two isoenzymes differ slightly in their amino-acid compositions and in their reactions to a rabbit antiserum prepared against the negative component, but the negative form is appreciably more stable at $50°$ than the neutral fraction.

Caeruloplasmin (copper oxidase)

Caeruloplasmin, the blue copper-containing α-globulin of human and animal sera, has enzymic properties and behaves as an oxidase, especially with *p*-phenylenediamine and certain phenols (Holmberg and Laurell, 1948, 1951). The serum caeruloplasmin level is markedly reduced or even absent in most patients with Wilson's disease (hepatolenticular degeneration), and determination of oxidase activity is an important diagnostic test for this disease (Scheinberg and Gitlin, 1952; Bearn and Kunkel, 1954; Cumings, Goodwin and Earl, 1955).

Since the enzyme molecule contains eight copper atoms, four of which remain in the cuprous form while four are successively oxidized and reduced (Broman, Malmström, Aasa and Vänngard, 1962), it has been suggested that the enzyme might be an octomer, comprising subunits each with a molecular weight of about 20,000. Another possibility is that the enzyme may consist of two pairs of dissimilar polypeptide chains of low molecular weight (α, β and γ) and one pair of high molecular weight (δ) (Poulik, 1968). Caeruloplasmin is strongly inhibited by azide in such a manner as to suggest the formation of an enzyme–azide complex (Curzon, 1966).

The heterogeneity of caeruloplasmin has been demonstrated by a number of investigators. When human serum is subjected to gradient-elution chromatography on a hydroxylapatite column the eluates contain one major and one minor component exhibiting oxidase activity with p-phenylenediamine as substrate (Broman, 1958; Richterich, Gautier, Stillhart and Rossi, 1960; Curzon and Vallet, 1960). Similar results may be obtained by electrophoresis (Uriel, 1958; Sankar, 1959), and Morell and Scheinberg (1960) have detected the presence in human sera of three chromatographically distinct forms, having the same colour, copper content and catalytic activity, but which could be differentiated into four electrophoretic variants.

Combined chromatographic and electrophoretic studies have demonstrated the presence in the sera of normal human adults of two caeruloplasmin components, which differ considerably in their electrophoretic mobilities in acetate buffer at pH 5·5, but only slightly in borate buffer at pH 8·6 (Sass-Kortsak, Jackson and Charles, 1960; Hirschman, Morell and Scheinberg, 1961). The sera of new-born infants (cord blood) also contain two fractions, but these migrate somewhat faster than the corresponding components of adult sera. Although the total amount is much reduced in comparison with normal subjects, two components may also be detected in the sera of patients with Wilson's disease. Their electrophoretic mobilities resemble those of the corresponding fractions in new-born sera (Hirschman *et al.*, 1961).

Poulik and Bearn (1962) found several zones of oxidase activity when highly purified stored specimens of human caeruloplasmin were subjected to two-dimensional electrophoresis on paper and starch gel. From the results of their electrophoretic, immunological and ultracentrifugal studies they suggest that the multiplicity of zones might be due to polymerization and that such polymers may be present in circulating blood. However, the possibility remains that such polymerization occurs during storage, a view which receives support from the observation of Uriel (1963), who obtained from a human serum caeruloplasmin preparation two catalytically similar, electrophoretic fractions, only one of which was immunologically identical with native serum caeruloplasmin.

Genetic variants of caeruloplasmin have been observed by Shreffler, Brewer, Gall and Honeyman (1967) and Poulik (1968). Shreffler *et al.* (1967) detected three bands of activity [A (fast), B (intermediate) and C (slow)] by starch-gel electrophoresis of the sera of American Negroes.

Five phenotypes, A, AB, B, AC and BC, were recognized, and family studies suggest that the A and B components are determined by a pair of autosomal co-dominant alleles, while the C form is possibly controlled by a third allele at the same locus.

Poulik (1968) also found the A and B variants in the sera of American Negroes by starch-gel electrophoresis, but reported other variants which were apparently not genetically controlled.

Catalase

Variant forms of catalase have been discovered during the course of investigations into the nature of the genetic defect, acatalasia, which was first recognized in Japan by Takahara (Takahara and Miyamoto, 1949; Takahara, 1952) and which has since been reported in Sweden, Switzerland and Israel. Most of the affected individuals are asymptomatic, but some suffer from a progressive oral gangrene due to their inability to destroy hydrogen peroxide produced by *Streptococcus haemolyticus*. The deficiency of catalase, however, does not appear to be complete: amounts ranging from 0·1% to about 0·2% of the normal activity have been detected in the erythrocytes of acatalasic persons. Aebi and his co-workers have shown that the abnormal enzyme differs from the normal in several respects.

Two different fractions of red-cell catalase were eluted from column chromatographs of normal (A_1, A_2), heterozygous (B_1, B_2) and acatalasics (C_1, C_2) haemolysates on calcium phosphate–DEAE-cellulose gels, one with 0·15M-sodium chloride and the other with 0·2M-sodium phosphate (Matsubara, Suter and Aebi, 1966). On subsequent electrophoresis on a mixture of starch, agar and acrylamide gels, the two normal fractions, A_1 and A_2, showed the same mobilities, as did the C_1 and C_2 fractions from the acatalasic cells. C_1 and C_2, however, were much less mobile than A_1 and A_2. The two heterozygous fractions B_1 and B_2 differed considerably in their mobilities; B_2 migrated as far as the normal catalases, while B_1 resembled the acatalasic fractions.

The normal and variant enzymes could not be differentiated by gel filtration with Sephadex G-100, by starch-gel electrophoresis, by their sensitivites to azide or by precipitation with anticatalase serum in the Ouchterlony double-diffusion test (Aebi, Baggliolini, Dewald, Lauber, Suter, Michell and Frei, 1964). However, the variant form is much more stable to heat than the normal form (Matsubara, Suter and Aebi, 1967). Similar differences in heat stability have been shown by the enzyme from acatalasic and normal mice (Feinstein, Seaholm, Howard and Russell, 1964).

A second familial catalase abnormality (allocatalasia) has been observed in six members of a family of Scandinavian–British extraction by Baur (1963, 1964), who found that the atypical enzyme differed from normal catalase in its electrophoretic mobility, but had similar catalytic activity. The individuals concerned had no clinical symptoms. An excellent survey of acatalasia has recently been given by Aebi (1967).

Purified catalase preparations from normal human and rat liver and erythrocytes have been shown to contain several components distinguishable by column chromatography, sucrose density gradient centrifugation and immuno-electrophoresis (Nishimura, Carson and Kobara, 1964; King and Gutmann, 1964; Thorup, Carpenter and Howard, 1964). Human red-cell catalase has been separated by chromatography on DEAE-cellulose columns into three electrophoretically distinct fractions, A, B and C, whose relative distribution is related to the age of the erythrocytes. Fraction A is increased and fraction C is decreased in young cells (Thorup *et al.*, 1964).

Catalase polymorphism has also been observed in plants. The diseased tissue of bean plants infected with *Pseudomonas phaseolica* has been shown to contain bands of catalase activity additional to those found in healthy tissue (Rudolph and Stahmann, 1964).

Scandalios (1965, 1968) has recently summarized his investigations into the genetic variants of this enzyme in the endosperm of maize. Six electrophoretic variants have been recognized, and since any pair can be subjected to *in vitro* hybridization by the freezing and thawing technique to produce a set of five isoenzymes, it is concluded that each is a tetramer. Two other variants, however, when crossed gave only two hybrid isoenzymes in addition to the parental types, an observation which Scandalios (1968) suggests may indicate the occurrence of a trimeric enzyme molecule.

Multiple forms of catalase have also been described in various maize tissues by Alexandrescu and Popov (1968) and Alexandrescu and Mihăilescu (1968).

Carbonic anhydrase

The occurrence of multiple forms of carbonic anhydrase was first demonstrated by Lindskog (1960), who separated the bovine erythrocytic enzyme into two forms by DEAE-cellulose chromatography and cellulose column electrophoresis. Three carbonic anhydrase isoenzymes, A, B and C, have since been separated from human erythrocytes by a variety of chromatographic and electrophoretic techniques

(Nyman, 1961; Rickli and Edsall, 1962; Laurent, Charrel, Castay, Nahan, Marriq and Derrien, 1962; Rickli, Ghazanfar, Gibbons and Edsall, 1964; Nyman and Lindskog, 1964; Laurent, Charrel, Luccioni, Autran and Derrien, 1965; Furth, 1968; McIntosh, 1969).

Multiple forms of the enzyme have also been observed in tissues other than erythrocytes, e.g. bovine lens (Sen, Drance and Woodford, 1963), rat brain, kidney and lens (Korhonen and Korhonen, 1965) and rat prostate (McIntosh, 1969).

It is difficult to detect carbonic anhydrase isoenzymes by specific staining after electrophoresis, and several indirect procedures have been employed. Partial purification by column chromatography, followed by protein staining, has been used by Tashian and Shaw (1962), while other workers have taken advantage of the carboxylesterase activity of carbonic anhydrase. Difficulty arises because esterases are also inhibited by acetazolamide and other carbonic anhydrase inhibitors (Tappan, Jacey and Boyden, 1964). Duplicate electropherograms, one stained for protein and the other eluted for enzyme determination, appear to offer the most specific means for the detection of carbonic anhydrase isoenzymes. With this procedure, Tappan *et al.* (1964) have demonstrated differences in the isoenzyme patterns of erythrocytes of different species.

Genetic variants of human erythrocytic carbonic anhydrase B, also known as carbonic anhydrase I, have been reported by Tashian (1965), Shows (1967) and Tashian, Shreffler and Shows (1968), but isoenzyme C (II) has hitherto been found to exhibit a single component only. Shows (1967) isolated a slower-migrating variant (Id) by DEAE-cellulose column chromatography followed by starch-gel electrophoresis, and demonstrated that it was more stable to heat than the normal Ia enzyme. The variant form appears to contain a lysine group in place of a threonine residue of the normal enzyme.

Variants of carbonic anhydrase I, and more rarely of carbonic anhydrase II, have also been detected in the erythrocytes of other primates by Tashian *et al.* (1968), who conclude that isoenzyme II is an evolutionary older form of the enzyme.

Amylase

Serum and tissue amylases

The main component of human serum amylase has an electrophoretic mobility similar to that of the γ-globulins. Earlier reports that it was also associated with albumin were discounted when it was demonstrated that the starch–iodine reaction employed for its detection is not specific for amyloclastic measurement and that serum albumin also takes up iodine in significant amounts (Wilding, 1963). When saccharogenic methods are employed no activity is found in the albumin region (Searcy, Ujihira, Hayashi and Berk, 1964), but there is abundant evidence that the serum amylase is not homogeneous. Two peaks of saccharogenic activity, both of which migrate with the γ-globulins on electrophoresis, have been separated by DEAE-cellulose chromatography from rabbit serum (Berk, Kawagushi, Zeineh, Ujihira and Searcy, 1963).

Human salivary and pancreatic amylases have also been shown to have slightly different electrophoretic mobilities, though both are found in the γ-globulin region (Nørby, 1964; Muus and Vnenchak, 1964; Berk, Searcy, Hayashi and Ujihira, 1965; Kamarýt and Laxová, 1965; Aw, 1966; Berk, Hayashi, Searcy and Hightower, 1966), and to be separable by gel filtration on DEAE-Sephadex (Aw and Hobbs, 1966). The two bands of activity in human serum have electrophoretic mobilities almost identical with those of the salivary and pancreatic amylases.

The relative rates of digestion of different starches have been reported to aid the differentiation of pancreatic amylase from that of the saliva. Meites and Rogols (1968) found the amylase of normal serum, urine and saliva to digest potato starch more rapidly than corn starch, whereas the converse held for pancreatic extracts, pancreatitis serum and secretin-stimulated duodenal juice.

Up to six amylolytic components have been detected in human colostrum. Two fractions with amylase activity can be eluted from a Sephadex G-100 column after other proteins have been removed, and on subsequent acrylamide-gel electrophoresis the second peak gives six peaks, while the first gives only two (Got, Bertagnolio, Pradal and Frot-Coutaz, 1968).

Two amylase isoenzymes have recently been reported in homogenates of chick pancreas by Heller and Kulka (1968), who overlaid acrylamide-gel electropherograms with a starch gel, subsequently treated with iodine. Both bands were observed throughout embryonic development. A study of individual chicks suggested a genetic control mechanism, since three phenotypes having either the fast band only, the slow band only or both bands were observed. Their distribution was consistent with the idea that the fast and slow components are the products of two allelic genes.

Rowe, Wakim and Thoma (1968) separated porcine pancreatic amylase into two fractions by DEAE-cellulose chromatography and acrylamide-gel electrophoresis. Since both isoenzymes have the same specific activity and identical catalytic properties, it seems that they have different but related structures, though probably the same active centres.

Two components have also been detected in rat serum after distal ligation of the choledochal duct, though not after proximal ligation (Docter-Hünicke and Goetze, 1968).

Urinary amylase

Human serum and tissue amylases have molecular weights of about 45,000, and consequently are readily filtered by the kidney glomerulus (Blainey and Northam, 1967). They are not reabsorbed by the renal tubules (McGeachin and Hargan, 1956). Considerable attention has recently been paid to the electrophoretic heterogeneity of the urinary amylase.

Using fractional elution of paper electropherograms, Franzini (1965*a*) studied the amylase distribution in normal urines concentrated by ultrafiltration. Two distinct fractions, a major component with the mobility of a)'-globulin and a minor component which migrated with the β-globulins, were detected by means of the maltosazone reaction. Urine from a patient with pancreatitis showed the presence of both fractions, but the)'-globulin peak was greatly increased above that found in normal urine, whereas the β-globulin peak was practically the same as that of normal urine. Similar separation of urinary amylases was obtained by gel filtration on Sephadex G-200 (Franzini, 1965*b*; Walravens and Estas, 1967).

Since they found the β-globulin peak to be increased in acute hepatitis, Franzini and Moda (1965) have suggested that at least part of this fraction might originate in the liver. The occurrence of a hepatic amylase has been the subject of controversy; some investigators (Somogyi, 1941; Aw and Hobbs, 1966) were unable to detect its presence, while others claim to have demonstrated starch-splitting enzymes in the liver (McGeachin and Potter, 1960; Brosemer and Rutter, 1961; Torres and Olavarrjia, 1964).

It has been suggested that the liver may regulate the serum amylase (Gray, Probstein and Heifetz, 1941), but this also has not been confirmed (Cummins and Bockus, 1951). Joseph, Olivero and Ressler (1966), however, have reported the presence in liver extracts of an amylase more cationic than that of the pancreas. It was detected by applying the *o*-dianisidine reaction to a coupled enzyme system containing maltase, glucose oxidase and peroxidase, but it failed to prevent the starch–iodine reaction. Rosenfeld (1964) has observed the presence of a)'-amylase in human liver extracts, and McGeachin and Reynolds (1961) reported that though the pancreatic and salivary amylases are immunologically identical, liver amylase is different from these (also see McGeachin, 1968). Whatever the status of the liver amylase ultimately proves to be, there appears to be little evidence of a β-globulin component in the liver, and the possibility of a hepatic contribution to the urinary enzyme remains doubtful at the present time.

The urinary amylase isoenzymes have recently been investigated in chronic pancreatitis by Aw, Hobbs and Wootton (1967), who concentrated the urines by ultrafiltration and separated the isoenzymes by agar-gel electrophoresis. Imprints of the pattern were transferred on cellulose acetate strips to a starch substrate plate. After incubation the strips were removed and the starch plate was stained with iodine (Aw, 1966). The dried plates were then scanned in a microdensitometer. The digested bands indicative of amylase activity were easily seen as pale zones against a purple background. The writer is indebted to Prof Hobbs for the further information that the surface of the starch plate is etched by contact with the amylase zones and that there is exact concordance between the etching and decolorization of the starch–iodine stain.

Three zones of pancreatic (P_1, P_2, P_3) and three zones of salivary (S_1, S_2, S_3) were detected, the pancreatic isoenzymes being more cationic. In normal subjects three distinct patterns are seen: type 1 (87%) containing P_2 and S_1; type 2 (8%) containing P_1, P_2, S_1 and S_2; and type 3 (5%) containing P_2, P_3, S_1 and S_2. The ratio of pancreatic to salivary amylase ranged from 0·9 to 3·4. Seven of eight patients with pancreatic

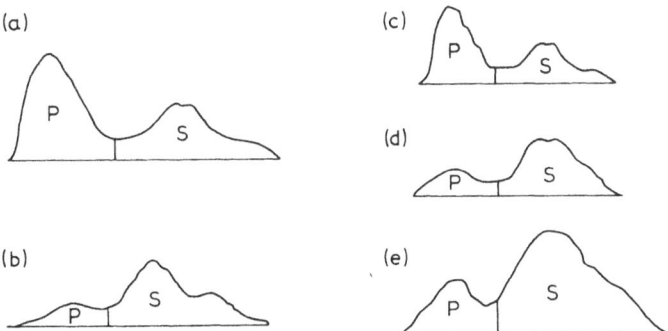

Fig. 63. Electrophoretic separation of 24-hr. urinary amylase on agar-gel. P = pancreatic fraction. S = salivary fraction. (*a*) Normal adult (type 1); (*b*) patient with chronic pancreatitis; (*c*) normal adult (type 1); (*d*) normal infant (6 months); (*e*) boy (4 years) with mumps. (Aw, Hobbs and Wootton, 1967.)

disease gave ratios of 0·6 or less – some typical patterns are illustrated in Fig. 63, which also shows that the salivary fraction is markedly increased in patients with mumps (Aw *et al.*, 1967; Hobbs and Aw, 1968).

Macroamylasaemia

There have been several reports of patients with markedly elevated serum amylase levels, but normal urinary output (Sachar and Weinhaus, 1956; Sweeney and Finkel, 1963; Wilding, Cooke and Nicholson, 1964). A probable explanation for the failure of the kidney to excrete the excess serum amylase has recently been advanced by Berk, Kizu, Wilding and Searcy (1967), who studied three patients with high serum amylase levels but low renal clearance values for the enzyme. When the serum from each of these patients was subjected to gel filtration on Sephadex G-200 two peaks of amylase activity were detected in the eluate. The lesser of the two was found in a fraction free from serum proteins and proved to be similar to the amylase of normal serum, but the other component was eluted in the same fractions as the 7S globulins.

On acrylamide-gel electrophoresis the minor component behaved like normal serum amylase, but the major fraction was significantly more cationic than the normal serum and normal parotid enzyme. The authors therefore conclude that the abnormal amylase is a macromolecular form too large to be excreted via the kidney, and have suggested the term '*macroamylasaemia*' to describe the state. There is insufficient evidence to define the nature of the macroamylase molecule at present. Berk *et al.* (1967) suggest three possibilities: it might be a complex of amylase with a serum globulin component, but if so the protein to which it is bound must have been fully saturated, since when the authors added normal parotid and pancreatic amylases to the serum the increase in activity was confined to the normal peak. Another possibility is the formation of a polymer or some other form of molecular aggregation, while a third explanation might be based upon an auto-immune reaction involving the development of an antibody against endogenous amylase.

Further publications from Berk's laboratory are awaited with interest, but meanwhile Levitt, Goetzl and Cooperband (1968) have reported a second type of macroamylasaemia in which there appears to be an association of amylase with normal serum immunoglobulins. The sera of two elderly female patients with malabsorption and villous atrophy exhibited on sucrose density gradient ultracentrifugation an 11S amylase which appeared to be a compound of normal 4·5S amylase with IgA globulin. In three other patients without evidence of malabsorption a 7S serum amylase, similar to that reported by Berk *et al.* (1967), was detected.

Other enzymes occurring in multiple forms

Glycogen phosphorylase
Yunis and Arimura (1968) demonstrated the occurrence of three different forms of glycogen phosphorylase in extracts of human muscle, platelets and leucocytes. Muscle contains a single component which is also found as a major band in the platelets and as a minor band in the white cells. Platelets also contain a major faster band and a minor slower band, whereas in leucocytes the intensities of these bands are reversed. The authors suggest that the muscle-type phosphorylase in the platelets may be related to the presence of thrombosthenin, a contractile protein similar to the actomyosin of muscle.

Adenylate kinase
Adenylate kinase polymorphism has been demonstrated in human erythrocytes, heart and skeletal muscle by starch-gel electrophoresis (Fildes and Harris, 1966). The common homozygote (AK1) in an English population, on electrophoresis at pH 7·0, exhibited a major band, which remained near the origin, along with a strong second band migrating towards the anode and two minor fast bands. The rare homozygous phenotype (AK2) was detected in only one individual of the 960 studied. In this phenotype the major band was cationic, and was accompanied by a second strong band which remained at the origin and three faint anionic bands. The heterozygote (AK2-1), which comprised about 10% of the population studied, exhibited three major bands corresponding in mobilities to those of both homozygous phenotypes. The two minor bands found in the AK1 phenotype were also detected.

Adenosine deaminase
Multiple forms of adenosine deaminase were first observed in crude and purified preparations of various animal tissues (Brady and O'Connell, 1962; Brady and O'Donovan, 1965; Cory, Weinbaum and Suhadolnik, 1967).

A more specific method, based upon the conversion of the inosine formed into hypoxanthine with nucleoside phosphorylase and thence into xanthine with the aid of xanthine oxidase ocupled to the tetrazolium system, has recently been developed by Spencer, Hopkinson and Harris

(1968). This technique has enabled these authors to detect three genetically determined phenotypes in human haemolysates. The homozygotes exhibited sets of three electrophoretically separable fast or slow isoenzymes, while the heterozygote gives a pattern similar to that obtained with a mixture of the homozygous haemolysates. The three phenotypes appear to be determined by two autosomal alleles.

Tyrosinase
Several electrophoretically distinct forms of tyrosinase have been reported, and it seems that they may play different roles in the synthesis of different types of melanin (Shimao, 1962; Pomerantz, 1963; Burnett and Seiler, 1966; Holstein, Burnett and Quevido, 1967). A maximum of three components can be separated from the follicular melanocytes of the mouse, and population studies have indicated that their synthesis is controlled by specific alleles at separate *a* and *b* loci (Holstein *et al.*, 1967).

Multiple forms of tyrosinase have also been demonstrated in *Neurospora* (Fling, Horowitz and Heinemann, 1963).

Ribonuclease
The separation of recrystallized bovine pancreatic ribonuclease into two enzymically active forms by partition chromatography on kieselguhr columns (Martin and Porter, 1951; Hirs, Moore and Stein, 1953) was probably one of the first demonstrations of heterogeneity in a purified enzyme. The occurrence of multiple forms of ribonuclease has been confirmed by zone electrophoresis (Raacke and Li, 1954) and by chromatography on a CM-cellulose column (Taborsky, 1959). Altogether about five or six fractions have been distinguished in the bovine pancreatic enzyme, but as many as eight ribonucleases have been isolated from sheep pancreas (Aqvist and Anfinsen, 1959).

The principal fractions are designated 'ribonuclease *A*' and 'ribonuclease *B*', the primary structures of which appear to resemble each other very closely (Tanford and Hauenstein, 1956; Shapira and Parker, 1960), though they differ considerably in their interactions with antibodies (Carter, Cinader and Ross, 1961).

Human serum and tissue ribonucleases have also been shown to exist in multiple forms. Of the two fractions separated from serum by starch-block electrophoresis, Leeper (1963) found one to be selectively increased in hypothyroidism. Ressler, Olivero, Thompson and Joseph (1966) identified up to three components in various human tissues and serum by starch-gel electrophoresis.

The development of a chemical method for the assay of ribonuclease by Crook, Matthias and Rabin (1960), with cyclic cytidine 2',3'-phosphate as substrate, has enabled a possible mechanism for the catalytic action of the enzyme to be postulated (Findlay, Herries, Matthias, Rabin and Ross, 1962), but this technique does not yet appear to have been applied to isoenzyme studies.

Desoxyribonucleic acid (DNA) polymerase
DNA polymerase has also been shown to occur in multiple forms (Hori, Fujiki and Takagi, 1966; Lezius, Hennig, Menzel and Metz, 1967; Cavalieri and Carroll, 1968). Cavalieri and Rosenberg (1963) had previously suggested that DNA polymerase might be a dimer, since the simultaneous synthesis of two chains would require two functional groups. This has now been confirmed by fractionation of the *E. coli* enzyme by Sephadex G-100 chromatography and acrylamide-gel electrophoresis. Three components having molecular weights about 120,000, 60,000 and 30,000 were isolated, and it is suggested that these represent the tetramer, dimer and monomer respectively (Cavalieri and Carroll, 1968).

Peroxidase
Six separate fractions exhibiting peroxidase activity may be separated by starch-gel electrophoresis from corn-leaf sheath preparations. These components differ from each other in their substrate specificities with respect to guaiacol and pyrogallol and, in the presence of manganous chloride and 2,4-dichlorophenol, with indoleacetic acid (McCune, 1961).

Starch-gel electrophoresis has also been employed in the study of peroxidase isoenzymes in the root nodules of various leguminous plants (Moustafa, 1963; Ockerse, Ziegel and Galston, 1966) and in commercial preparations of horseradish peroxidase (Klapper and Hackett, 1965). The peroxidase patterns of various plants are modified by the action of the growth hormone, indole-3-acetic acid (Ockerse *et al.*, 1966), *Ceratocystis* and *Pseudomonas phaseolica* infections (Weber and Stahmann, 1964; Rudolph and Stahmann, 1964) and by γ-ray irradiation (Endo, 1967).

As many as eight cationic and five anionic peroxidase isoenzymes have recently been reported in various tissues of the wild rice, *Oryza perennis*. The perennial plants from India and Africa show a variety of peroxidase patterns, but annual plants from India and the Amazon basin are not polymorphic (Chu, 1967).

Hyaluronate lyase (Hyaluronidase)

The heterogeneity of staphylococcal hyaluronate lyase was first observed by Abramson and Friedman (1964), who subjected crude extracts and partially purified preparations to agar-gel electrophoresis. By column chromatography on DEAE-cellulose, Greilung, Stuhlsatz and Eberhard (1965) separated three components, but Abramson (1967) has recently extended his studies, using agar-gel electrophoresis and Sephadex G-100 column chromatography, to obtain four distinct fractions, two of which were anionic and two cationic at pH 8·6.

α-Glycosidases

Evidence has recently been obtained indicating that a single gene of certain strains of Saccharomyces can produce α-glycosidases which differ in their substrate specificities as well as in their electrophoretic mobilities on starch gel. Yau and Lindegren (1967) observed three different forms: a cationic isoenzyme, A, active with melizitose; an intermediate component, B, active with *p*-nitrophenyl-α-glucopyranoside, turanose, maltose and sucrose; and an anionic form, C, which is active with α-methylglycoside and palatinose. Any inducer to which the gene responds was found to lead to the production of all three isoenzymes.

Other enzymes

Among other enzymes which have been reported to occur in multiple forms are glutaminase (Roberts, 1960; Katunama, Tomino and Nichino, 1966), asparaginase (Suld and Herbut, 1965), iodotyrosine desiodase (Kusukabe and Miyake, 1966), *β*-glucuronidase (Sadihiro, Takanashi and Kowada, 1965) and phenolase (Smith and Krueger, 1962; Bouchilloux, McMahill and Mason, 1963; Jolley and Mason, 1965).

REFERENCES

ABRAMSON, C. (1967), *Arch. Biochem. Biophys.* **121**, 103.

ABRAMSON, C. and FRIEDMAN, H. (1964), *Federation Proc.* **23**, 191.

AEBI, H. (1967), *Proc. 3rd Int. Congr. Human Genet., Chicago*, ed. CROW, J. F. and NEEL, J. V., Baltimore: Johns Hopkins Press, p. 189.

AEBI, H., BAGGLIOLINI, M., DEWALD, B., LAUBER, E., SUTER, H., MICHELL, A. and FREI, J. (1964), *Enzymol. biol. clin.* **4**, 121.

AEBI, U., RICHTERICH, R., COLOMBO, J. P. and ROSSI, E. (1961), *Enzymol. biol. clin.* **1**, 61.

AEBI, U., RICHTERICH, R., STILLHART, H., COLOMBO, J. P. and ROSSI, E. (1961), *Helv. Paed. Acta* **16**, 543.

ALEXANDRESCU, V. and MIHĂILESCU, F. (1968), *Rev. roum. Biochim.* **5**, 189.

ALEXANDRESCU, V. and POPOV, D. (1968), *Rev. roum. Biochim.* **5**, 97.

AQVIST, S. E. G. and ANFINSEN, C. B. (1959), *J. biol. Chem.* **234**, 1112.

ARTURSON, G., BECKMAN, L. and PERSSON, B. H. (1967), *Nature, Lond.* **214**, 1252.

ASKONAS, B. A. (1951), *Nature, Lond.* **167**, 933.

AW, S. E. (1966), *Nature, Lond.* **209**, 298.

AW, S. E. and HOBBS, J. R. (1966), *Biochem. J.* **99**, 16P.

AW, S. E., HOBBS, J. R. and WOOTTON, I. D. P. (1967), *Gut* **8**, 402.

BAUR, E. W. (1963), *Science,* **140**, 816.

BAUR, E. W. (1964), *Clin. chim. Acta* **9**, 252.

BEARN, A. G. and KUNKEL, H. G. (1954), *J. clin. Invest.* **33**, 400.

BECKMAN, L. (1967), *Isozyme Variations in Man,* Basel: Karger, p. 43.

BECKMAN, L., BECKMAN, G. and DAHLGREN, S. (1968), *Scand. J. Gastroenterol.* **3**, 241.

BECKMAN, L., BJORLING, G. and CHRISTODOULOU, C. (1966), *Acta Genet., Basel* **16**, 223.

BECKMAN, L. and JOHNSON, F. M. (1964), *Hereditas* **51**, 221.

BECKMAN, L. and WETTERBERG, L. (1967), *Acta Genet., Basel* **17**, 314.

BENESCH, R. E., LARDY, H. A. and BENESCH, R. (1955), *J. biol. Chem.* **216**, 663.

BERK, J. E., HAYASHI, S., SEARCY, R. L. and HIGHTOWER, N. C. (1966), *Amer. J. dig. Dis.* **11**, 695.

BERK, J. E., KAWAGUCHI, M., ZEINEH, R., UJIHIRA, I. and SEARCY, R. L. (1963), *Science,* **141**, 1182.

BERK, J. E., KIZU, H., WILDING, P. and SEARCY, R. L. (1967), *New Engl. J. Med.* **277**, 941.

BERK, J. E., SEARCY, R. L., HAYASHI, S. and UJIHIRA, I. (1965), *J. Am. med. Assoc.* **192**, 389.

BIGAZZI, P. L. (1962). *Therapeutikon* **3**, 185.

BIGAZZI, P. L. and CIAMPI, G. P. (1963), *Prog. Med.* **19**, 590.

BLAINEY, J. D. and NORTHAM, B. E. (1967), *Clin. Sci.* **32**, 377.

BLETHEN, S. L. and KAPLAN, N. O. (1968), *Biochemistry* **7**, 2123.

BOUCHILLOUX, S., MCMAHILL, P. and MASON, H. S. (1963), *J. biol. Chem.* **238**, 1699.

BRADY, T. G. and O'CONNELL, W. (1962), *Biochim. biophys. Acta* **62**, 216.

BRADY, T. G. and O'DONOVAN, C. I. (1965), *Comp. Biochem. Physiol.* **14**, 101.

BROMAN, L. (1958), *Nature, Lond.* **182**, 1655.

BROMAN, L., MALMSTRÖM, B. G., AASA, R. and VANNGARD, T. (1962), *J. molec. Biol.* **5**, 301.

BROSEMER, R. W. and RUTTER, W. J. (1961), *J. biol. Chem.* **236**, 1253.

BURNETT, J. B. and SEILER, H. (1966), *Federation Proc.* **25**, 294.

CARTER, B. G., CINADER, B. and ROSS, C. A. (1961), *Ann. N.Y. Acad. Sci.* **94**, 1004.

CAVALIERI, L. F. and CARROLL, E. (1968). *Proc. Natl. Acad. Sci., Wash.* **59**, 951.

CAVALIERI, L. F. and ROSENBERG, B. H. (1963), *Prog. Nucleic Acid Res.* **2**, 1.

CHU, Y. (1967), *Jap. J. Genet.* **42**, 233.

CORY, J. G., WEINBAUM, G. and SUHADOLNIK, R. J. (1967), *Arch. Biochem. Biophys.* **118**, 428.

CROOK, E. M., MATTHIAS, A. P. and RABIN, B. R. (1960), *Biochem. J.* **74**, 234.

324 · Isoenzymes

CUMINGS, J. N., GOODWIN, H. J. and EARL, C. J. (1955), *J. clin. Path.* **8**, 60.

CUMMINS, A. J. and BOCKUS, H. L. (1951), *Gastroenterology* **18**, 518.

CURZON, G. (1966), *Biochem. J.* **100**, 295.

CURZON, G. and VALLET, L. (1960), *Biochem. J.* **74**, 279.

DANCE, N. and WATTS, D. C. (1962), *Biochem. J.* **84**, 114P.

DAWSON, D. M., EPPENBERGER, H. M. and KAPLAN, N. O. (1967), *J. biol. Chem.* **242**, 210.

DEUL, D. H. and VAN BREEMAN, J. F. L. (1964a), Abstracts, 1st Meeting, Federation of European Biochemical Societies, p. 52.

DEUL, D. H. and VAN BREEMAN, J. F. L. (1964b), *Clin. chim. Acta* **10**, 276.

DIOGUARDI, N., AGOSTONI, A., FIORELLI, G., TITTOBELLO, A., CIRLA, E. and SCHWEIZER, M. (1961), *Enzymol. biol. clin.* **1**, 204.

DOCTER-HÜNICKE, G. and GOETZE, T. (1968), *Acta biol. med. Germ.* **21**, 495.

DREYFUS, J.-C., SCHAPIRA, G. and DEMOS, J. (1960), *Rev. franç. d'Études Clin. Biol.* **5**, 384.

DREYFUS, J.-C., SCHAPIRA, G., RESNAIS, J. and SCEBAT, L. (1960), *Rev. franç. d'Études Clin. Biol.* **5**, 386.

EBASHI, S., TOYOKURA, Y., MOMOI, M. and SUGITA, H. (1959), *J. Biochem. (Japan)* **46**, 103.

ENDO, T. (1967), *Rad. Bot.* **7**, 35.

ENNOR, A. H. and ROSENBERG, H. (1954), *Biochem. J.* **57**, 203.

ENNOR, A. H. and STOCKEN, L. A. (1948), *Biochem. J.* **42**, 557.

EPPENBERGER, H. M., EPPENBERGER, M., RICHTERICH, R. and AEBI, H. (1964), *Develop. Biol.* **10**, 1.

FEINSTEIN, R. N., SEAHOLM, J. E., HOWARD, J. B. and RUSSELL, W. L. (1964), *Proc. Natl. Acad. Sci., Wash.* **52**, 661.

FILDES, R. A. and HARRIS, H. (1966), *Nature, Lond.* **209**, 261.

FINDLAY, D., HERRIES, D. G., MATTHIAS, A. P., RABIN, B. R. and ROSS, C. A. (1962), *Biochem. J.* **85**, 152.

FLING, M., HOROWITZ, N. H. and HEINEMANN, S. F. (1963), *J. biol. Chem.* **238**, 2045.

FORSTER, G. and ESCHER, J. (1961), *Helv. med. Acta* **28**, 513.

FRANZINI, C. (1965a), *J. clin. Path.* **18**, 664.

FRANZINI, C. (1965b), *Boll. Soc. Ital. Biol. sper.* **42**, 55.

FRANZINI, C. and MODA, S. (1965), *J. clin. Path.* **18**, 775.

FUJIE, Y. (1960), *Seitai no Kagaku,* **11**, 207 |quoted by Bigazzi (1962)|.

FURTH, A. J. (1968), *J. biol. Chem.* **243**, 4832.

GOEBELSMAN, V. and BELLER, F. K. (1965), *Z. klin. Chem.* **3**, 49.

GOLDBARG, J. A. and RUTENBURG, A. M. (1958), *Cancer* **11**, 283.

GOT, R., BERTAGNOLIO, G., PRADAL, M. B. and FROT-COUTAZ, J. (1968), *Clin. chim. Acta* **22**, 545.

GRAIG, F. A. and ROSS, G. (1963), *Metabolism* **12**, 57.

GRAIG, F. A. and SMITH, J. C. (1966), *Science,* **156**, 254.

GRAY, S. H., PROBSTEIN, J. G. and HEIFETZ, C. J. (1941), *Arch. intern. Med.* **67**, 805.

GREILING, H., STUHLSATZ, H. W. and EBERHARD, T. (1965), *Hoppe-Seyler's Z. physiol. Chem.* **340**, 243.

GRIFFITHS, P. D. (1963), *Lancet* i, 894.

HANSON, H., BOHLEY, P. and MANNSFELDT, H. G. (1963), *Clin. chim. Acta* **8**, 555.

HELLER, H. and KULKA, R. G. (1968), *Biochim. biophys. Acta* **165**, 393.

HIRS, C. H. W., MOORE, S. and STEIN, W. H. (1953), *J. biol. Chem.* **200**, 493.

HIRSCHMAN, S. Z., MORELL, A. G. and SCHEINBERG, I. H. (1961), *Ann. N.Y. Acad. Sci.* **94**, 960.

HOBBS, J. R. and AW, S. E. (1968), in *Enzymes in Urine and Kidney*, ed. DUBACH, U. C., Berne: Huber, p. 281.

HOLMBERG, C. G. and LAURELL, C. B. (1948), *Acta chem. Scand.* **2**, 550.

HOLMBERG, C. G. and LAURELL, C. B. (1951), *Acta chem. Scand.* **5**, 476.

HOLSTEIN, T. J., BURNETT, J. B. and QUEVIDO, W. C. (1967), *Proc. Soc. exp. Biol., N.Y.* **126**, 415.

HORI, K., FUJIKI, H. and TAKAGI, Y. (1966), *Nature, Lond.* **210**, 604.

HUGHES, B. P. (1962), *Clin. chim. Acta* **7**, 597.

HUGHES, B. P. (1963), *Proc. Roy. Soc. Med.* **56**, 179.

JOHNSON, F. M., KANAPI, C. G., RICHARDSON, R. H. and SAKAI, R. K. (1967), *Biochem. Genet.* **1**, 35.

JOHNSON, F. M. and SAKAI, R. K. (1964), *Nature, Lond.* **203**, 373.

JOLLEY, R. L. and MASON, H. S. (1965), *J. biol. Chem.* **240**, PC1489.

JOSEPH, R. R., OLIVERO, E. and RESSLER, N. (1966), *Gastroenterology* **51**, 377.

KALBAG, R. M., PARK, D. C. and PENNINGTON, R. J. (1966), *Proc. Assoc. clin. Biochem.* **4**, 88.

KAMARÝT, J. and LAXOVÁ, R. (1965), *Humangenetik* **1**, 579.

KATUNAMA, N., TOMINO, I. and NISHINO, H. (1966), *Biochem. biophys. Res. Commun.* **22**, 321.

KING, C. M. and GUTMANN, H. R. (1964), *Cancer Res.* **24**, 770.

KLAPPER, M. H. and HACKETT, D. P. (1965), *Biochim. biophys. Acta* **96**, 272.

KOKOT, F. and KUSKA, J. (1968), *Enzymol. biol. clin.* **9**, 59.

KONTTINEN, A. and HALONEN, P. I. (1963), *Cardiologia* **43**, 56.

KORHONEN, L. K. and KORHONEN, E. (1965), *Histochemie* **5**, 279.

KOWLESSAR, O. D., HAEFFNER, L. J. and RILEY, E. M. (1961), *Ann. N.Y. Acad. Sci.* **94**, 836.

KOWLESSAR, O. D., HAEFFNER, L. J. and SLEISENGER, M. H. (1960), *J. clin. Invest.* **39**, 671.

KRETSCHMER, K. and HANSON, H. (1968), *Hoppe-Seyler's Z. physiol. Chem.* **349**, 831.

KUBY, S. A., NODA, L. and LARDY, H. A. (1954a), *J. biol. Chem.* **209**, 191.

KUBY, S. A., NODA, L. and LARDY, H. A. (1954b), *J. biol. Chem.* **210**, 65.

KUSUKABE, T. and MIYAKE, T. (1966), *J. clin. Endocrin.* **26**, 615.

LAURENT, G., CHARREL, M., CASTAY, M., NAHAN, D., MARRIQ, C. and DERRIEN, Y. (1962), *Compt. rend. Soc. Biol., Paris* **156**, 461.

LAURENT, G., CHARREL, M., LUCCIONI, F., AUTRAN, M. F. and DERRIEN, Y. (1965), *Bull. Soc. chim. biol.* **47**, 1101.

LAWRENCE, S. H., MELNICK, P. J. and WEIMER, H. E. (1960), *Proc. Soc. exp. Biol., N.Y.* **105**, 572.

LEEPER, R. A. (1963), *J. clin. Endocrin.* **23**, 426.

LEHMANN, H. (1935a), *Naturwiss.* **21**, 337.

LEHMANN, H. (1935b), *Biochem. Z.* **281**, 271.

LEHMANN, H. (1936), *Biochem. Z.* **286**, 336.

LEHMANN, H. and POLLAK, L. (1942), *Biochem. J.* **36**, 672.

LEVITT, M. D., GOETZL, E. J. and COOPERBAND, S. R. (1968), *Lancet* **i**, 957.

LEZIUS, A. G., HENNIG, S. B., MENZEL, C. and METZ, E. (1967), *Europ. J. Biochem.* **2**, 90.

LINDSKOG, S. (1960), *Biochim. biophys. Acta* **39**, 218.

LOHMANN, K. (1934), *Biochem. Z.* **271**, 264.

MCCUNE, D. C. (1961), *Ann. N.Y. Acad. Sci.* **94**, 723.

MCGEACHIN, R. L. (1968), *Ann. N.Y. Acad. Sci.* **151**, 208.

MCGEACHIN, R. L. and HARGAN, L. A. (1956), *J. appl. Physiol.* **9**, 129.

MCGEACHIN, R. L. and POTTER, B. A. (1960), *J. biol. Chem.* **235**, 1354.

MCGEACHIN, R. L. and REYNOLDS, J. M. (1961), *Ann. N.Y. Acad. Sci.* **94**, 996.

MCINTOSH, J. E. A. (1969), *Biochem. J.* **114**, 463.

MAHOWALD, T. A. and KUBY, S. A. (1960), *Federation Proc.* **19**, 46.

MÄKINEN, P.-L. and RAEKALLIO, J. (1967), *Acta chem. Scand.* **21**, 761.

MARTIN, A. J. P. and PORTER, R. R. (1951), *Biochem. J.* **49**, 215.

MATSUBARA, S., SUTER, H. and AEBI, H. (1966), *Experientia* **22**, 428.

MATSUBARA, S., SUTER, H. and AEBI, H. (1967), *Humangenetik* **4**, 29.

MEADE, B. W. and ROSALKI, S. B. (1964), *J. clin. Path.* **17**, 61.

MEITES, S. and ROGOLS, S. (1968), *Clin. Chem.* **14**, 1176.

MELTZER, H. (1968), *Science,* **159**, 1368.

MORELAND, B. and WATTS, D. C. (1967), *Nature, Lond.* **215**, 1092.

MORELL, A. G. and SCHEINBERG, I. H. (1960), *Science,* **131**, 930.

MOUSTAFA, E. (1963), *Nature, Lond.* **199**, 1189.

MUUS, J. and VNENCHAK, J. M. (1964), *Nature, Lond.* **204**, 283.

NISHIMURA, E. T., CARSON, N. and KOBARA, T. Y. (1964), *Arch. Biochem. Biophys.* **108**, 452.

NØRBY, S. (1964), *Exptl. Cell Res.* **36**, 663.

NYMAN, P.-O. (1961), *Biochim. biophys. Acta* **52**, 1.

NYMAN, P.-O. and LINDSKOG, S. (1964), *Biochim. biophys. Acta* **85**, 141.

OCKERSE, R., ZIEGEL, B. Z. and GALSTON, A. W. (1966), *Science,* **151**, 452.

OLIVER, I. T. (1955), *Biochem. J.* **61**, 116.

ONO, T., ETO, K. and ARAKAWA, K. (1968), *Clin. chim. Acta* **19**, 257.

ORLOWSKI, M. and SZCZEKLIK, A. (1967), *Clin. chim. Acta* **15**, 387.

PAGE, E. W., TITUS, M. A., MOHUN, G. and GLENDENING, M. B. (1961), *Am. J. Obstet. Gyn.* **82**, 1090.

PERKOFF, G. T. (1968), *Ann. intern. Med.* **122**, 326.

PETERS, J. E., REHFELD, N., BEIER, L. and HASCHEN, R. J. (1968), *Clin. chim. Acta* **19**, 277.

PETERS, J. E., REHFELD, N. and HASCHEN, R. J. (1967), *Clin. chim. Acta* **17**, 516.

PETERS, J. E., REHFELD, N., SCHNEIDER, I., BATTKE, H. and HASCHEN, R. J. (1969), *Clin. chim. Acta* **24**, 314.

PINEDA, E. P., GOLDBARG, J. A., BANKS, B. M. and RUTENBURG, A. M. (1960), *Gastroenterology* **38**, 698.

POMERANTZ, S. H. (1963), *J. biol. Chem.* **238**, 2351.

POULIK, M. D. (1968), *Ann. N.Y. Acad. Sci.* **151**, 476.

POULIK, M. D. and BEARN, A. G. (1962), *Clin. chem. Acta* **7**, 374.

RAACKE, I. D. and LI, C. H. (1954), *Biochim. biophys. Acta* **14**, 290.

RAEKALLIO, J. (1960), *Nature, Lond.* **188**, 234.

RAEKALLIO, J. and MÄKINEN, P.-L. (1967a), *Nature, Lond.* **213**, 1037.

RAEKALLIO, J. and MÄKINEN, P.-L. (1967b), *Ann. Med. exp. Fenn.* **45**, 224.

RESSLER, N., OLIVERO, E., THOMPSON, G. R. and JOSEPH, R. R. (1966), *Nature, Lond.* **210**, 695.

RICHARDSON, R. H. and JOHNSON, F. M. (1967), *Biochem. Genet.* **1**, 73.

RICHTERICH, R., GAUTIER, E., STILLHART, H. and ROSSI, E. (1960), *Helv. Paed. Acta* **5**, 424.

RICKLI, E. E. and EDSALL, J. T. (1962), *J. biol. Chem.* **237**, PC258.

RICKLI, E. E., GHAZANFAR, S. A. S., GIBBONS, B. H. and EDSALL, J. T. (1964), *J. biol. Chem.* **239**, 1065.

ROBERTS, E. (1960), in *The Enzymes*, ed. BOYER, P. D., LARDY, H. and MYRBÄCK, K., New York: Academic Press, vol. 4, p. 285.

ROSALKI, S. B. (1965), *Nature, Lond.* **207**, 414.

ROSENFELD, E. L. (1964), in *Control of Glycogen Metabolism*, Ciba Foundation Symposium, ed. WHELAN, W. J. and CAMERON, M. P., London: Churchill, p. 176.

ROWE, J. J. M., WAKIM, J. and THOMA, J. A. (1968), *Anal. Biochem.* **25**, 206.

RUDOLPH, K. and STAHMANN, M. A. (1964), *Nature, Lond.* **204**, 744.

SACHAR, L. A. and WEINHAUS, R. (1956), *Arch. Surg.* **73**, 305.

SADIHIRO, R., TAKANASHI, S. and KOWADA, N. (1965), *Neurology* **58**, 104.

SANKAR, S. (1959), *Federation Proc.* **18**, 441.

SASS-KORTSAK, A., JACKSON, S. J. and CHARLES, A. F. (1960), *Vox Sanguinis* (*N.S.*) **5**, 87.

SCANDALIOS, J. G. (1965), *Proc. Natl. Acad. Sci., Wash.* **53**, 1035.

SCANDALIOS, J. G. (1968), *Ann. N.Y. Acad. Sci.* **151**, 274.

SCHAPIRA, F. (1966), *Compt. rend. Acad. Sci., Paris* **262**, 2291.

SCHAPIRA, F., DREYFUS, J.-C. and ALLARD, D. (1968), *Clin. chim. Acta* **20**, 439.

SCHEINBERG, I. H. and GITLIN, D. (1952), *Science*, **116**, 484.

SEARCY, R. L., UJIHIRA, I., HAYASHI, S. and BERK, J. E. (1964), *Clin. chim. Acta* **9**, 505.

SEN, M., DRANCE, S. M. and WOODFORD, V. R. (1963), *Canad. J. Biochem. Physiol.* **41**, 1235.

SHAPIRA, R. and PARKER, S. (1960), *Biochem. biophys. Res. Commun.* **3**, 200.

SHIMAO, K. (1962), *Biochim. biophys. Acta* **62**, 205.

SHOWS, T. B. (1967), *Biochem. Genet.* **1**, 171.

SHREFFLER, D. C., BREWER, G. J., GALL, J. C. and HONEYMAN, M. S. (1967), *Biochem. Genet.* **1**, 101.

SJÖVALL, K. and VOIGT, A. (1964), *Nature, Lond.* **201**, 701.

SMITH, E. E., PINEDA, E. P. and RUTENBURG, A. M. (1962), *Proc. Soc. exp. Biol., N.Y.* **110**, 683.

SMITH, E. E. and RUTENBURG, A. M. (1963), *Nature, Lond.* **197**, 800.

SMITH, E. E. and RUTENBURG, A. M. (1966), *Science*, **152**, 1256.

SMITH, J. L. and KRUEGER, R. C. (1962), *J. biol. Chem.* **237**, 1121.

SOMOGYI, M. (1941), *Arch. intern. Med.* **67**, 665.

SPENCER, N., HOPKINSON, D. A. and HARRIS, H. (1964), *Nature, Lond.* **204**, 742.

SULD, H. M. and HERBUT, P. A. (1965), *J. biol. Chem.* **240**, 2234.

SWEENEY, D. F. and FINKEL, M. (1963), *Gastroenterology* **45**, 756.

TABORSKY, G. (1959), *J. biol. Chem.* **234**, 2652.

TAKAHARA, S. (1952), *Lancet* ii, 1101.

TAKAHARA, S. and MIYAMOTO, H. (1949), *Oto-Rhino-Laryng.* (*Japan*) **21**, 7.

TANFORD, C. and HAUENSTEIN, J. D. (1956), *Biochim. biophys. Acta* **19**, 535.

TANZER, M. L. and GILVARG, C. (1959), *J. biol. Chem.* **234**, 3201.

328 · *Isoenzymes*

TAPPAN, D. V., JACEY, M. J. and BOYDEN, H. M. (1964), *Ann. N.Y. Acad. Sci.* **121**, 589.

TASHIAN, R. E. (1965), *Am. J. Human Genet.* **17**, 257.

TASHIAN, R. E. and SHAW, M. W. (1962), *Am. J. Human Genet.* **14**, 295.

TASHIAN, R. E., SHREFFLER, D. C. and SHOWS, T. B. (1968), *Ann. N.Y. Acad. Sci.* **151**, 64.

THORUP, O. A., CARPENTER, J. T. and HOWARD, P. (1964), *Br. J. Haematol.* **10**, 542.

TORRES, H. N. and OLAVARRJIA, J. M. (1964), *J. biol. Chem.* **239**, 2427.

URIEL, J. (1958), *Nature, Lond.* **181**, 999.

URIEL, J. (1963), *Ann. N.Y. Acad. Sci.* **103**, 956.

VAN DER VEEN, K. J. and WILLEBRANDS, A. F. (1966), *Clin. chim. Acta* **13**, 312.

VIRDEN, R. and WATTS, D. C. (1964), *Comp. Biochem. Physiol.* **13**, 161.

WALRAVENS, P. and ESTAS, A. (1967), *Clin. chim. Acta* **18**, 82.

WATTS, D. C., RABIN, B. R. and CROOK, E. M. (1961), *Biochim. biophys. Acta* **48**, 380.

WATTS, D. C., RABIN, B. R. and CROOK, E. M. (1962), *Biochem. J.* **82**, 412.

WEBER, D. J. and STAHMANN, M. A. (1964), *Science*, **146**, 929.

WILDING, P. (1963), *Clin. chim. Acta* **8**, 918.

WILDING, P., COOKE, W. T. and NICHOLSON, G. I. (1964), *Ann. intern. Med.* **60**, 1053.

WOOD, T. (1963a), *Biochem. J.* **87**, 453.

WOOD, T. (1963b), *Biochem. J.* **89**, 210.

YAU, T. M. and LINDEGREN, C. C. (1967), *Biochem. biophys. Res. Commun.* **27**, 305.

YUNIS, A. A. and ARIMURA, G. K. (1968), *Biochem. biophys. Res. Commun.* **33**, 119.

CHAPTER 12

The Biological Significance
of Isoenzymes

The biological significance of the occurrence of enzymes in multiple forms is a problem which has greatly interested biochemists during the past decade. At the present time it is impossible to do more than comment upon a number of more or less isolated observations on their metabolic roles and the conclusions which have been drawn from them. Most of these have been mentioned in earlier chapters, and this concluding section is therefore restricted to a brief summary of some of the more important findings concerning the possible physiological functions of the heteropolymeric isoenzymes, particularly lactate dehydrogenase, the significance of genetically determined enzymes and finally the role of isoenzymes in metabolic regulation.

The physiological functions of heteropolymeric isoenzymes

The difference in the sensitivity of LD_1 and LD_5 to substrate inhibition suggests a possible function of lactate dehydrogenase isoenzymes. As pointed out in Chapter 6, LD_1 is strongly inhibited by the high pyruvate concentration optimal for LD_5, whereas at the lower concentrations optimal for LD_1, LD_5 is only feebly active. This led Cahn, Kaplan, Levine and Zwilling (1962) to suggest that the H and M sub-units have different functions: isoenzymes containing major proportions of the M sub-unit (LD_5 and LD_4) can permit the rapid accumulation of lactate and are therefore to be found in skeletal muscle, in which anaerobic glycolysis is a normal metabolic route. In tissues rich in LD_1, such as the heart and brain, on the other hand, lactate does not accumulate and pyruvate is oxidized via the tricarboxylic acid cycle. Several examples of tissues in which LD_5 is associated with anaerobic glycolysis have been mentioned, e.g. the flight muscles of birds, the white fibres of skeletal muscle and tumour tissues (pp. 144–169).

Although this theory has been criticized on the grounds that it is difficult to demonstrate pyruvate inhibition of LD_1 under physiological conditions (Vesell, 1965; Stambaugh and Post, 1966), and liver metabolism does not correlate with the preponderance of LD_5 in this tissue, there is much circumstantial evidence in support. There are also reports that certain tricarboxylic acid cycle intermediates, e.g. oxaloacetate and citrate, may activate LD_5 (Fritz, 1965, 1967; Orleans-Harding and Mahler, 1968). If these observations are confirmed it seems likely that over-production of such metabolites might divert pyruvate from the citric acid cycle to lactate. The accumulation of lactate during vigorous exercise may in part be mediated by this means.

Further confirmation of the possible role of lactate dehydrogenase isoenzymes as regulators of oxidative and glycolytic processes has been obtained from tissue culture and chick-embryo studies under aerobic

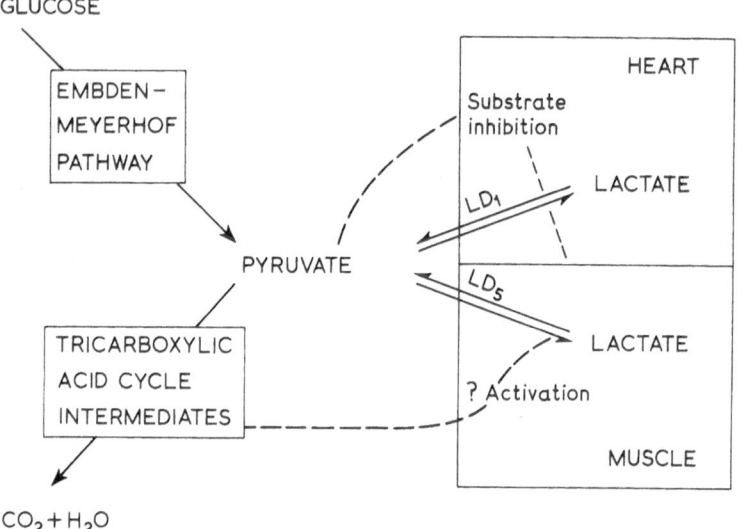

GLUCOSE

EMBDEN–MEYERHOF PATHWAY

PYRUVATE

TRICARBOXYLIC ACID CYCLE INTERMEDIATES

$CO_2 + H_2O$

HEART

Substrate inhibition

LD_1

LACTATE

LD_5

LACTATE

? Activation

MUSCLE

Fig. 64. Some of the factors governing the metabolic activity of lactate dehydrogenase isoenzymes.

and anaerobic conditions (Goodfriend and Kaplan, 1963; Cahn, 1963, 1964; Dawson, Goodfriend and Kaplan, 1964; Goodfriend, Sokol and Kaplan, 1966; Lindy and Rajasalmi, 1966; Güttler and Clausen, 1969). It has been demonstrated that under anaerobic conditions synthesis of the M sub-unit is increased, while in the presence of increasing oxygen tension there is a progressive switch towards synthesis of the H sub-unit.

Some of the factors which may play a part in the control of the activity of LD_1 and LD_5 are illustrated in Fig. 64.

There is some evidence of hormonal control of the synthesis of the M and H sub-units of lactate dehydrogenase. During pregnancy the proportion of the slower isoenzymes increases in the uterus of the rat

(Allen, 1961), the rabbit (Biron, 1964) and of women (Richterich, Schafroth and Aebi, 1963). A similar effect on the uterine muscle resulting from the administration of oestradiol to immature female rats has been reported by Goodfriend and Kaplan (1964), who suggest that in altering the isoenzyme pattern, oestradiol prepares the uterus for the anaerobic conditions associated with prolonged contraction during labour.

Thyroid hormones also affect the lactate dehydrogenase isoenzyme pattern. In the rat injection of tri-iodothyronine in sufficient amounts to produce a 30% increase in the basal metabolic rate causes a loss of LD_5 in the liver (Allison, Gerszten and Sanchez, 1964). A reduction in the synthesis of the M sub-unit is also produced in thyroxine-treated tadpoles (Kim, D'Ioro and Paik, 1966).

Lactate dehydrogenase isoenzymes may also play a major part in regulating the $NAD^+/NADH$ ratio (Markert, 1963). A similar role has been proposed for the cytoplasmic and mitochondrial forms of α-glycerophosphate dehydrogenase (Chapter 7). It is suggested that these isoenzymes may aid the build-up of mitochondrial NADH if dihydroxy-acetone phosphate is reduced in the cytoplasm and α-glycerophosphate is oxidized in the mitochondria (Bücher and Klingenberg, 1958; Sacktor, 1958). Kaplan (1961, 1963) has extended this concept to malate dehydrogenase. He proposes that cytoplasmic oxaloacetate is reduced to malate, which passes through the mitochondrial membrane and is then oxidized back to oxaloacetate. This process would have the advantage that the NADH produced by the mitochondrial enzyme could be oxidized by the ADP-dependent respiratory enzymes, so leading to the production of ATP.

Some of the multiple forms of enzymes of the Embden–Meyerhof pathway appear to be associated with specific functions. Mention has already been made (p. 96) of the relationship postulated by Katzen (1967) between the multiplicity of the hexokinases and the action of insulin on glucose transport across cell membranes. Rutter, Rajkumar, Penhoet and Kochman (1968) have reported that the catalytic properties of aldolase A are more suitable for the cleavage of fructose 1,6-diphosphate, while those of aldolase B favour the synthesis of this metabolite. They therefore suggest that isoenzyme A is associated with glycolysis, while isoenzyme B is concerned with gluconeogenesis.

Genetically determined isoenzymes
Several examples of genetically determined isoenzymes have been described in the preceding chapters. The glycolytic enzymes of the red cell

have been extensively studied and have proved of immense interest to geneticists and structural biochemists, but the biological advantages of such mutations, if any, remain obscure. It is generally assumed that mutant forms persist only if they confer some advantage. Thus there is evidence to suggest that the presence of the sickle trait (haemoglobin S) increases resistance to the malarial parasite, and consequently individuals with this trait are better fitted for life in areas where malaria is endemic. Homozygotes, however, are at a serious disadvantage, since they develop sickle-cell anaemia. It is possible that certain enzyme anomalies, e.g. the common A+ variant of glucose 6-phosphate dehydrogenase, may similarly aid adaptation to environment. Studies so far, however, suggest that while most individuals with the variant phenotypes are symptomless, they do not appear to have gained any particular advantage from it, and in a minority the variant is associated with severe haemolytic anaemia.

Several of the glycolytic enzymes discussed in Chapter 5, e.g. erythrocytic 6-phosphogluconate dehydrogenase, occur in variant forms which are apparently without detectable effect on the individuals concerned. No physiological role has yet been suggested for the variants of alkaline phosphatase in the placenta. The presence of such isoenzymes, however, certainly explains biochemical individuality.

Isoenzymes as metabolic regulators

Perhaps the most definitive proposal for the physiological role of isoenzymes is that put forward by Umbarger and his co-workers. These investigators deduced that when an enzyme controlling a given reaction, common to more than one metabolic pathway, is subject to inhibition by the end-products of one pathway, a second form of the enzyme is required to maintain supplies of substrates for succeeding steps in other pathways (Umbarger, 1961). This principle has been elaborated by Stadtman and his associates, who studied the role of the three aspartate kinase isoenzymes of *Escherichia coli* in the biosynthesis of lysine, methionine and threonine. Each of these isoenzymes is selectively inhibited by one of the end-products (Stadtman, Cohen and Le Bras, 1961; Stadtman, 1968). It was therefore proposed that the three aspartate kinases might occupy different sites within the cell and that end-product inhibition by one amino acid would lead to diversion of aspartyl phosphate to another branch according to the scheme illustrated in Fig. 65. This indicates secondary control at the first stage in each branch. There are also two distinct homoserine dehydrogenases, one of which is inhibited by methionine and the other by threonine. Thus an excess of threonine would not only diminish the total supply of aspartyl phosphate, and hence of aspartate semialdehyde, but would also divert these intermediates towards the synthesis of lysine and methionine by inhibiting homoserine dehydrogenase and the phosphorylation of homoserine.

Several other examples are known where isoenzymes appear to be involved in the regulation of branched metabolic pathways in bacteria.

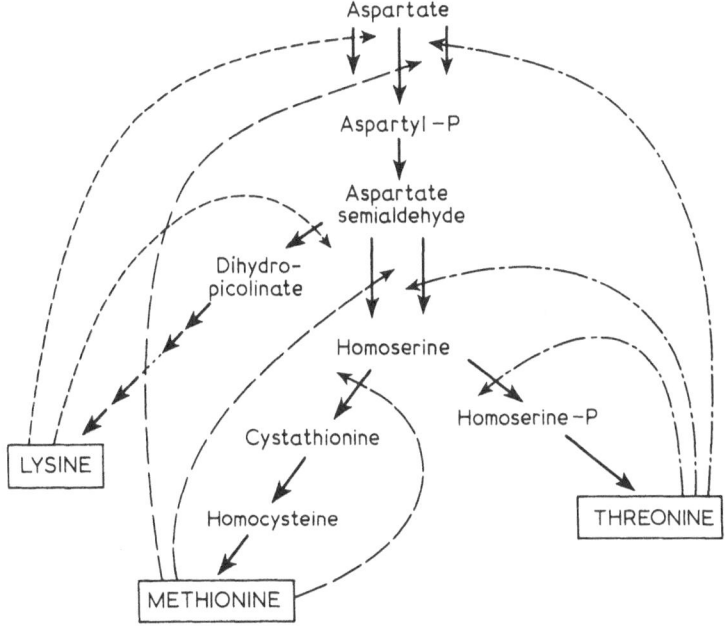

Fig. 65. The biosynthetic pathways for lysine, methionine and threonine in *E. coli*. Feedback control is indicated by the broken lines. (Stadtman, 1968.)

The first step in the formation of isoleucine from threonine in *E. coli*, the conversion of threonine into α-oxobutyric acid, is catalysed by two separate forms of threonine deaminase. One of these is directly concerned with the isoleucine pathway and is subject to feedback inhibition by this amino acid, whereas the second isoenzyme has a degradative function. When grown under anaerobic conditions with an adequate supply of amino acids, *E. coli* obtains energy from fermentation processes. The first form of threonine deaminase is inhibited, and threonine is utilized by the second, but under aerobic conditions fermentation is not needed as a source of energy, and only the first isoenzyme is produced (Umbarger and Brown, 1957).

Another instance is the selective inhibition of the three isoenzymes of 3-deoxy-D-arabinoheptulonic acid 7-phosphate synthetase in *E. coli* and *Aerobacter aerogenes* by phenylalanine, tyrosine and tryptophan. Such inhibition or repression of isoenzyme synthesis, coupled with feedback controls at the first divergent stage, regulates the biosynthesis of these three amino acids (Doy and Brown, 1965; Brown and Doy, 1966). The part played by isoenzymes in the regulation of branched metabolic pathways has recently been reviewed by Stadtman (1968), and a more general account is given by Datta (1969).

REFERENCES

ALLEN, J. M. (1961), *Ann. N.Y. Acad. Sci.* **94**, 937.

ALLISON, M. J., GERSZTEN, E. and SANCHEZ, B. (1964), *Endocrinology* **74**, 87.

BIRON, P. (1964), *Rev. Canad. Biol.* **23**, 497.

BROWN, K. D. and DOY, C. H. (1966), *Biochim. biophys. Acta* **118**, 157.

BÜCHER, T. and KLINGENBERG, M. (1958), *Angew. Chem.* **70**, 552.

CAHN, R. D. (1963), *J. Cell. Biol.* **19**, 12A.

CAHN, R. D. (1964), *Develop. Biol.* **9**, 327.

CAHN, R. D., KAPLAN, N. O., LEVINE, L. and ZWILLING, E. (1962), *Science*, **136**, 962.

DATTA, P. (1969), *Science, N.Y.* **165**, 556.

DAWSON, D. M., GOODFRIEND, T. L. and KAPLAN, N. O. (1964), *Science*, **143**, 929.

DOY, C. H. and BROWN, K. D. (1965), *Biochim. biophys. Acta* **104**, 377.

FRITZ, P. J. (1965), *Science*, **150**, 364.

FRITZ, P. J. (1967), *Science*, **156**, 82.

GOODFRIEND, T. L. and KAPLAN, N. O. (1963), *J. Cell Biol.* **19**, 28A.

GOODFRIEND, T. L. and KAPLAN, N. O. (1964), *J. biol. Chem.* **239**, 130.

GOODFRIEND, T. L., SOKOL, D. M. and KAPLAN, N. O. (1966), *J. molec. Biol.* **15**, 18.

GÜTTLER, F. and CLAUSEN, J. (1969), *Biochem. J.* **114**, 839.

KAPLAN, N. O. (1961), in *Mechanism of Action of Steroid Hormones*, ed. VILLEE, C. A. and ENGEL, A. A., Oxford: Pergamon Press, p. 247.

KAPLAN, N. O. (1963), *Bact. Rev.* **27**, 155.

KATZEN, H. M. (1967), in *Advances in Enzyme Regulation*, ed. WEBER, G., Oxford: Pergamon Press, Vol. 5, p. 335.

KIM, H. C., D'IORO, A. D. and PAIK, W. K. (1966), *Canad. J. Biochem. Physiol.* **44**, 103.

LINDY, S. and RAJASALMI, M. (1966), *Science*, **153**, 1401.

MARKERT, C. L. (1963), in *21st Symposium, Soc. Study Develop. Growth*, ed. RUDNICK, D., New York: Academic Press, p. 65.

ORLEANS-HARDING, J. G. and MAHLER, R. (1968), *Biochem. J.* **107**, 31P.

RICHTERICH, R., SCHAFROTH, P. and AEBI, H. (1963), *Clin. chim. Acta* **8**, 178.

RUTTER, W. J., RAJKUMAR, T., PENHOET, E. and KOCHMAN, M. (1968), *Ann. N.Y. Acad. Sci.* **151**, 102.

SACKTOR, B. (1958), *Proc. 4th Int. Congr. Biochem., Vienna* **12**, 138.

STADTMAN, E. R. (1968), *Ann. N.Y. Acad. Sci.* **151**, 516.

STADTMAN, E. R., COHEN, G. N. and LE BRAS, G. (1961), *Ann. N.Y. Acad. Sci.* **94,** 952.

STAMBAUGH, R. and POST, D. (1966), *J. biol. Chem.* **241,** 1462.

UMBARGER, H. E. (1961), in *Control Mechanisms in Cellular Processes,* ed. BONNER, D. M., New York: Ronald Press, p. 67.

UMBARGER, H. E. and BROWN, B. (1957), *J. Bacteriol.* **73,** 195.

VESELL, E. S. (1965), *Science,* **150,** 1590.

Author Index

Boulter, D., 50, 55
Bourne, G. H., 249, 278
Bourne, J. G., 289, 298
Bowbeer, D. R., 98, 127
Bowman, H. S., 124, 127
Bowman, J. E., 22, 33, 102, 113, 115, 127, 128
Bowman, M., 259, 263, 278
Boyd, J. W., 3, 6, 19, 28, 33, 150, 196, 225, 226, 227, 228, 229, 236
Boyde, T. R. C., 23, 33, 50, 54, 226, 227, 231, 233, 234, 236, 237
Boyden, H. M., 27, 35, 314, 328
Boyer, P. D., 102, 117, 124, 127
Boyer, S. H., 21, 22, 33, 103, 104, 107, 108, 109, 127, 131, 170, 171, 196, 243, 244, 263, 265, 274
Brady, T. G., 319, 323
Brahen, L., 193
Braid, P. E., 291, 299
Bratton, A. C., 45, 54
Braun, A. D., 135, 199
Brehmeyer, B. A., 215, 222
Brewbaker, J. L., 284, 299
Brewer, G. J., 23, 34, 38, 92, 93, 98, 99, 113, 114, 127, 128, 131, 132, 311
Briere, R. O., 17, 37, 50, 56, 176, 188, 195, 200
Brinson, A. G., 129
Brinster, R. L., 141, 195
Brock, M. J., 258, 274
Brody, I. A., 33, 67, 87, 154, 169, 196
Brody, S., 293, 294, 298
Brok, F., 103, 131
Broman, L., 31, 33, 310, 311, 323
Brosemer, R. W., 316, 323
Brosteaux, J., 141, 198
Brown, B., 333, 335
Brown, J., 91, 92, 94, 97, 98, 127
Brown, K. D., 334
Browne, E. A., 107, 128
Brownlow, E. K., 19, 33
Brudenell-Woods, J., 18, 35, 37, 247, 256, 274, 276, 278
Brunelli, F., 273, 277
Brunner, A. L., 229, 237
Bücher, T., 82, 85, 204, 207, 221, 331, 334
Buck, F., 3, 7
Buck, G. M., 119, 120, 121, 127, 128
Buckley, J., 71, 73, 87, 149, 201
Builder, J. E., 181, 195
Bukowsky, J., 3, 6, 228, 237
Burger, A., 19, 32, 33, 37, 152, 181, 201
Burk, D., 295, 299

Burka, E. R., 111, 127
Burke, J. O., 239, 275
Burnett, J. B., 320, 323
Burnett, R. W., 320, 325
Burston, D., 30, 38
Burstone, M. S., 54, 55
Butterworth, P. J., 77, 78, 84, 246, 248, 256, 257, 275
Byrne, R., 268, 278

Cahn, R. D., 23, 35, 60, 62, 65, 84, 144, 145, 159, 160, 164, 165, 166, 169, 196, 198, 203, 329, 330, 334
Cammarata, P. S., 227, 237
Campbell, D. M., 21, 22, 33, 36, 40, 45, 53, 56, 213, 214, 216, 221, 243, 251, 277
Campos, J. O., 125, 126, 127
Cao, A., 235, 237
Carlin, L., 218, 222
Carlson, C., 254, 276
Carlström, A., 29, 33
Carmichael, R., 232, 236
Carpenter, J. T., 313, 328
Carroll, E., 321, 323
Carson, N., 313, 326
Carson, P. E., 22, 33, 102, 111, 113, 115, 116, 127, 128
Carter, B. G., 320, 323
Carter, N. D., 23, 33, 111, 116, 117, 128
Casida, J., 294, 298
Castay, M., 314, 325
Castellino, F. J., 118, 128
Castro-Sierra, E., 218, 221
Cather, J. N., 169, 197
Cavalieri, L. F., 321, 323
Cawley, L. P., 19, 39, 176, 203
Cenciotti, L., 180, 196
Center, E. M., 285, 298
Chan, W., 75, 84, 118, 119, 128
Changeux, J.-P., 82, 86
Chardonnens, H., 77, 86
Charles, A. F., 311, 327
Charrel, M., 83, 86, 314, 325
Chen, K. K. W., 180, 200
Chen, R. F., 146, 203
Cheng, C. S., 49, 56
Cheng, T., 118, 129
Cherrick, G. R., 219, 222
Chiancone, E., 86
Chiandussi, L., 21, 25, 33, 53, 54, 81, 243, 244, 245, 246, 259, 275
Childs, B., 104, 107, 128, 267, 278
Childs, V. A., 167, 196
Chilson, C. A., 62, 84

Subject Index